Frank Schwierz (Ed.)

Two-Dimensional Electronics - Prospects and Challenges

MDPI

This book is a reprint of the Special Issue that appeared in the online, open access journal, *Electronics* (ISSN 2079-9292) from 2015–2016 (available at: http://www.mdpi.com/journal/electronics/special_issues/2d_electronics).

Guest Editor
Frank Schwierz
Institut für Mikro- und Nanotechnologien
Technische Universität Ilmenau
Germany

Editorial Office
MDPI AG
St. Alban-Anlage 66
Basel, Switzerland

Publisher
Shu-Kun Lin

Senior Assistant Editor
Xiaoyan Chen

1. Edition 2016

MDPI • Basel • Beijing • Wuhan • Barcelona

ISBN 978-3-03842-249-5 (Hbk)
ISBN 978-3-03842-250-1 (PDF)

Table of Contents

V

List of Contributors

Nicolas Agraït Instituto Madrileño de Estudios Avanzados en Nanociencia (IMDEA-nanociencia), Campus de Cantoblanco, E-18049 Madrid, Spain; Condensed Matter Physics Center (IFIMAC); Dpto. de Física de la Materia Condensada, Universidad Autónoma de Madrid, 28049 Madrid, Spain.

Deji Akinwande Microelectronics Research Center, Department of Electrical and Computer Engineering, The University of Texas at Austin, Austin, TX 78758, USA.

Andreas Bablich Department of Electrical Engineering and Computer Science, University of Siegen, Hölderlinstr. 3, 57076 Siegen, Germany.

Yaser M. Banadaki College of Engineering, Southern University, Baton Rouge, LA 80813, USA; Division of Electrical and Computer Engineering, Louisiana State University, Baton Rouge, LA 70803, USA.

Rudolf Bratschitsch Center for Nanotechnology; Institute of Physics, University of Münster, D-48149 Münster, Germany.

Andres Castellanos-Gomez Instituto Madrileño de Estudios Avanzados en Nanociencia (IMDEA-nanociencia), Campus de Cantoblanco, E-18049 Madrid, Spain.

David Perez de Lara Instituto Madrileño de Estudios Avanzados en Nanociencia (IMDEA-nanociencia), Campus de Cantoblanco, E-18049 Madrid, Spain.

Francesco Driussi DIEGM—University of Udine via delle Scienze, 208 22100 Udine, Italy.

Sébastien Frégonèse CNRS and University of Bordeaux, IMS UMR 5218, Talence 33400, France.

Georgy Fursey Department of Physics, University of Telecommunications, St. Petersburg 198504, Russia.

Zhansong Geng Institut für Mikro-und Nanoelektronik, Technische Universität Ilmenau, PF 100565, 98684 Ilmenau, Germany.

Ralf Granzner Institut für Mikro-und Nanoelektronik, Technische Universität Ilmenau, PF 100565, 98684 Ilmenau, Germany.

Ruben Guerrero Instituto Madrileño de Estudios Avanzados en Nanociencia (IMDEA-nanociencia), Campus de Cantoblanco, E-18049 Madrid, Spain.

Thomas Heine Department of Physics and Earth Sciences, Jacobs University Bremen, Campus Ring 1, 28759 Bremen, Germany; Wilhelm-Ostwald-Institut für Physikalische und Theoretische Chemie, Universität Leipzig, Linnéstr. 2, 04103 Leipzig, Germany.

Fahim Ferdous Hossain Department of Electrical and Electronic Engineering, Bangladesh University of Engineering and Technology (BUET), Dhaka 1205, Bangladesh.

Satender Kataria Department of Electrical Engineering and Computer Science, University of Siegen, Hölderlinstr. 3, 57076 Siegen, Germany.

Asir Intisar Khan Department of Electrical and Electronic Engineering, Bangladesh University of Engineering and Technology (BUET), Dhaka 1205, Bangladesh.

Pavel Konorov Department of Solid State Electronics, St. Petersburg State University, St. Petersburg 193232, Russia.

Agnieszka Kuc Wilhelm-Ostwald-Institut für Physikalische und Theoretische Chemie, Universität Leipzig, Linnéstr. 2, 04103 Leipzig, Germany.

Jongho Lee Microelectronics Research Center, Department of Electrical and Computer Engineering, The University of Texas at Austin, Austin, TX 78758, USA.

Max C. Lemme Department of Electrical Engineering and Computer Science, University of Siegen, Hölderlinstr. 3, 57076 Siegen, Germany.

Jiantong Li KTH Royal Institute of Technology, School of Information and Communication Technology, Electrum 229, SE-16440 Kista, Sweden.

Vikas Mittal Department of Chemical Engineering, The Petroleum Institute, Abu Dhabi 2533, United Arab Emirates.

Cedric Nanmeni Bondja Institut für Mikro-und Nanoelektronik, Technische Universität Ilmenau, PF 100565, 98684 Ilmenau, Germany.

Toshiaki Natsuki Institute of Carbon Science and Technology; Faculty of Textile Science and Technology, Shinshu University, 3-15-1 Tokida, Ueda-shi 386-8567, Japan.

Ishtiaque Ahmed Navid Department of Electrical and Electronic Engineering, Bangladesh University of Engineering and Technology (BUET), Dhaka 1205, Bangladesh.

Maliha Noshin Department of Electrical and Electronic Engineering, Bangladesh University of Engineering and Technology (BUET), Dhaka 1205, Bangladesh.

Mikael Östling KTH Royal Institute of Technology, School of Information and Communication Technology, Electrum 229, SE-16440 Kista, Sweden.

Saungeun Park Microelectronics Research Center, Department of Electrical and Computer Engineering, The University of Texas at Austin, Austin, TX 78758, USA.

Kristen N. Parrish Microelectronics Research Center, Department of Electrical and Computer Engineering, The University of Texas at Austin, Austin, TX 78758, USA.

Boris Pavlov Center for Theoretical Chemistry and Physics, New Zealand Institute of Advanced Study, Massey University, Auckland 0632, New Zealand.

Jörg Pezoldt Institut für Mikro-und Nanoelektronik, Technische Universität Ilmenau, PF 100565, 98684 Ilmenau, Germany

Jorge Quereda Dpto. de Física de la Materia Condensada, Universidad Autónoma de Madrid, 28049 Madrid, Spain.

Gabino Rubio-Bollinger Condensed Matter Physics Center (IFIMAC); Dpto. de Física de la Materia Condensada, Universidad Autónoma de Madrid, 28049 Madrid, Spain.

Frank Schwierz Institut für Mikro-und Nanoelektronik, Technische Universität Ilmenau, PF 100565, 98684 Ilmenau, Germany.

Krishna Kumar Singh Department of Physics, Birla Institute of Technology and Science Pilani, Dubai Campus, Dubai 345055, United Arab Emirates.

Ashok Srivastava Division of Electrical and Computer Engineering, Louisiana State University, Baton Rouge, LA 70803, USA.

Samia Subrina Department of Electrical and Electronic Engineering, Bangladesh University of Engineering and Technology (BUET), Dhaka 1205, Bangladesh.

Sundaram Swaminathan Department of Electrical and Electronics Engineering, Birla Institute of Technology and Science Pilani, Dubai Campus, Dubai 345055, United Arab Emirates.

Li Tao Microelectronics Research Center, Department of Electrical and Computer Engineering, The University of Texas at Austin, Austin, TX 78758, USA.

H. M. Ahsan Uddin Department of Electrical and Electronic Engineering, Bangladesh University of Engineering and Technology (BUET), Dhaka 1205, Bangladesh.

Luis Vaquero-Garzon Instituto Madrileño de Estudios Avanzados en Nanociencia (IMDEA-nanociencia), Campus de Cantoblanco, E-18049 Madrid, Spain.

Saino Hanna Varghese Department of Chemistry, Catholicate College, Pathanamthitta, 689645 Kerala, India; Department of Electrical and Electronics Engineering, Birla Institute of Technology and Science Pilani, Dubai Campus, Dubai 345055, United Arab Emirates; Department of Chemical Engineering, The Petroleum Institute, Abu Dhabi 2533, United Arab Emirates.

Stefano Venica DIEGM — University of Udine via delle Scienze, 208 22100 Udine, Italy.

Adil Yafyasov Department of Solid State Electronics, St. Petersburg State University, St. Petersburg 193232, Russia.

Maruthi N. Yogeesh Microelectronics Research Center, Department of Electrical and Computer Engineering, The University of Texas at Austin, Austin, TX 78758, USA.

Thomas Zimmer CNRS and University of Bordeaux, IMS UMR 5218, Talence 33400, France.

About the Guest Editor

Frank Schwierz received the Dr.-Ing. and Dr. habil. degrees from the Technical Institute (TU) Ilmenau, Germany, in 1986 and 2003, respectively. Presently he serves as a private lecturer at TU Ilmenau and is Head of the Radio Frequency and Nano Device Research Group. His research interests include semiconductor device physics, ultra-high speed transistors, and novel device and material concepts for future transistor generations. Currently he is particularly interested in two-dimensional electronic materials, such as graphene and transition metal dichalcogenides. Dr. Schwierz is conducting research projects funded by the European Community, German government agencies, and the industry. In conjunction with German universities and industrial companies he was involved in the development of the fastest Si based transistors worldwide in the late 1990s, of Europe's smallest MOSFETs (metal-semiconductor field-effect transistors) in the early 2000s, and of the world's fastest GaN HEMT on Si this decade. His work on two-dimensional materials has been a major contribution to the current understanding of the merits and drawbacks of graphene transistors. Dr. Schwierz has published more than 260 journal and conference papers including 40 invited papers. He has published two books, Modern Microwave Transistors - Theory, Design, and Performance (J. Wiley and Sons 2003) and Nanometer CMOS (Pan Stanford Publishing 2010), and is a Senior Member of the Institute of Electrical and Electronics Engineers (IEEE). He also serves as a Distinguished Lecturer of the IEEE Electron Devices Society and as an editor of the IEEE Transactions on Electron Devices.

Preface to "Two-Dimensional Electronics - Prospects and Challenges"

In the beginning, there was graphene. For a little more than a decade now, 2D (two-dimensional) materials have represented one of the hottest trends in solid-state research. The rise of 2D materials began in 2004 when the Novoselov–Geim group from the University of Manchester and the group of Berger and de Heer from Georgia Tech published their pioneering papers on graphene, a 2D material consisting of a single layer of carbon atoms arranged in a honeycomb lattice [1–2]. Since graphene shows outstanding properties, e.g., very high carrier mobilities, excellent heat conductivity, and superior mechanical strength, this new material attracted enormous attention from various scientific communities, ranging from chemists and physicists to material scientists and electronic device engineers. An impression of the unabatedly strong interest in graphene can be obtained by counting the papers listed in the database Web of Science [3] under the search term "graphene". For 2004, one finds 183 entries compared to over 7,000 for 2010, about 29,000 for 2014, and more than 34,800 for the year 2015 up to the date of publishing, i.e., it seems that the peak of publications on graphene had not been reached in 2015.

From graphene to beyond graphene. Many of the early papers on graphene envisaged its use in electronics as most promising [2,4]. This has raised expectations that graphene could become perfect electronic material and possibly replace the conventional semiconductors. Consequently, many groups worldwide started working on graphene transistors. The first graphene MOSFET was demonstrated in 2007 by M. Lemme (one of the authors of the present book) and co-workers, at the Advanced Microelectronic Center Aachen, Germany,) [5], and integrated graphene circuits were successfully fabricated [6]. However, the prospects of graphene for electronic applications, in particular transistors, were assessed less optimistically. The main problem being that graphene does not possess a bandgap. As a consequence, graphene MOSFETs cannot be switched off and therefore are not suitable for digital CMOS (complementary MOS) which comprise around 70% of the overall chip market. Moreover, the missing gap degrades the performance of graphene transistors for RF (radio frequency) circuits, another important branch of semiconductor electronics. Yet, research on graphene transistors is still underway and the use of graphene transistors for specific applications is still on the agenda.

Possibly even more important than the work on graphene transistors, is the fact that advances in graphene research have motivated scientists to extend their work to 2D materials beyond graphene. A milestone in this direction was the demonstration of the first MOSFETs with single-layer MoS_2 channels in 2011 [9].

Meanwhile hundreds of 2D materials other than graphene have been discovered—many of them possess sizeable bandgaps and therefore are potentially useful for electronics [10-13]. This has led to intensive research on the application of 2D materials beyond graphene in semiconductor electronics.

Organization of the book. The present book is divided into two parts. Part I comprises four chapters which review the current state of the art in different areas of research on 2D materials and devices. Chapter 1 provides an excellent overview of the status of 2D material synthesis and puts special emphasis on the scalability of the discussed techniques and the attainable 2D material quality. Chapter 2 presents an overview on recent progress in the field of flexible graphene devices and describes a flexible graphene-based radio frequency receiver operating at 2.4 GHz. In Chapter 3, a thorough overview on the application of 2D materials in optoelectronics is given, and Chapter 4 comprehensively discusses gas sensors made of 2D materials.

The eight chapters of Part II deal with specific important aspects of 2D materials and devices. Chapter 5 describes theoretical investigations on the stability and electronic structure of monolayer TMD (transition metal dichalcogenide) alloys and, in Chapter 6, the thermal conductivity of graphene nanoribbons is studied by molecular dynamics simulations. Chapter 7 presents the results of experimental investigations on the visibility of exfoliated TMD and black phosphorus flakes. The following three chapters are devoted to the simulation of advanced graphene transistors: Chapter 8 deals with the effects of band-to-band tunneling and edge roughness on the behavior of graphene nanoribbon MOSFETs. In Chapter 9, recent results on the steady-state and RF operation of graphene nanoribbon transistors are presented, and in Chapter 10 a compact modeling approach for a novel graphene-based transistor type (called graphene base transistor) is elaborated. The last two chapters of the book deal with several aspects of the application of graphene in non-transistor devices: Chapter 11 presents the analysis of graphene nanomechanical mass sensors and Chapter 12 provides a discussion on graphene-based field emitters.

Outlook. Compared to conventional semiconductors such as Si and III-V compounds, 2D materials are still in their infancy and many problems have still to be solved before they can be used in commercial electronic devices and circuits. This makes it currently difficult to identify the most promising applications for 2D materials. On the other hand, within only a very few years, substantial progress has been achieved. This makes me confident that 2D materials will find their applications, in particular given the fact that electronics comprise much more than transistors. A few examples are given below.

Two dimensional materials are bendable and stretchable and therefore ideally suited for the emerging field of flexible electronics [14]. Applications for the 2D materials beyond transistors are, for example, touch screens and batteries. In November 2013, the Chinese smart phone maker, AWIT, announced the shipment of 2000 AT26 phones equipped with a graphene touch screen [15] and in March 2015, a consortium of two Chinese companies, the graphene maker Moxi and the tablet maker Galapad, announced the shipment of 30,000 Android Settler α smartphones which use graphene for the screens, batteries, and heat conduction [16]. Also Samsung, one of the big players in the smartphone business, is working intensively on the application of graphene in mobile phones.

Research on 2D materials, both graphene and beyond graphene, will remain an exciting field for many years to come. According to Kroemer's Lemma of New Technology [17], which reads *"The principal applications of any sufficiently new and innovative technology always have been—and will continue to be—applications created by that technology"*, we should be prepared to see a great many new applications for 2D materials, which at least in part are not yet envisaged today.

Acknowledgements. First of all, I would like to thank all the authors of this book for their excellent contributions. I am very grateful to Mostafa Bassiouni, the Editor-in-Chief of *Electronics*, and Xiaoyan Chen of MDPI for giving me the opportunity to edit this book, for their continuous encouragement and support, and for their patience when I missed deadlines. Furthermore, I acknowledge financial support of my own research on 2D materials from the Technical University, Ilmenau, in the frame of several Excellence Research Grants and from the German Research Foundation (Deutsche Forschungsgemeinschaft/DFG) under contract number SCHW 729/16-1.

Frank Schwierz
Guest Editor

Scalable Fabrication of 2D Semiconducting Crystals for Future Electronics

Jiantong Li and Mikael Östling

Abstract: Two-dimensional (2D) layered materials are anticipated to be promising for future electronics. However, their electronic applications are severely restricted by the availability of such materials with high quality and at a large scale. In this review, we introduce systematically versatile scalable synthesis techniques in the literature for high-crystallinity large-area 2D semiconducting materials, especially transition metal dichalcogenides, and 2D material-based advanced structures, such as 2D alloys, 2D heterostructures and 2D material devices engineered at the wafer scale. Systematic comparison among different techniques is conducted with respect to device performance. The present status and the perspective for future electronics are discussed.

Reprinted from *Electronics*. Cite as: Li, J.; Östling, M. Scalable Fabrication of 2D Semiconducting Crystals for Future Electronics. *Electronics* **2015**, *4*, 1033–1059.

1. Introduction

Two-dimensional (2D) layered materials, such as graphene, boron nitride (BN) and transition metal dichalcogenides (TMDCs), have attracted tremendous interest in extensive research fields [1–3], due to their general excellent electronic, optical and mechanical properties. It is widely accepted that 2D materials will play an important role in the applications of next-generation nanoelectronics [4–6], optoelectronics [7,8], emerging (flexible, organic, printed and stretchable) electronics [9], energy conversion and storage [10], sensing [11] and medicine and biology [12]. As for the application in the upcoming high-end nanoelectronics, 2D materials possess unique electronic performance as compared with their bulk form (conventional three-dimensional (3D) materials) [13] and easy manipulation for complex structures as compared with the one-dimensional materials [14]. Recently, 2D semiconducting crystals, such as TMDCs and black phosphorous [15], have received increasing attention in electronic applications. At present, the majority of 2D semiconducting crystals lie in the family of TMDCs with the formula MX_2, where M is a transition metal element from Group IV (e.g., Ti, Zr or Hf), Group V (e.g., V, Nb or Ta) or Group VI (e.g., Mo or W), and X is a chalcogen (S, Se or Te) [2]. With a simple mechanical cleavage technique, monolayer (sub-nm thick) MoS_2 has been fabricated and used to demonstrate high-performance field-effect transistors (FETs) with room-temperature mobility ~200 cm^2 V^{-1} s^{-1} and a current on/off ratio ~10^8 [14], verifying the promise

1

of 2D materials for nanoelectronics. However, in order to realize the transfer from lab-scale fabrication to industrial-scale manufacturing for future electronics, scalable synthesis techniques of 2D semiconducting crystals are indispensable.

A variety of scalable synthesis techniques for 2D TMDCs have been developed in the recent literature. In this review, we systematically introduce those techniques that produce highly-crystalline large-area 2D crystals and that have great potential for manufacturing in future nanoelectronics. There are also some other scalable synthesis techniques for nanosheets of 2D materials, such as solvent exfoliation [16] and inkjet printing [17,18]. Despite their promising applications in other fields, e.g., printed electronics and energy storage, they may not be suitable for high-end nanoelectronics and are not included in this review. Section 2 introduces scalable synthesis techniques for large-area (typically wafer-scale) monolayer or few-layer TMDCs. Section 3 introduces the advanced engineering of 2D TMDCs, including 2D TMDC alloys, 2D heterostructures and wafer-scale fabrication of 2D TMDC devices. In Section 4, we benchmark the performance of TMDC FETs based on different scalable synthesis techniques against those based on mechanical cleavage, and offer an outlook for the research tendency in the near future.

2. Scalable Synthesis Techniques for TMDCs

Here, we briefly introduce the recently-developed synthesis techniques for wafer-scale TMDCs, including vapor phase deposition, thermal decomposition, magnetron sputtering and molecular beam epitaxial.

2.1. Vapor Phase Deposition

Currently, the predominant and most promising scalable synthesis technique for large-area atomically-thin TMDCs is vapor phase deposition (VPD). In general, VPD relies on the chemical reaction or physical transport (often with inert gas as the carrier) of vaporized precursors to deposit TMDCs onto the substrate surface. In terms of the initial precursor state, VPD can be classified into solid-precursor VPD and gas-precursor VPD.

2.1.1. Solid-Precursor VPD

So far, most of the VPD techniques are based on solid-state precursors. Usually, transition metal, metal oxides or chlorides, such as MoO_3, $MoCl_5$ and WO_3, act as the metal precursors, while S/Se powders as the chalcogen precursors [19]. Solid-precursor VPD has been widely employed to synthesize not only pure 2D TMDC crystals, but also TMDC alloys and heterostructures, to be discussed in Section 3. As illustrated in Figure 1, three classes of solid-precursor VPD techniques have been demonstrated in the literature [20]: (1) simple chalcogenization of predefined metal (or metal oxide) film (Figure 1a), where metal or metal oxide

2

thin films are first deposited onto the substrates, then chalcogen powers are heated to be vapors and transported to the substrates and, finally, the metal/metal oxide films are annealed at the atmosphere of chalcogen vapors to produce TMDCs; (2) chemical vapor deposition (CVD) based on the reaction between metal precursors and chalcogen precursors (Figure 1b), where both solid metal precursors and chalcogen powders are heated separately into the vapor phase, and are transported to and react on the substrate surface to form 2D TMDC crystals; and (3) vapor-solid growth based on vapor phase transport and recrystallization of TMDCs (Figure 1c), where TMDC powders serve as the precursors which are vaporized and transported to the cool substrate region and recrystallize to 2D crystals.

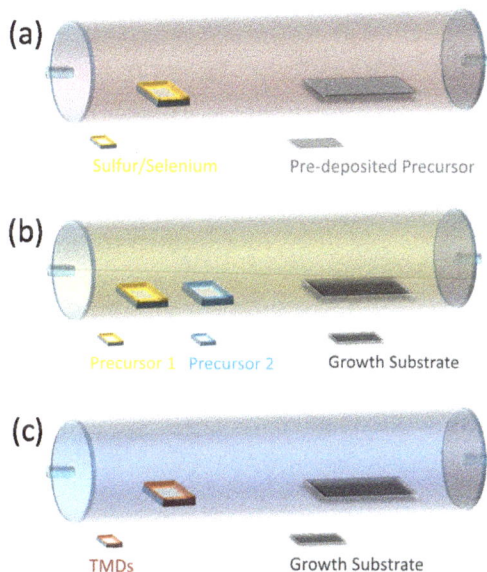

Figure 1. Schematic of three types of solid-precursor vapor phase deposition techniques for scalable synthesis of large-area transition metal dichalcogenides (TMDCs). (**a**) Vapor phase chalcogenization of pre-deposited precursor (metal, metal oxide, *etc.*) film; (**b**) chemical vapor deposition; (**c**) physical vapor deposition. Adapted with permission from [20]. Copyright 2015, The Royal Society of Chemistry.

The principle and methodology of the solid-precursor VPD techniques have been elaborated in a recent relevant review by Shi *et al.* [20]. In general, TMDCs synthesized through chalcogenization of predefined metal/metal oxide film (Class (1)) give rise to low carrier mobility (<0.1 cm^2 V^{-1} s^{-1}) [20,21]. In contrast, the chemical reaction-based synthesis (Class (2)) is more favorable because of the technique's simplicity, preference for monolayer growth and typically high mobility over

10 cm^2 V^{-1} s^{-1} [20,22]. For all of the techniques, a critical challenge is the limited spatial uniformity. In most cases, spatially inhomogeneous mixtures of monolayer, multi-layer and no-growth regions are obtained [23]. In this section, we focus on the introduction of recent innovative strategies that lead to higher uniformity and/or better controllability.

Pre-treated substrates: Most studies on VPD synthesis use SiO$_2$ as the growth substrates. However, because of the amorphous nature of the SiO$_2$ substrate and its relatively high surface roughness, the obtained TMDCs usually suffer from high-density grain boundaries and random orientation among domains. This inevitably generates severe non-uniformity and hinders the scalable growth of large-area high-quality 2D TMDCs. One effective solution is to use an atomically-smooth crystalline substrate [22] to control the crystallographic orientation of TMDC domains during growth and to attain a uniform layer with reduced grain boundary density.

Sapphire is a suitable substrate for CVD growth of TMDCs [24,25]. In particular, Dumcenco *et al.* [24] have recently achieved good control over lattice orientation (Figure 2) during CVD growth (based on the reaction between vaporized MoO$_3$ and sulfur) and obtained high-quality monolayer MoS$_2$ with centimeter-scale uniformity (Figure 2a,b). The key to the lattice orientation is the introduction of terraces on the sapphire surface (Figure 2e) via annealing in air at an elevated temperature of 1000 °C prior to the growth process. The CVD growth produces well-defined equilateral triangular single-crystal domains (Figure 2d) that merge into a continuous monolayer film with a typical coverage area of 6 mm × 1 cm in the center of the sapphire substrate (Figure 2b). In addition, over 90% of domains (Figure 2c,f) are well aligned with the relative edge orientation at multiples of 60°. A striking merit of the technique is that the relatively weak van der Waals interaction between sapphire and MoS$_2$, on the one hand, effectively induces lattice alignment, while on the other hand, allows easy transfer of the grown MoS$_2$ from the sapphire substrate to silicon wafers, which facilitates the fabrication of high-performance FETs. As a result, a high mobility of about 43 cm^2 V^{-1} s^{-1} has been attained for devices based on single grains of the grown MoS$_2$. Even for devices based on the large-area MoS$_2$ films, which contain grain boundaries, the mobility still retains about 25 cm^2 V^{-1} s^{-1}. In particular, there is no evident mobility degradation when the channel length increasing from 4 μm to ~80 μm.

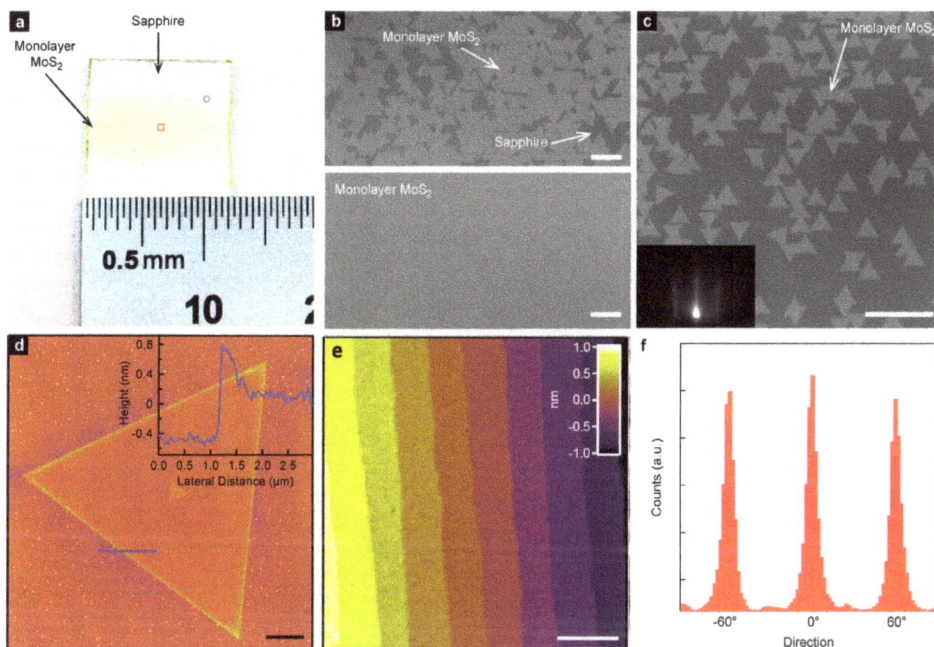

Figure 2. Monolayer MoS$_2$ growth on sapphire with controlled lattice orientation. (a) Photograph of centimeter-scale MoS$_2$ film on sapphire; (b) optical microscope images of the sample in (a) at the edge region (top, the circle region in (a), scale bar: 20 μm) and the center region (bottom, the square region in (a), scale bar: 10 μm); (c) optical microscope images of the grown monolayer MoS$_2$ grains, scale bar: 50 μm; the inset is a reflection a high-energy electron diffraction (RHEED) pattern showing a long-range structure order; (d) atomic force microscope (AFM) image of a monolayer MoS$_2$ grain (inset: thickness profile along the blue line; scale bar: 2 μm); (e) AFM image of annealed sapphire as the growth substrates, where atomically-smooth terraces occurs on the surface (scale bar: 100 nm); (f) the orientation histogram based on the area shown in (c) confirms good control over lattice orientation. Adapted with permission from [24]. Copyright 2015, American Chemical Society.

In addition to sapphire substrates, Najmaei *et al.* [21] developed a CVD process to grow centimeter-scale MoS$_2$ film directly on patterned SiO$_2$/Si wafers. Based on the observation that MoS$_2$ triangular domains and films are commonly nucleated and formed in the vicinity of substrates' edges, scratches, dust particles or rough surfaces, they used conventional lithography to strategically create step edges by patterning SiO$_2$/Si substrates with a uniform distribution of square SiO$_2$ pillars, as shown in Figure 3a. The pillars facilitate a high density of domain nucleation, and the continued growth allows the formation of large-area continuous film. The MoS$_2$ films, grown on both the pillar surface and the valley space in between (Figure 3b,c), are predominantly single layered (multiple layers typically accumulate at the preferred nucleation sites). The synthesized MoS$_2$ films can be readily transferred to other substrates (Figure 3d) or directly applied to fabricate devices. FETs based on these MoS$_2$ films can exhibit average mobility of about 4.8 cm^2 V^{-1} s^{-1} and a maximum on/off current ratio approaching 6 × 10^6.

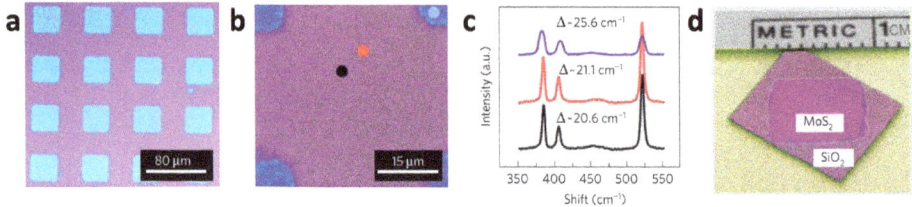

Figure 3. Controlled nucleation for CVD growth of large-area MoS$_2$ growth films on patterned SiO$_2$/Si substrates. (**a**) Optical image of a large-area continuous MoS$_2$ film on substrates with patterned square SiO$_2$ pillars; (**b**) the close-up view indicating monolayer and bilayer MoS$_2$ films in between pillars and a thicker MoS$_2$ film on top of pillars; (**c**) Raman spectra acquired at different regions in (**b**) confirming the thickness of the sample; (**d**) centimeter-scale MoS$_2$ film transferred to a new substrate. Adapted with permission from [21]. Copyright 2013, Nature Publishing Group.

Seeding promoters: In order to improve the uniformity of CVD-grown TMDC films directly on SiO$_2$/Si substrates, a simple and effective strategy is to use seeding promoters (Figure 4) [22,26]. In 2012, Lee *et al.* [26] obtained uniform, large-area monolayer MoS$_2$ by spinning aqueous solutions of graphene-like molecules, such as reduced graphene oxides (rGO), perylene-3,4,9,10-tetra-carboxylic acid tetra-potassium salt (PTAS) and perylene-3,4,9,10-tetracarboxylic dianhydride (PTCDA), as the seeding promoters onto the SiO$_2$/Si substrates prior to growth. Later on, Ling *et al.* [22] identified more aromatic molecules, such as copper phthalocyanine (CuPc) and bathocuproine (BCP), as effective promoters. In particular, in contrast to the previous seeding promoters deposited via aqueous solution [26], the newly-identified seeding promoters can be deposited by thermal evaporation to various substrates, including hydrophobic substrates, to allow direct growth of a variety of hybrid structures (e.g., MoS$_2$/Au, MoS$_2$/BN and MoS$_2$/graphene). Besides, PTAS has also been used as seeding promoters for scalable synthesis of WS$_2$ [27]. However, possibly because of the presence of grain boundaries, back-gate FETs based on such TMDCs only attain a carrier mobility up to 1.2 cm^2 V^{-1} s^{-1} for MoS$_2$ and 0.01 cm^2 V^{-1} s^{-1} for WS$_2$, in spite of a high on/off ratio above 10^5 [27].

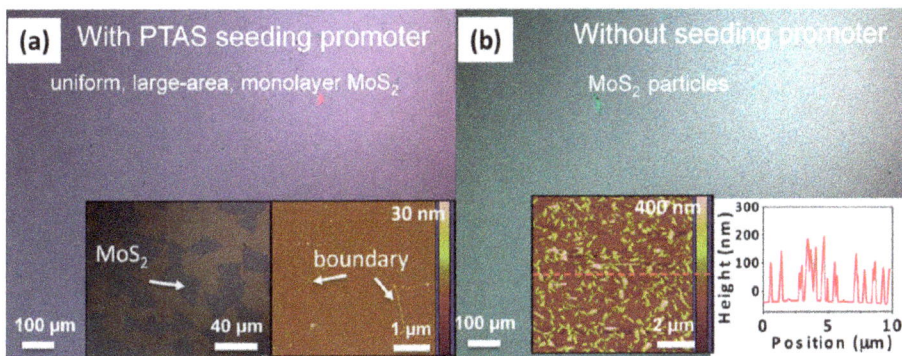

Figure 4. The effects of seeding promoter on the CVD growth of MoS$_2$ films. (**a**) Optical image of MoS$_2$ film grown on SiO$_2$/Si substrate with perylene-3,4,9,10-tetra-carboxylic acid tetra-potassium salt (PTAS) seeding promoter; (**b**) optical image of MoS$_2$ film grown on bare SiO$_2$/Si substrate without seeding promoter. The insets show a close-up view through optical or AFM images. Adapted with permission from [22], Copyright 2014, American Chemical Society.

Atomic layer-deposited precursors: In order to obtain layer-controlled wafer-scale WS$_2$ atomic layers on SiO$_2$/Si substrates, Song *et al.* [28] introduced an innovative synthesis technique through sulfurizing the WO$_3$ film prepared by atomic layer deposition (ALD). The precursor WO$_3$ film is first deposited on SiO$_2$/Si substrates by the ALD process and then sulfurized through the VDP technique (Figure 5a). Benefiting from the excellent controllability of the ALD process, the synthesis of WS$_2$ films exhibits systematic layer controllability, good reproducibility, wafer-level thickness uniformity and high conformity. In particular, the number of final WS$_2$ layers can be well controlled by controlling the number of the ALD cycles during WO$_3$ deposition. With 20, 30 and 50 ALD cycles, mono-, bi- and tetra-layer WS$_2$ are obtained, respectively (Figure 5b–g). Wafer-scale (2 cm × 13 cm) thickness uniformity has been demonstrated for all of the mono-, bi- and tetra-layer WS$_2$ films (Figure 5h,i). Top-gate FETs based on the WS$_2$ films exhibit high mobility of about 3.9 cm^2 V^{-1} s^{-1} with an on/off current ratio around 10^3.

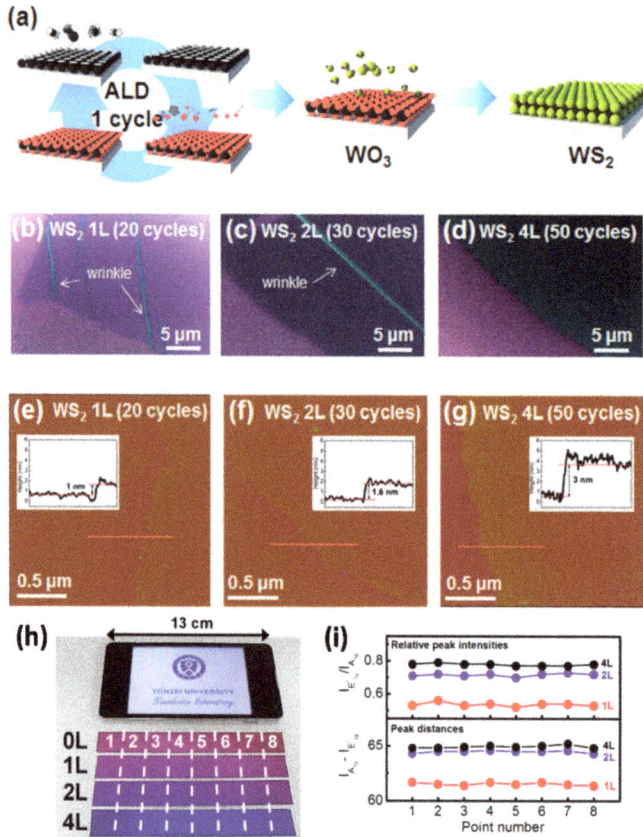

Figure 5. Atomic layer deposition (ALD)-based layer-controlled CVD growth of WS$_2$ films with precursors. (**a**) Schematic of the synthesis procedure; (**b–d**) optical microscope images and (**e–g**) AFM images and height profile (insets) of the mono-, bi- and tetra-layer WS$_2$ films transferred onto SiO$_2$ substrates, respectively; (**h**) wafer-sale (about 13 cm, comparable to the size of a cellular phone display screen) mono-, bi- and tetra-layer WS$_2$ films on SiO$_2$ substrates; (**i**) Raman peak intensity ratio (top) and peak distance (bottom) of the E_{2g}^1 and A_{1g} bands for the WS$_2$ films at eight different measurement positions in (h) indicating the wafer-scale uniformity. Adapted with permission from [28]. Copyright 2013, American Chemical Society.

Oxygen plasma treatment: Besides the layer-controlled synthesis by ALD-deposited precursors, Jeon *et al.* [29] have recently developed another facile layer-controlled CVD growth technique for wafer-scale MoS$_2$ film on silicon substrates. Prior to the growth, the SiO$_2$/Si substrates are treated with oxygen plasma. With the low pressure CVD process, uniform MoS$_2$ film can be obtained in a large area

(Figure 6a). In contrast, on untreated SiO_2/Si substrates, only small-scale (<100 nm) triangular MoS_2 nanoparticles are synthesized. Besides, by changing the duration of the oxygen plasma treatments, different MoS_2 layers can be adjusted in a controllable manner. With treatment durations of 90 s, 120 s and 300 s, mono-, bi- and tri-layer MoS_2 films have been obtained, respectively (Figure 6b–g). All back-gate FETs based on these MoS_2 show a high on/off current ratio between 10^5 and 10^6. The carrier mobility increases with the layer number, which is 3.6 cm^2 V^{-1} s^{-1}, 8.2 cm^2 V^{-1} s^{-1} and 15.6 cm^2 V^{-1} s^{-1} for the mono-, bi- and tri-layer MoS_2 transistors, respectively. However, one may notice that the residual oxygen caused by the plasma treatment may impact the quality of the grown MoS_2 [30]. Further cleaning of the plasma-treated SiO_2/Si wafers (e.g., through purging with Ar at high temperature [29]) might be necessary to obtain high crystallinity.

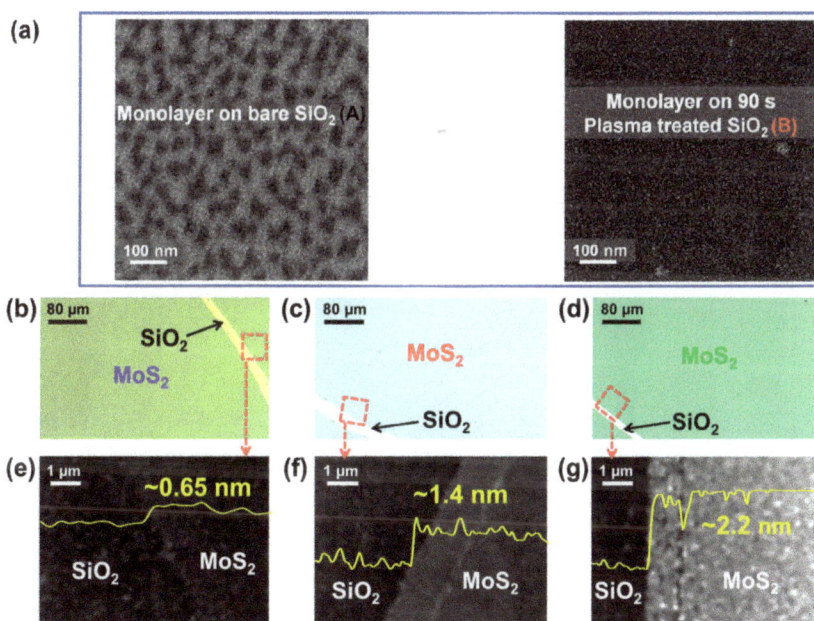

Figure 6. Layer-controlled CVD growth of MoS_2 film on plasma-treated SiO_2 substrates. (**a**) Scan electron microscope (SEM) images of the grown MoS_2 film on bare (left) and plasma-treated (right) SiO_2 substrates; (**b**–**d**) optical microscope images and (**e**–**g**) AFM images for the uniform mono-, bi- and tri-layer MoS_2 films, respectively. Adapted with permission from [29]. Copyright 2015, The Royal Society of Chemistry.

Self-limiting CVD: Yu *et al.* [31] developed a simple self-limiting CVD process to grow centimeter-scale MoS_2 films with precisely-controlled layer number ranging from 1–4. Highly uniform MoS_2 films (Figure 7) are synthesized on various substrates, including silicon oxide, sapphire and graphite. The MoS_2 films are grown at a high temperature (>800 °C) with $MoCl_5$ and sulfur as the precursors. At high temperature, precursors react to produce MoS_2 species, which then precipitate onto the substrates to form MoS_2 films. The layer number of the MoS_2 films is precisely controlled by the amount of $MoCl_5$ used in the source or the total pressure during the growth process. The greater the amount of $MoCl_5$ or the higher the total pressure used, the larger the layer number of MoS_2 films obtained. The obtained large-area, highly-uniform MoS_2 thin films suggest that the growth is a self-limiting process, *i.e.*, the growth automatically stops once the formation of each individual layer finishes. The self-limiting mechanism may lie in a thermodynamic balance between the partial pressure of gaseous MoS_2 species (P_1) and the vapor pressure of MoS_2 thin films on the substrate (P_2). P_2 could increase with the layer number. The initial force of $P_1 > P_2$ drives the formation of monolayer MoS_2. When the growth of the monolayer finishes, P_2 increases, and the driving force $P_1 > P_2$ may vanish, so that the growth ceases. To grow more layers, higher P_1 should be introduced. Therefore, the layer number can be well controlled by the precursor amount and environmental pressure. However, for thicker MoS_2 films, precise control becomes more difficult. Back-gate FETs based on monolayer MoS_2 films exhibit carrier mobility ranging from 0.003–0.03 cm^2 V^{-1} s^{-1}.

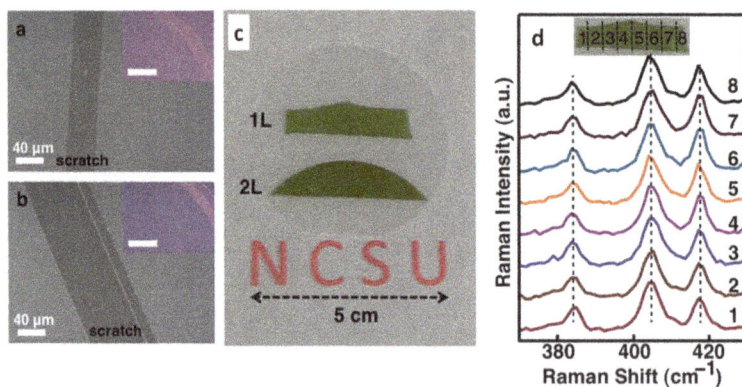

Figure 7. Controlled synthesis of monolayer and bilayer MoS_2 films by self-limiting CVD. Optical microscope images of highly uniform (**a**) monolayer and (**b**) bilayer MoS_2 films (scale bars in insets: 80 μm); (**c**) photograph of as-grown large-area monolayer and bilayer MoS_2 films; (**d**) Raman spectra acquired at different locations indicating the homogeneity. Adapted with permission from [31]. Copyright 2013, Nature Publishing Group.

Immediate-state precursor: Usually, CVD synthesis for MoS_2 is based on MoO_3 as the initial metal precursor. During growth, however, the conversion from MoO_3 to MoS_2 involves one intermediate step. MoO_3 is first partially reduced by the sulfur to a volatile state MoO_{3-x}, and then, MoO_{3-x} is further reduced to MoS_2. The incomplete reaction (partial reduction) may produce an uncontrollable intermediate phase, which provides extra binding sites to chemisorb oxygen. The absorbed oxygen can generate defects/vacancies in the final MoS_2 flakes and significantly degrade their electrical performance. To diminish the impacts arising from the intermediate chemistry, Bilgin *et al.* [30] developed a CVD process with MoO_2 powders as the metal precursor and directly sulfurized MoO_2 in the vapor phase. With this technique, monolayer, as well as few-layer MoS_2 with an edge-size (Figure 8) ranging from 10–50 µm have been grown in various substrates, including amorphous SiO_2, crystalline Si wafer, transparent quartz and silicon nitride, as well as conductive graphene. Through transferring the as-grown monolayer MoS_2 onto SiO_2/Si substrates and building up back-gate device structures, the obtained FETs exhibit excellent performance of high mobility of about 35 cm^2 V^{-1} s^{-1}, and a large on/off current ratio about 10^8.

2.1.2. Gas-Precursor VPD

In addition to the solid-precursor VPD, conventionally with sulfur (or selenium) powder as the precursor, recent studies have also employed gas state precursors to improve the controllability during the course of VPD growth.

Cycle-based epitaxy on gold: Large-area epitaxial monolayer MoS_2 films have been recently grown on a Au(111) surface [32] under ultrahigh vacuum with H_2S gas and e-beam evaporated Mo as the precursors (Figure 9a). The synthesis is based on growth cycles, each of which consists of two sequent steps, Mo evaporation and sulfurization by H_2S gases. Upon the finishing of one growth cycle, the H_2S gases are pumped out, and the chamber is prepared for a second growth cycle. The first growth cycle is to nucleate MoS_2 nano-islands (Figure 9c), and the following cycles gradually increase the island size (Figure 9d,e). The growth cycles are repeated until a continuous film forms with coverage approaching unity. Without such a cycle-based synthetic process, it is challenging to obtain high-coverage uniform MoS_2, since a single long Mo deposition in a H_2S atmosphere only produces low quality sub-monolayer films.

11

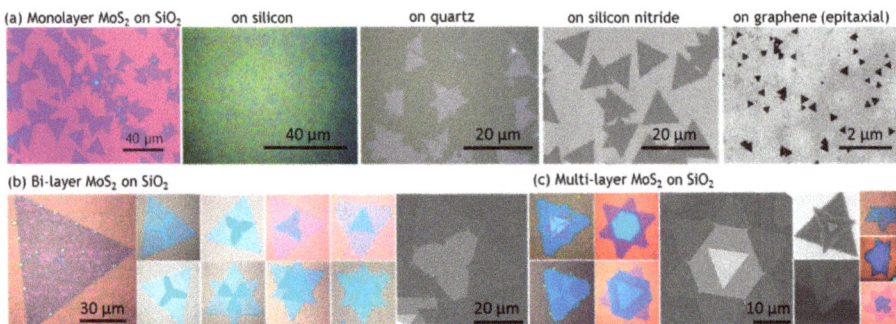

Figure 8. Optical (colored) and SEM (grey) images of (**a**) monolayer, (**b**) bilayer and (**c**) multi-layer MoS_2 films on various substrates grown by CVD with MoO_2 as the metal precursor. Reprinted with permission from [30]. Copyright 2015, American Chemical Society.

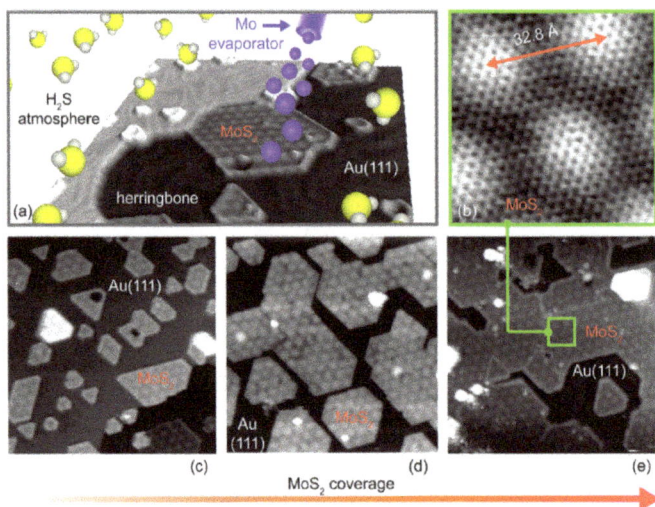

Figure 9. (**a**) Schematic of cycle-based epitaxy of MoS_2 on gold; (**b**) high resolution scan tunneling microscope (STM) image of the moiré lattice of MoS_2; (**c–e**) STM images acquired at different growth times. Reprinted with permission from [32]. Copyright 2015, American Chemical Society.

All-gas precursor CVD: Kang *et al.* [23] have developed a metal-organic chemical vapor deposition (MOCVD) technique to grow monolayer MoS_2 and WS_2 films with safer-scale homogeneity. Remarkably, only gas-phase precursors (Figure 10a) are used in the technique, including $Mo(CO)_6$ or $W(CO)_6$ as the metal precursors, $(C_2H_5)S$ as the sulfide precursors and H_2 to remove carbonaceous species generated during the MOCVD growth. All of the gases are diluted in the carrier gas, argon, so that their concentration can be precisely controlled throughout the growth period by regulating the corresponding partial pressure. In this way, the MoS$_2$ film is grown layer by layer, which is ideal for uniform layer control over a large scale. Figure 10b reveals the controllable growth process at different times: initial nucleation on the SiO$_2$ surface ($t = 0.5t_0$), monolayer growth ($0.8t_0$), maximum monolayer coverage (t_0), secondary nucleation mainly at grain boundaries ($1.2t_0$) and bilayer growth ($2t_0$). In addition, the average grain size and inter-grain connection can be well controlled by the concentrations of H$_2$, (C$_2$H$_5$)$_2$S and residual water (Figure 10c). The wafer-scale growth of a high-quality monolayer MoS$_2$ film allows mass production of 8100 FETs (Figure 10d) on a four-inch SiO$_2$/Si wafer with a yield as high as 99% and excellent electrical performance of the on/off current ratio of ~10^6 and mobility of ~29 cm^2 V^{-1} s^{-1}. Similar devices based on a CVD-grown WS$_2$ monolayer also exhibit a high on/off current ratio of ~10^6 and mobility of ~18 cm^2 V^{-1} s^{-1}.

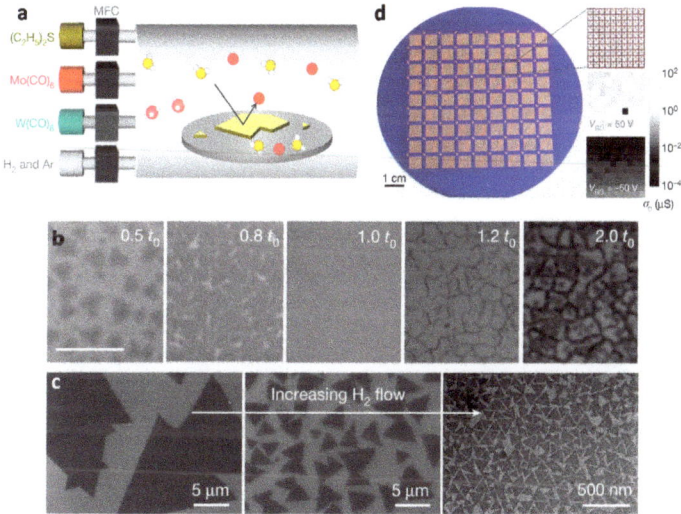

Figure 10. All-gas precursor metal-organic chemical vapor deposition (MOCVD) growth of continuous large-area MoS$_2$ film and device fabrication. (**a**) Diagram of the MOCVD setup; (**b**) optical images of the grown MoS$_2$ films at different growth times, where t_0 is the optimal growth time for full monolayer coverage (scale bar: 10 μm); (**c**) the effects on the H$_2$ flow rate on the grown MoS$_2$ grains; (**d**) mass production of 8100 MoS$_2$ FETs on a four-inch SiO$_2$/Si wafer. Adapted with permission from [23]. Copyright 2015, Nature Publishing Group.

2.2. Thermal Decomposition

In addition to the VPD techniques, other techniques have also been developed for scalable synthesis of 2D TMDC materials.

In 2011, Liu *et al.* [33] developed a thermal decomposition technique to produce large-area MoS_2 thin layers on insulating substrates (Figure 11). First, a thin and uniform film of the precursor, ammonium thiomolybdates, $(NH_4)_2MoS_4$, is coated onto insulating substrates (sapphire or SiO_2/Si wafer) through dip coating. Then, the precursor is annealed and converts to MoS_2. The annealing process comprises a first low-pressure (1 Torr) annealing at 500 °C in the presence of H_2 for the decomposition of $(NH_4)_2MoS_4$ to MoS_2 and a second high-pressure (500 Torr) annealing at 1000 °C in an atmosphere of Ar (or a mixture of Ar and sulfur) to improve crystallinity and/or increase the grain domain size. Typically, uniform and continuous bi- or tri-layer MoS_2 films are obtained throughout the substrates (Figure 11). Bottom-gate FETs based on the synthesized MoS_2 annealed under the atmosphere of Ar and S exhibit a current ratio up to 1.6×10^5 and a mobility up to 4.7 cm^2 V^{-1} s^{-1}.

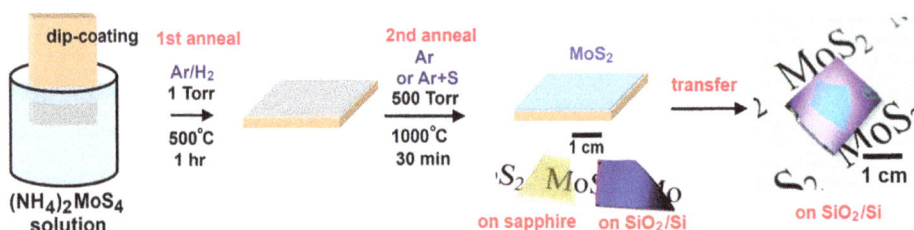

Figure 11. Schematic of the thermal decomposition process for the growth of MoS_2 films on insulating substrates. Adapted with permission from [33]. Copyright 2012, American Chemical Society.

2.3. Magnetron Sputtering

Tao *et al.* [34] recently developed a one-step magnetron sputtering technique to synthesize wafer-scale MoS_2 atomic layers on various substrates. As illustrated in Figure 12b, the growth is based on the reaction between vaporized S and sputtered Mo. Sulfur is first vaporized and then transported to the chamber, where the metal target is sputtered in an Ar atmosphere by a low power. The sputtered Mo is more reactive than Mo or MoO_3 in CVD processes and can easily react with the vaporized S. After the reaction, the product is deposited onto the hot substrate to form MoS_2 layers. Because of the low sputtering power, the growth rate is very low. Therefore, mono- or few-layer MoS_2 can be obtained in a controllable manner by adjusting the sputtering power or deposition time. Since magnetron sputtering is compatible with large-scale production, highly-homogeneous MoS_2 flakes with dimensions up to centimeter scale (Figure 12a) are obtained with a controllable layer number of 1, 2, 3 and more. The back-gate FETs based on the grown MoS_2 exhibit a p-type behavior with an on/off current ratio of about 10^3 and an average mobility of about 7 cm^2 V^{-1} s^{-1}.

2.4. Molecular Beam Epitaxy

Recently, Yue *et al.* [35] developed a molecular beam epitaxy (MBE) to grow one 2D TMDC (HfSe$_2$) on another 2D material (highly-ordered pyrolytic graphite (HOPG) or MoS$_2$). Although MBE was introduced for hetero-epitaxial growth of TMDC (e.g., NbSe$_2$) three decades ago, few studies have demonstrated the MBE growth of 2D HfSe$_2$ on another 2D material to fabricate an all-2D heterostructure. Hf-based TMDCs have a small band gap, a large work function and reasonable mobility and, hence, are promising for a variety of applications in nanoelectronics and optoelectronics. In Yue's processing, the HfSe$_2$ films are grown on mechanically-exfoliated HOPG and MoS$_2$ through a VG-Semicon V80H molecular beam epitaxy system, where the growth chamber is equipped with a vertical e-beam evaporator to allow the growth of high melting-point metals (Hf, Ti, Mo, W, *etc.*) in addition to effusion cell evaporation of the chalcogen. Prior to growth, the substrates are cleaned by heating for 2 h in the growth chamber, and the Hf and S sources are cleaned by outgassing. During the growth, the Se and Hf shutters were opened and closed simultaneously, and the Se:Hf flux was maintained at a ratio of 5:1. As shown in Figure 13, the interface is atomically sharp, and there are no detectable misfit dislocations or strains between the MBE-grown HfSe$_2$ and the HOPG or MoS$_2$ underneath, in spite of the large lattice mismatch (41% and 17%, respectively). The grains are typically larger than 100 nm × 100 nm, and the layer number depends on the growth time. The grown HfSe$_2$ is slightly n-type with a band gap of about 1.1 eV.

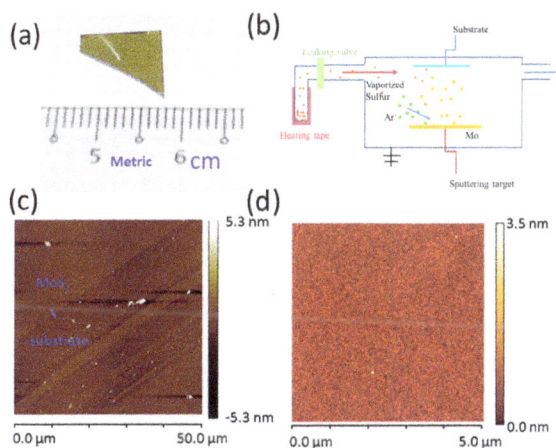

Figure 12. Large-area uniform MoS$_2$ film grown on sapphire by magnetron sputtering. (**a**) Photograph of a centimeter-scale film; (**b**) schematic of the magnetron sputtering growth process; (**c**) AFM image of as-grown MoS$_2$ film showing a clear film edge; (**d**) a close-up view of the AFM image showing good uniformity. Adapted with permission from [34]. Copyright 2015, The Royal Society of Chemistry.

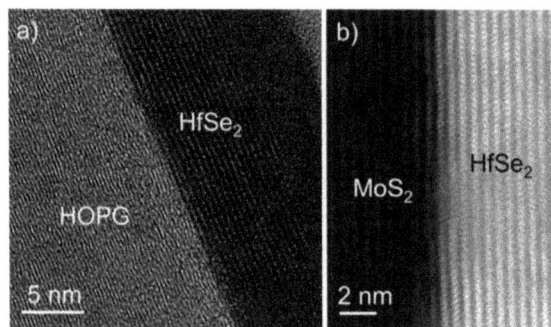

Figure 13. Transmission electron microscope (TEM) images of epitaxial HfSe$_2$ on (a) highly-ordered pyrolytic graphite (HOPG) and (b) MoS$_2$ with a sharp interface and layered crystallinity. Reprinted with permission from [35]. Copyright 2015, American Chemical Society.

3. Advanced Engineering for Materials, Heterostructures and Devices

As the scalable synthesis of 2D TMDC becomes mature, more and more techniques have been developed to allow advanced engineering in TMDC-based electronics at different levels, from synthesis of the TDMC alloys (material level), to fabrication of various 2D heterostructures, to controllable and batch production of electronic devices at the wafer scale.

3.1. TMDC Alloys

Both theoretical calculation and experiments have demonstrated that the band gap of 2D TMDC alloys can be modulated in a wide spectrum range by changing the composition [36]. This is a desired property for many applications in electronic and optoelectronics and has stimulated great efforts in controllable fabrication of various 2D TMDC alloys. So far, two types of alloys have been synthesized through scalable VDP techniques. One is a mixture of chalcogens, e.g., MoS$_{2(1-x)}$Se$_{2x}$ alloys, and the other is a mixture of metals, e.g., Mo$_{1-x}$W$_x$S$_2$ alloys.

3.1.1. MoS$_{2(1-x)}$Se$_{2x}$ Alloys

Two CVD processes have been developed for scalable synthesis of MoS$_{2(1-x)}$Se$_{2x}$ alloys. One [37,38] uses MoSe$_2$ and MoS$_2$ powders as the precursors, which are placed separately, as illustrated in Figure 14a. The precursors are directly vaporized at a high temperature (940–975 °C), while the SiO$_2$/Si substrate is put in a relatively low temperature (600–700 °C) region. In order to optimize the supersaturation for the nucleation process, a moderately high temperature gradient of about 50 °C/cm in the substrate (deposition) region is applied [37,38]. Too low of a temperature gradient (<30 °C/cm) may suffer from insufficient supersaturation and nucleation for the alloy

16

growth [37], whereas too high of a temperature gradient (>80 °C/cm) may cause surplus nucleation, which greatly reduces the domain size of the monolayer alloy [38] or even changes the morphology to a fiber-like structure [37]. By finely adjusting the evaporation temperature, the temperature gradient and the H_2 flow rate in the carrier gas, high-quality centimeter-scale $MoS_{2(1-x)}Se_{2x}$ monolayer alloy can grow on the substrate (Figure 14b). By fixing the evaporation temperature of MoS_2 at 940 °C and increasing the evaporation temperature of $MoSe_2$ from 940–975 °C, the composition in the $MoS_{2(1-x)}Se_{2x}$ alloy can be finely controlled with x ranging from 0–0.4. However, at an even higher evaporation temperature (>975 °C), $MoSe_2$ tends to decompose, and it becomes hard to further increase x in the alloy. Nevertheless, if extra Se vapor is introduced into the upstream of the furnace, more Se-rich alloy with $0.4 < x < 1$ can be obtained [38]. Atomic-resolution high-angle annular dark filed (HAADF) scanning transmission electron microscopy (STEM) imaging (Figure 14c,d) shows the distribution of S and Se atoms in the monolayer alloys. Raman and photoluminescence spectra (Figure 14e) also indicate the shift of characteristic peaks induced by composition change.

The other process [36,39] uses mixed S and Se fine powders as the chalcogen source and MoO_3 as the metal source (Figure 14f). By changing the ratio of the S and Se powders in the source, the fraction in the alloy can be well controlled with x ranging from 0–0.75. FETs based on the synthesized $MoS_{2(1-x)}Se_{2x}$ alloy exhibit a high on/off current ratio of about 10^6 and a mobility up to 15.3 cm^2 V^{-1} s^{-1} [39].

3.1.2. $Mo_{1-x}W_xS_2$ Alloys

Song *et al.* [40] recently reported a synthesis method of $Mo_{1-x}W_xS_2$ alloys by sulfurization of super-cycle ALD $Mo_{1-x}W_xO_y$ alloy thin films. First, 10 cycles of WO_3 were pre-deposited through ALD to address the nucleation delay issues of ALD-grown WO_3. Then, one super-cycle ALD process (Figure 15a) comprising n cycles of ALD MoO_x and m cycles of ALD-grown WO_3 is performed to grow $Mo_{1-x}W_xO_y$ alloy thin films, followed by sulfurization to obtain 2D $Mo_{1-x}W_xS_2$ alloys. By adjusting the n/m ratio, the fraction of W in the final alloys, x, can be well controlled. Typically, one super-cycle gives rise to uniform monolayer $Mo_{1-x}W_xS_2$ alloys with a thickness of ~1 nm (Figure 15b), and two and three super-cycles produce bi- or tri-layer 2D alloys, respectively. In other words, the alloy composition and layer number can be well controlled during the ALD-based synthesis process. Consequently, the band gaps of the alloys can also be precisely controlled. Furthermore, vertically-composition-controlled (VCC) $Mo_{1-x}W_xS_2$ multilayer alloys (Figure 15c) have also been synthesized by using a sequential super-cycle ALD process, which consists of five continuous super-cycles with different n/m ratios for each super-cycle. The VCC $Mo_{1-x}W_xS_2$ multilayer alloys exhibit a broadband light absorption property and are very promising for efficient photodetector applications. In particular, the ALD-based super-cycle process is anticipated to apply also to other 2D TMDC alloys than $Mo_{1-x}W_xS_2$.

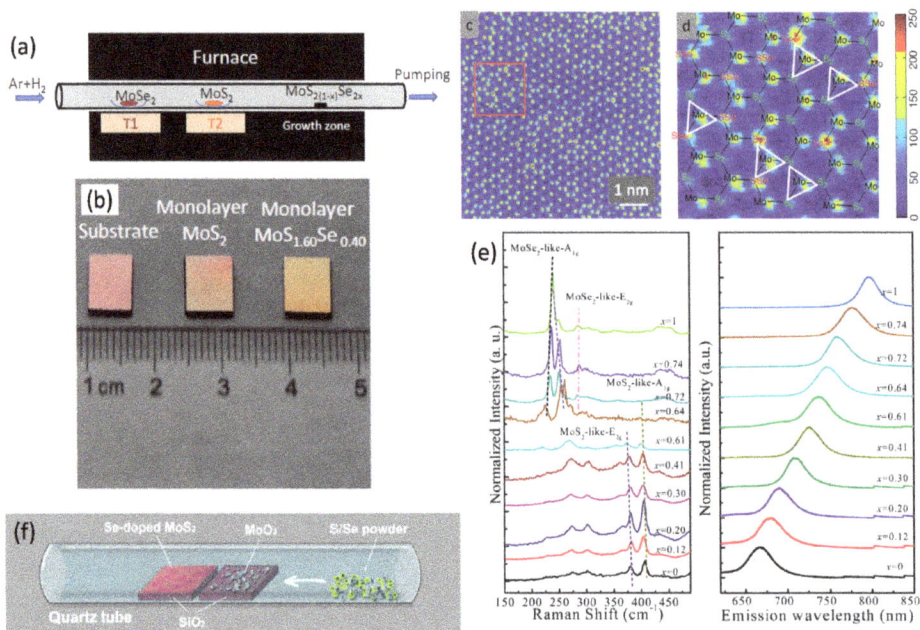

Figure 14. Two CVD processes for the synthesis of $MoS_{2(1-x)}Se_{2x}$ alloys. (**a**) Schematic of CVD with separate $MoSe_2$ and MoS_2 precursors; (**b**) photograph of bare SiO_2/Si substrate and centimeter-scale MoS_2 and $MoS_{1.60}Se_{0.4}$ monolayer films on substrates; (**c,d**) high-angle annular dark filed (HAADF)-STEM image of $MoS_{1.60}Se_{0.4}$ monolayer film in false color; (**e**) Raman (left) and photo-luminescence (right) spectra of $MoS_{2(1-x)}Se_{2x}$ alloys with x ranging from 0 to 1; (**f**) schematic of CVD with separate mixtures of S and Se powders as precursors. (**a–d**) Adapted with permission from [37]. Copyright 2014, WILEY-VCH Verlag GmbH. (**e**) Adapted with permission from [38]. Copyright 2015, American Chemical Society. (**f**) Adapted with permission from [39]. Copyright 2014, American Chemical Society.

Meanwhile, Zhang *et al.* [36] also developed a CVD process to synthesize 2D $Mo_{1-x}W_xS_2$ alloys. They used a mixture of WO_3 and MoO_3 powders as the metal precursor and sulfur powder as the chalcogen precursor. The band gaps of the grown $Mo_{1-x}W_xS_2$ alloys are tuned through adjusting the Mo/W ratio in the precursors.

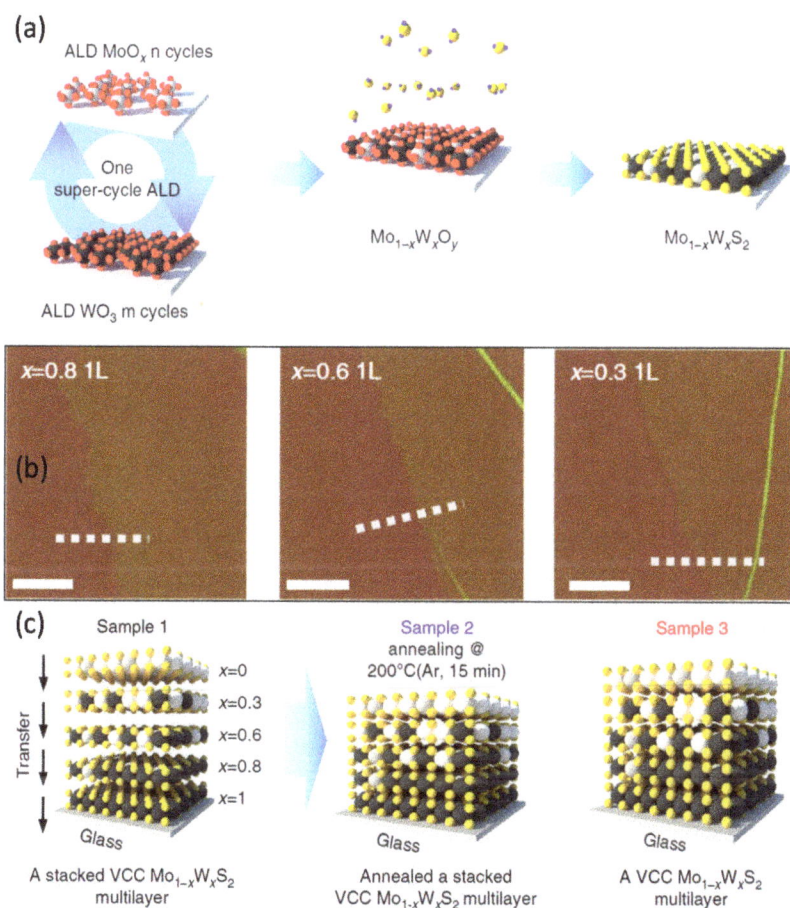

Figure 15. ALD-based synthesis of $Mo_{1-x}W_xO_y$ alloy. (**a**) Schematic of the synthesis procedure with the super-cycle of the ALD processes; (**b**) AFM images of the alloys with different x (scale bars: 500 nm); (**c**) schematics of three vertically-composition-controlled (VCC) $Mo_{1-x}W_xO_y$ multilayers. Adapted with permission from [40]. Copyright 2015, Nature Publishing Group.

3.2. 2D Heterostructures

One of the greatest interests in recent research is to reassemble isolated 2D materials into heterostructures in a layer-by-layer form and in a controllable sequence, just like building with atomic-scale Lego blocks [13]. The assembled 2D heterostructures, sometimes called van der Waals heterostructures, often reveal unusual properties and new phenomena and are anticipated to play an important role in future electronics and optoelectronics [8]. The van der Waals heterostructures can be readily fabricated by stacking different 2D materials through mechanical transfer techniques.

19

However, the techniques suffer from uncontrolled stacking orientation, contaminated interfaces and significant challenges for massive production. As a matter of fact, researchers have also begun to develop scalable synthesis techniques for various 2D heterostructures, including semiconductor/conductor (TMDC/graphene), semiconductor/insulator (TMDC/BN) and semiconductor/semiconductor (TMDC/TMDC) 2D structures. Actually some techniques introduced above are already able to synthesize 2D TMDC heterostructures, such as the MBE-grown $HfSe_2$/HPOG and $HfSe_2$/MoS_2 heterostructures (Section 2.4, Figure 13, [35]) and the ALD-based CVD-grown VCC $Mo_{1-x}W_xO_y$ multilayers (Section 3.1.2, Figure 15c, [40]). Below, we briefly introduce other techniques for the 2D TMDC heterostructures.

3.2.1. TMDC/Graphene Heterostructures

In 2011, Shi *et al.* [41] introduced a method to synthesize MoS_2/graphene heterostructures (Figure 16). First, monolayer graphene is synthesized on copper foil through a CVD process. Then, the graphene/copper substrate is placed in a quartz tube chamber, and the $(NH_4)_2MoS_4$/DMF dispersion is carried by Ar gas to deposit onto the graphene surface as the MoS_2 precursor. Finally, after DMF evaporates, the samples are annealed at 400 °C in Ar/H_2 atmosphere to decompose $(NH_4)_2MoS_4$ and obtain the MoS_2/graphene heterostructures.

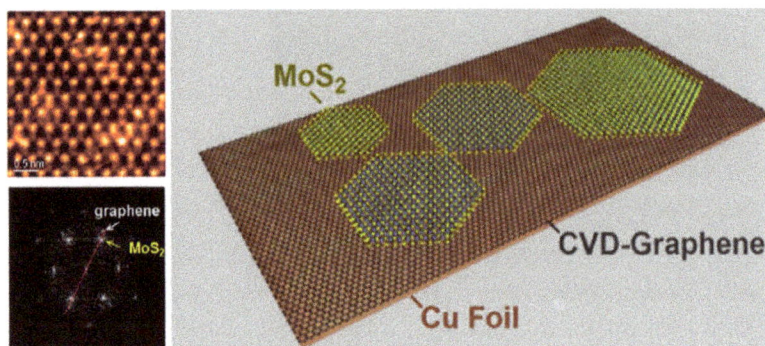

Figure 16. TEM analysis (left) and schematic (right) of the MoS_2/graphene heterostructure. Reprinted with permission from [41]. Copyright 2012, American Chemical Society.

Miwa *et al.* [42] recently demonstrated direct van der Waals epitaxy of MoS_2/graphene heterostructures on silicon carbide (SiC) substrates. The graphene/SiC substrate is first fabricated through direct current annealing of SiC under a mild flux of silicon atoms. Then, MoS_2 is synthesized in a similar way to the cycle-based MoS_2 growth on gold (Figure 9, [32]). The buffer layers coexisting with graphene on the substrates facilitate the MoS_2 growth. Due to the weak van der Waals interaction between graphene and MoS_2, the electronic structure of free-standing monolayer MoS_2 is retained in the heterostructure as confirmed by angle-resolved photoemission spectroscopy measurements.

3.2.2. TMDC/BN Heterostructures

Wang *et al.* [43] recently reported an all-CVD process to fabricate high-quality monolayer MoS_2/BN heterostructures with centimeter-scale MoS_2 domains. As illustrated in Figure 17a, first, few-layer hexagonal boron nitride (h-BN) is grown on copper substrates by CVD with an ammonia borane precursor. To suppress decomposition during MoS_2 growth, few-layer h-BN films (2–4 layers) are preferred to monolayer h-BN. Then, the h-BN film is transferred to SiO_2/Si wafer. Finally, MoS_2 is grown on top of h-BN films by CVD with MoO_3 and S powders as the precursors. Possibly because of the relatively strong interaction between MoS_2 and h-BN, the directly-grown MoS_2 on h-BN films exhibits smaller lattice strain, a lower doping level, cleaner and sharper interfaces and better interlayer contact, as compared with those directly grown on SiO_2/Si wafer (Figure 17b).

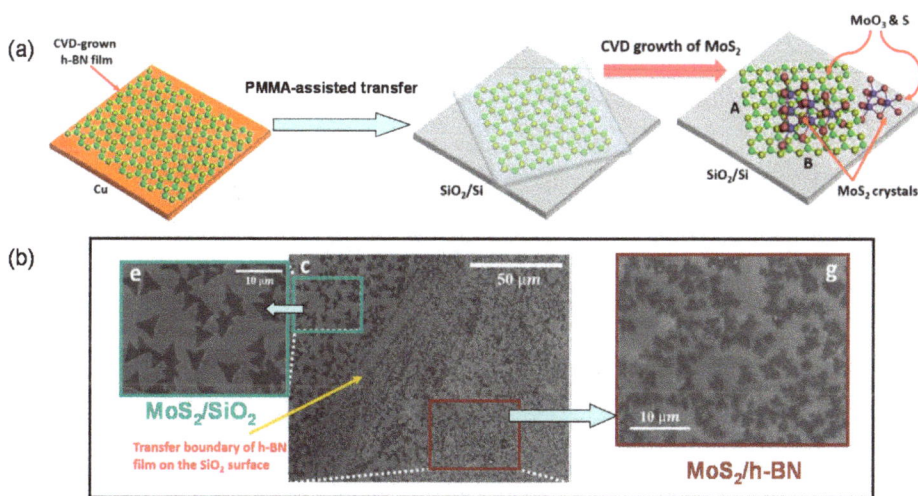

Figure 17. (**a**) Schematic of one fabrication process of the MoS_2/hexagonal boron nitride (h-BN) heterostructure; (**b**) SEM images of MoS_2 grown on the edge of the transferred h-BN indicating the difference in morphology of MoS_2 grown on SiO_2 and h-BN. Adapted with permission from [43]. Copyright 2015, American Chemical Society.

3.2.3. TMDC/TMDC Heterostructures

Gong *et al.* [44] report a scalable single-step VPD process to synthesize highly-crystalline MoS_2/WS_2 heterostructures with both a vertically-stacked bilayer structure (Figure 18a–d) and an in-plane inter-connected structure (Figure 18e–h). As shown in Figure 18i, in front of the SiO_2/Si substrate, MoO_3 powder is placed as the metal precursor for the growth of MoS_2. On top of the substrate, a mixture of tungsten (W) and tellurium (Te) powders is placed for the growth of WS_2. Here, the Te powder serves as the catalyst to accelerate the melting of W powder. Argon is used as the carrier gas to transport the S vapor from the upstream. Since MoS_2 and WS_2 have different nucleation and growth rates, this leads to sequential growth of the materials and, thereby, forms heterostructures, rather than growing the $Mo_{1-x}W_xS_2$ alloy. The structure of the final product is determined by the reaction temperature. High-temperature (~850 °C) growth produces mostly vertically-stacked bilayers, where WS_2 is grown epitaxially on top of monolayer MoS_2, while low-temperature (~650 °C) growth generates mostly in-plane lateral heterojunctions where MoS_2 and WS_2 locate within a single hexagonal monolayer lattice with seamless and atomically-sharp interfaces. Because of the clean interface, the vertically-stacked bilayers can retain the performance of individual monolayers and meanwhile generate an additional direct band gap via interlayer coupling. Remarkably, back-gate FETs based on the vertical MoS_2/WS_2 stacks exhibit an on/off current ratio over 10^6 at a high mobility ranging from 15–34 cm^2 V^{-1} s^{-1}. The in-plane MoS_2/WS_2 heterojunctions greatly enhance the localized photoluminescence and can serve as intrinsic monolayer p–n junctions with no need for external electrical tuning.

Huang *et al.* [45] also demonstrate a technique based on direct physical vapor deposition (PVD) to synthesize high-quality monolayer $MoSe_2$/WSe_2 lateral heterojunctions. The precursor, a mixture of WSe_2 and $MoSe_2$ powders, is vaporized at high temperature (~950 °C) and carried by hydrogen gas to the downstream SiO_2/Si substrate and, finally, crystallizes on the substrate to form the lateral heterojunctions (Figure 19). The growth mechanism may rely on the different evaporation rates of the two materials. In the beginning, $MoSe_2$ evaporates quickly, and pure monolayer $MoSe_2$ crystals predominate in the substrate. After some time, the $MoSe_2$ source has considerable surface depletion of Se, so that its evaporation slows down. Meanwhile, the evaporation of WSe_2 increases, leading to predominant growth of WSe_2 on the substrate. The different evaporation rates prevent the formation of $Mo_{1-x}W_xSe_2$ alloy. Due to their great similarity, WSe_2 can epitaxially grow on the $MoSe_2$ crystal edges to form the lateral heterojunctions. The monolayer heterojunctions retain an undistorted honeycomb lattice and exhibit enhanced photoluminescence.

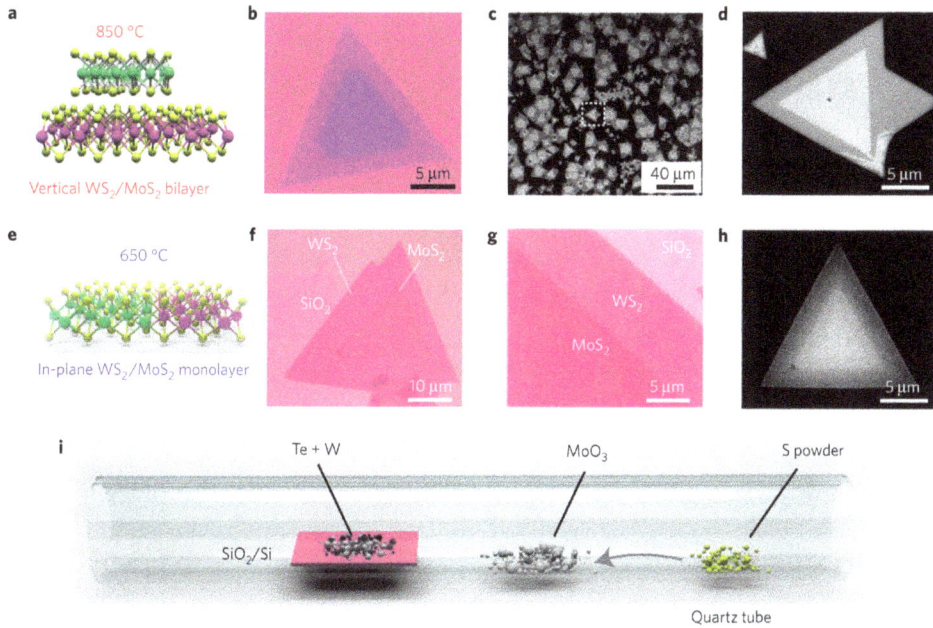

Figure 18. CVD growth of vertically-stacked and in-plane WS_2/MoS_2 heterostructures. (**a–d**) Schematic, optical and SEM images of vertically-stacked heterostructures; (**e–h**) schematic, optical and SEM images of in-plane heterostructures; (**i**) schematic of the synthesis process. Reprinted with permission from [44]. Copyright 2014, Nature Publishing Group.

3.3. Wafer-Scale Device Engineering

Despite the great progress in scalable and controllable synthesis of 2D TMDCs, with respect to device performance, however, most studies just explore a few devices in selected areas [23]. Large-scale device engineering still remains challenging. Fortunately, the increasing maturity of the VDP techniques has enabled some attempts in advanced engineering of 2D TMDC-based devices at the wafer scale.

3.3.1. Direct Growth of TMDC at Controlled Locations

Han *et al.* [46] use patterned Mo precursors to grow MoS_2 flakes at predefined locations at the resolution of micrometer scale. As illustrated in Figure 20a, first, lithography is employed to define the area for MoS_2 growth with photoresist. Then, the precursor, MoO_3 or ammonium heptamolybdate (AHM), is deposited into the area through thermal evaporation or spin coating. After the photoresist is stripped, a special aggregation step and chemical treatment by promoters, such as PTAS and PTCDA, are conducted to form spherical beads of precursors and to improve the uniformity of the following MoS_2 growth. Finally, the precursor is sulfurized by vaporized S in a nitrogen atmosphere. The grown MoS_2 shows high crystallinity and excellent optical properties. Back-gate FETs exhibit high carrier mobility between 8.2 and 14.4 cm^2 V^{-1} s^{-1} with the on/off current ratio around 10^6. In particular, the precise control over the location of discrete MoS_2 flakes (Figure 20b) facilitates direct integration and massive device production (Figure 20c) through conventional lithography without the need for an etching step.

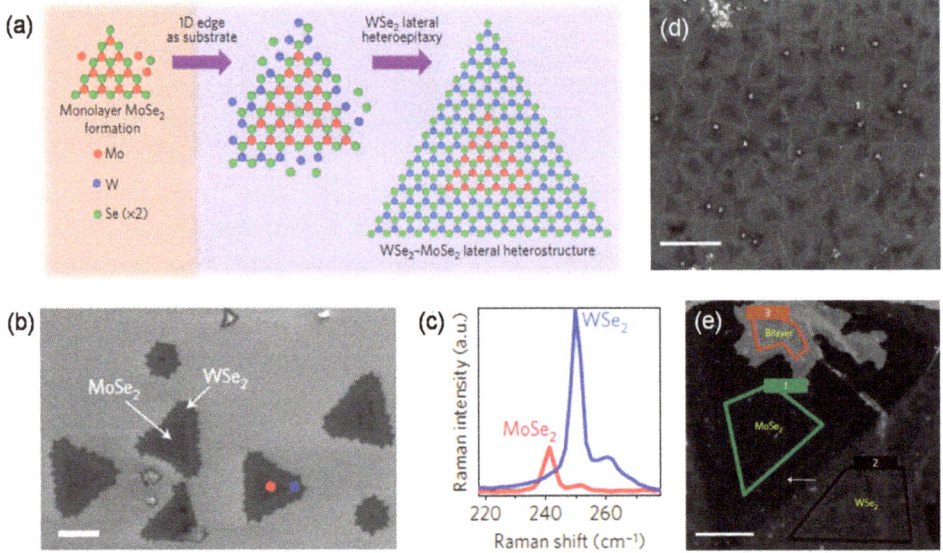

Figure 19. PVD growth of in-plane WSe_2/$MoSe_2$ heterostructures. (**a**) Schematic of the synthesis mechanism; (**b**) SEM image of the grown in-plane heterostructures (scale bar: 10 μm); (**c**) Raman spectra acquired at different regions in the heterostructure as indicated in (**b**); (**d**) SEM image of a film transfer film to a SiO_2 substrate (scale bar: 5 μm); (**e**) HAADF-STEM image of a triangle (scale bar: 100 nm). Adapted with permission from [45]. Copyright 2014, Nature Publishing Group.

3.3.2. Multi-Level Stacking of TMDC Devices

As introduced in Section 2.1.2, the all-gas precursor MOCVD process developed by Kang *et al.* [23] enables wafer-scale growth of high-quality monolayer MoS_2 film directly on SiO_2 substrates. This opens opportunities to fabricate multi-level stacked monolayer MoS_2 films (Figure 21a,b) and devices (Figure 21c–e). As illustrated in Figure 21c, the multi-level stacked structure can be fabricated in the following sequence: growing the first MoS_2 monolayer on the SiO_2/Si wafer, fabricating first-level FETs, depositing SiO_2 (~500 nm thick) by plasma-enhanced CVD, growing the second MoS_2 monolayer, fabricating second-level FETs, and so on. Figure 21d indicates an array of two vertical levels of MoS_2 FETs, both of which can be well modulated through the global back gate (Figure 21e). Note that as compared to the first-level device, which has an on-state sheet conductance of about 2.5 μS and a mobility of about 11.5 cm^2 V^{-1} s^{-1}, the second-level device only exhibits a little performance degradation with an on-state sheet conductance of about 1.5 μS and a mobility of about 8.8 cm^2 V^{-1} s^{-1}. This suggests that the technique can be employed for 3D device architecture based on TMDCs, which may provide a new solution to the most attractive, yet challenging field in recent electronics research, monolithic 3D integration [47].

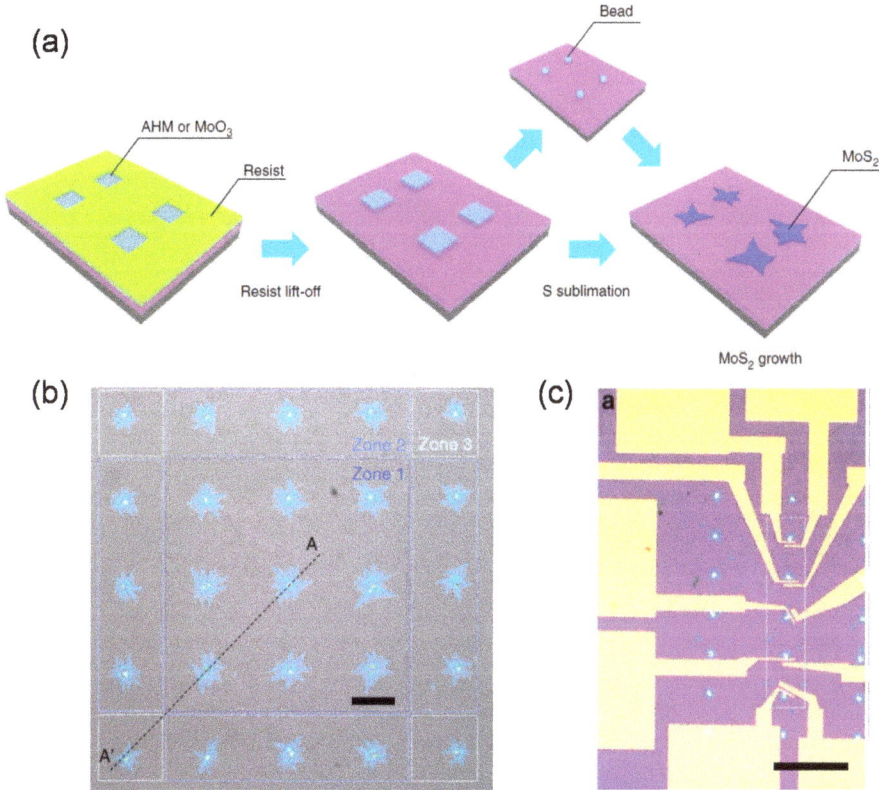

Figure 20. Selective CVD growth of MoS$_2$ using patterned precursors. (**a**) Schematic of growth process; (**b**) an array of CVD-grown MoS$_2$ monolayer flakes (scale bar: 50 μm); (**c**) batch production of FETs based on MoS$_2$ array (scale bar: 100 μm). Adapted with permission from [46]. Copyright 2015, Nature Publishing Group.

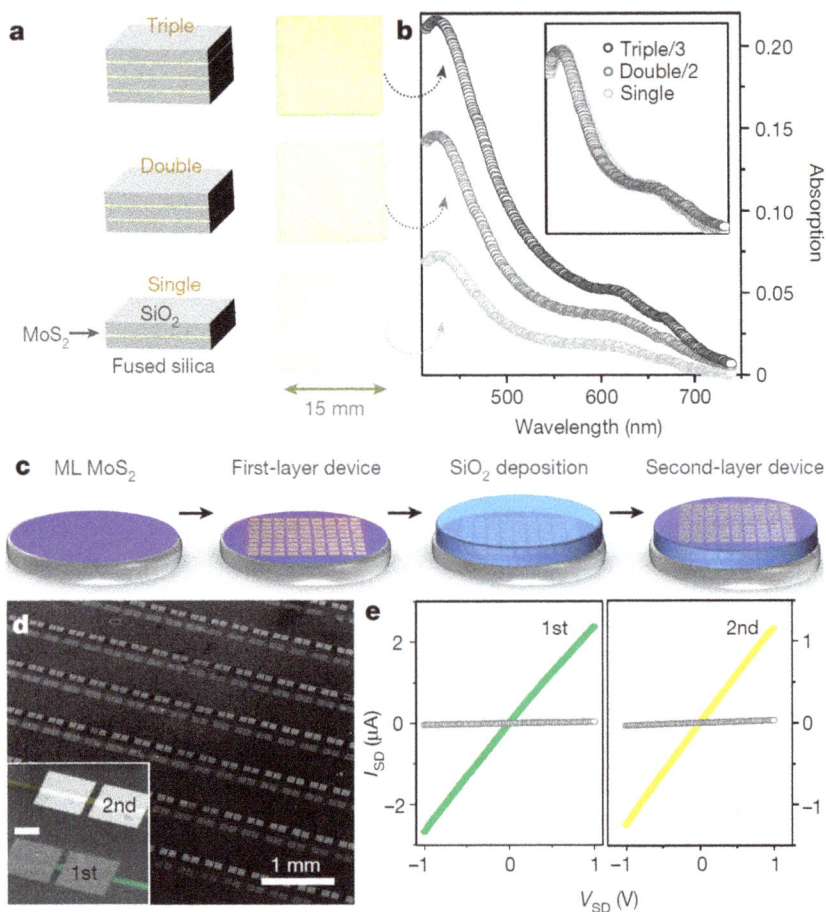

Figure 21. Multi-level stacking of MoS_2/SiO_2 structures. (**a**) Schematics (left) and optical images of 1–3 level stacking; (**b**) optical absorption of three levels of stacking; (**c**) schematic of the fabrication of multi-level stacking of MoS_2-based devices; (**d**) SEM image of MoS_2 FET arrays on the first (bottom) and second (top) layers (inset: close-up view; scale bar: 50 um); (**e**) current-voltage curves for a pair of devices on the first (left) and second (right) layers. The first-layer device is biased by the back gate of 50 V (green) and −50 V (grey), while the second-layer devices are at 100 V (yellow) and −100 V (grey). Reprinted with permission from [23]. Copyright 2015, Nature Publishing Group.

4. Discussions and Outlook

As introduced above, versatile scalable synthesis techniques for 2D semiconducting TMDCs have been developed in the literature, many of which have already led to commendable device performance (mobility >10 cm^2 V^{-1} s^{-1} with on/off current ratio $>10^6$) approaching the state-of-the-art performance acquired by the high-quality mechanically-exfoliated MoS_2 flakes (Figure 22) [14,48]. More importantly, the techniques tend to be mature and have been effective for advanced engineering of 2D TMDC alloys, various 2D heterostructures, wafer-scale device fabrication and vertical stacking of 2D transistors. These suggest that 2D TMDC-based devices have great potential for up-scaled manufacturing of high-end nanoelectronics in the near future. In particular, the unique properties of 2D TMDCs will enable innovative applications in emerging electronics and provide ideal solutions to the challenging research fields nowadays, such as monolithic 3D integration.

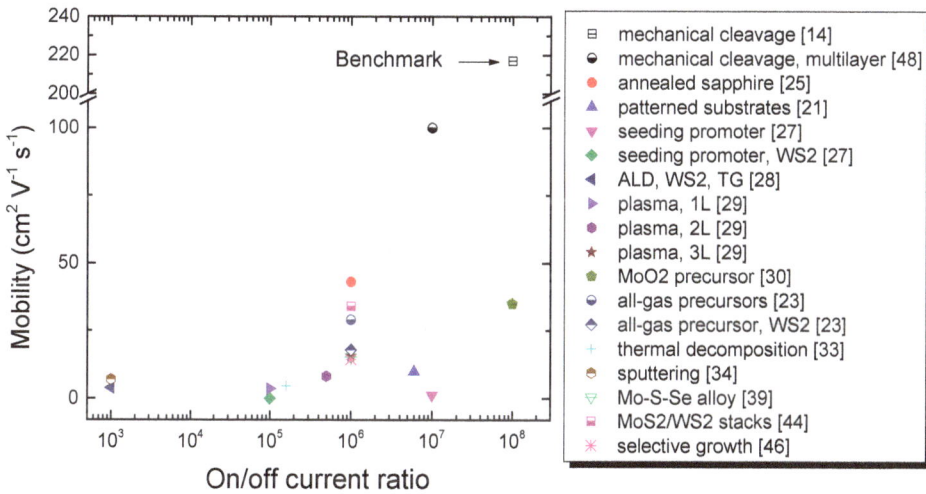

Figure 22. Plots of mobility against the on/off current ratio for FETs based on TMDCs synthesized by some techniques introduced above and the state-of-the-art performance acquired by the mechanically-exfoliated MoS_2 [14]. Unless specified in the legend, all devices are back-gate FETs with MoS_2 channels. "TG" in the legend means "top gate".

However, referring to Figure 22, one may see that despite the versatile scalable synthesis techniques demonstrated, none of them outperform the mechanically-exfoliated MoS_2 flakes with respect to device mobility and on/off current ratio. There is still great space for further advances in the fields of the scalable synthesis of 2D semiconducting materials. We foresee that the issues concerning material crystallinity, interface contamination, spatial uniformity, controlled layer number and integration with device engineering will remain critical for a relatively long time. Here, we envision several tendencies in the upcoming research: (1) The device performance boost may have to rely on effective integration among different synthetic techniques. Seemingly, one single technique can hardly make a breakthrough in addressing all of the issues. In contrast, a comprehensive technology integrating substrate treatment, seeding promoters, gas-phase precursors and growth cycle control will likely lead to higher crystallinity, better layer number controllability, higher uniformity at the wafer scale and, hence, substantially improved device performance. (2) Material synthesis will be integrated with device fabrication to enable advanced engineering at multiple levels. A good example has been demonstrated in [46], where the precursors are patterned before material growth to facilitate the following device fabrication. (3) More efforts will be transferred from synthesizing high-quality simple materials to fabricating complex materials and building advanced structures, such as the 2D alloys and van der Waals heterostructures, aiming to offer more flexibility to construct advanced device architectures. (4) The established synthesis techniques may extend to newly-discovered 2D semiconducting materials within or outside the family of TMDCs to seek a breakthrough through a new perspective.

It is worth mentioning that the final device performance is not merely determined by the material synthesis. The device engineering also plays an important role in the performance boost. For example, as shown in Figure 22, most TMDC-based FETs are of the back-gate device structure with exposed channels, which, although simplifying device fabrication, usually does not give rise to high device performance. This might be an important reason that they cannot outperform the state-of-the-art FETs based on mechanically-exfoliated MoS_2 where a high-k gate oxide is deposited on top of the channel to generate dielectric screening and to improve the mobility [14]. In most cases, the synthesized TMDCs have to be transferred from the growth substrates to the SiO_2/Si substrates for device fabrication. Versatile techniques (Figure 23) have been developed for wafer-scale transfer of 2D materials [49–51]. They provide more flexibility for material synthesis and device fabrication. However, a common issue is that the residual contamination induced by the transfer processing may also degrade the device performance [52]. Therefore, in addition to improving material synthesis techniques, properly addressing the issues arising from device engineering is also crucial for 2D material-based electronic devices.

Finally, there are considerable efforts on computational simulations in order to understand the growth dynamics of 2D materials [53–59]. These theoretical studies provide mechanisms and explanations for experimental observations and useful suggestions for material quality improvement [56,58]. They also indicate optimal conditions for nucleation [57] and edge reconstruction [53] during growth, as well as predict the structure and stability of 2D materials grown on different substrates [54–57,59]. Such guidance is also valuable for experimentalists to improve their growth processes or to explore innovative techniques.

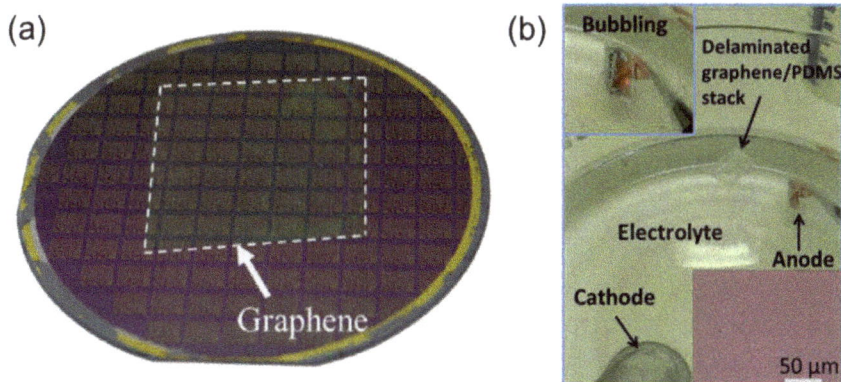

Figure 23. Wafer-scale transfer techniques for 2D materials. (**a**) Transferred graphene onto a 4-inch silicon wafer with poly(bisphenol A) carbonate (PC) as the carrier; (**b**) delamination of the graphene/polydimethylsiloxane (PDMS) stack using electrolysis (inset: optical microscope image of graphene transferred onto an SiO$_2$ layer). (**a**) Reprinted with permission from [49]. Copyright, 2014 IEEE. (**b**) Reprinted with permission from [50]. Copyright 2014, IEEE.

In summary, the review provides a systematic and comprehensive introduction of scalable synthesis techniques of 2D semiconducting materials for high-end applications in future nanoelectronics, most of which have shown promise in scaling up device fabrication. We hope the review is beneficial to the readers, either in establishing existing techniques for further research on 2D materials or in stimulating innovative ideas to enhance and extend the research on scalable synthesis techniques.

Acknowledgments: We acknowledge the financial support by the Swedish Research Council through the research project iGRAPHENE (No. 2013-5759), the Framework project (No. 2014-6160) and the Marie Skłodowska Curie International Career Grant (No. 2015-00395, co-funded by Marie Skłodowska-Curie Actions, through the Project INCA 600398), the European Research Council through the Proof of Concept Grant iPUBLIC (No. 641416), the Göran Gustafsson Foundation through the Young Researcher Prize (No. 1415B), the Olle Engkvist Byggmästare Foundation through the Research Project (No. 2014/799) and the Swedish Innovation Agency VINNOVA through the Strategic Innovation Program for Graphene, iEnergy (No. 2015-01337).

Author Contributions: Jiantong Li and Mikael Östling discussed the topic, surveyed the literature and wrote the manuscript.

Conflicts of Interest: The authors declare no conflict of interest.

References

1. Novoselov, K.S.; Fal'ko, V.I.; Colombo, L.; Gellert, P.R.; Schwab, M.G.; Kim, K. A roadmap for graphene. *Nature* **2012**, *490*, 192–200.

2. Wang, Q.H.; Kalantar-Zadeh, K.; Kis, A.; Coleman, J.N.; Strano, M.S. Electronics and optoelectronics of two-dimensional transition metal dichalcogenides. *Nat. Nanotechnol.* **2012**, *7*, 699–712.

3. Kim, S.J.; Choi, K.; Lee, B.; Kim, Y.; Hong, B.H. Materials for Flexible, Stretchable Electronics: Graphene and 2D Materials. *Annu. Rev. Mater. Res.* **2015**, *45*, 63–84.

4. Vaziri, S.; Lupina, G.; Henkel, C.; Smith, A.D.; Östling, M.; Dabrowski, J.; Lippert, G.; Mehr, W.; Lemme, M.C. A Graphene-Based Hot Electron Transistor. *Nano Lett.* **2013**, *13*, 1435–1439.

5. Lemme, M.C.; Li, L.-J.; Palacios, T.; Schwierz, F. Two-dimensional materials for electronic applications. *MRS Bull.* **2014**, *39*, 711–718.

6. Lemme, M.C.; Echtermeyer, T.J.; Baus, M.; Kurz, H. A Graphene Field-Effect Device. *IEEE Electron Device Lett.* **2007**, *28*, 282–284.

7. Lopez-Sanchez, O.; Lembke, D.; Kayci, M.; Radenovic, A.; Kis, A. Ultrasensitive photodetectors based on monolayer MoS_2. *Nat. Nanotechnol.* **2013**, *8*, 497–501.

8. Yim, C.; O'Brien, M.; McEvoy, N.; Riazimehr, S.; Schäfer-Eberwein, H.; Bablich, A.; Pawar, R.; Iannaccone, G.; Downing, C.; Fiori, G.; *et al.* Heterojunction Hybrid Devices from Vapor Phase Grown MoS_2. *Sci. Rep.* **2014**, *4*, 5458.

9. Li, J.; Lemme, M.C.; Östling, M. Inkjet Printing of 2D Layered Materials. *Chemphyschem* **2014**, *15*, 3427–3434.

10. Li, J.; Östling, M. Prevention of Graphene Restacking for Performance Boost of Supercapacitors—A Review. *Crystals* **2013**, *3*, 163–190.

11. Smith, A.D.; Niklaus, F.; Paussa, A.; Vaziri, S.; Fischer, A.C.; Sterner, M.; Forsberg, F.; Delin, A.; Esseni, D.; Palestri, P.; *et al.* Electromechanical Piezoresistive Sensing in Suspended Graphene Membranes. *Nano Lett.* **2013**, *13*, 3237–3242.

12. Shen, H.; Zhang, L.; Liu, M.; Zhang, Z. Biomedical applications of graphene. *Theranostics* **2012**, *2*, 283–294.

13. Geim, A.K.; Grigorieva, I.V. Van der Waals heterostructures. *Nature* **2013**, *499*, 419–425.

14. Radisavljevic, B.; Radenovic, A.; Brivio, J.; Giacometti, V.; Kis, A. Single-layer MoS_2 transistors. *Nat. Nanotechnol.* **2011**, *6*, 147–150.

15. Liu, H.; Du, Y.; Deng, Y.; Ye, P.D. Semiconducting black phosphorus: Synthesis, transport properties and electronic applications. *Chem. Soc. Rev.* **2015**, *44*, 2732–2743.

16. Nicolosi, V.; Chhowalla, M.; Kanatzidis, M.G.; Strano, M.S.; Coleman, J.N. Liquid Exfoliation of Layered Materials. *Science* **2013**, *340*.

17. Li, J.; Ye, F.; Vaziri, S.; Muhammed, M.; Lemme, M.C.; Östling, M. Efficient Inkjet Printing of Graphene. *Adv. Mater.* **2013**, *25*, 3985–3992.

18. Li, J.; Naiini, M.M.; Vaziri, S.; Lemme, M.C.; Östling, M. Inkjet Printing of MoS_2. *Adv. Funct. Mater.* **2014**, *24*, 6524–6531.

19. Kong, D.; Wang, H.; Cha, J.J.; Pasta, M.; Koski, K.J.; Yao, J.; Cui, Y. Synthesis of MoS_2 and $MoSe_2$ Films with Vertically Aligned Layers. *Nano Lett.* **2013**, *13*, 1341–1347.

20. Shi, Y.; Li, H.; Li, L.J. Recent advances in controlled synthesis of two-dimensional transition metal dichalcogenides via vapor deposition techniques. *Chem. Soc. Rev.* **2015**, *44*, 2744–2756.

21. Najmaei, S.; Liu, Z.; Zhou, W.; Zou, X.; Shi, G.; Lei, S.; Yakobson, B.I.; Idrobo, J.-C.; Ajayan, P.M.; Lou, J. Vapour phase growth and grain boundary structure of molybdenum disulphide atomic layers. *Nat. Mater.* **2013**, *12*, 754–759.

22. Ling, X.; Lee, Y.-H.; Lin, Y.; Fang, W.; Yu, L.; Dresselhaus, M.S.; Kong, J. Role of the Seeding Promoter in MoS$_2$ Growth by Chemical Vapor Deposition. *Nano Lett.* **2014**, *14*, 464–472.

23. Kang, K.; Xie, S.; Huang, L.; Han, Y.; Huang, P.Y.; Mak, K.F.; Kim, C.-J.; Muller, D.; Park, J. High-mobility three-atom-thick semiconducting films with wafer-scale homogeneity. *Nature* **2015**, *520*, 656–660.

24. Dumcenco, D.; Ovchinnikov, D.; Marinov, K.; Lazić, P.; Gibertini, M.; Marzari, N.; Sanchez, O.L.; Kung, Y.; Krasnozhon, D.; Chen, M.; *et al.* Large-Area Epitaxial Monolayer MoS$_2$. *ACS Nano* **2015**, *9*, 4611–4620.

25. Zhang, Y.; Zhang, Y.; Ji, Q.; Ju, J.; Yuan, H.; Shi, J.; Gao, T.; Ma, D.; Liu, M.; Chen, Y.; *et al.* Controlled Growth of High-Quality Monolayer WS$_2$ Layers on Sapphire and Imaging Its Grain Boundary. *ACS Nano* **2013**, *7*, 8963–8971.

26. Lee, Y.H.; Zhang, X.Q.; Zhang, W.; Chang, M.T.; Lin, C.T.; Chang, K.D.; Yu, Y.C.; Wang, J.T.W.; Chang, C.S.; Li, L.J.; *et al.* Synthesis of large-area MoS$_2$ atomic layers with chemical vapor deposition. *Adv. Mater.* **2012**, *24*, 2320–2325.

27. Lee, Y.; Yu, L.; Wang, H.; Fang, W.; Ling, X.; Shi, Y.; Lin, C.-T.; Huang, J.-K.; Chang, M.-T.; Chang, C.-S.; *et al.* Synthesis and Transfer of Single Layer Transition Metal Disulfides on Diverse Surfaces. *Nano Lett.* **2013**, *13*, 1852–1857.

28. Song, J.; Park, J.; Lee, W.; Choi, T.; Jung, H.; Lee, C.W.; Hwang, S.-H.; Myoung, J.M.; Jung, J.-H.; Kim, S.; *et al.* Layer-Controlled, Wafer-Scale, and Conformal Synthesis of Tungsten Disulfide Nanosheets Using Atomic Layer Deposition. *ACS Nano* **2013**, *7*, 11333–11340.

29. Jeon, J.; Jang, S.K.; Jeon, S.M.; Yoo, G.; Jang, Y.H.; Park, J.-H.; Lee, S. Layer-controlled CVD growth of large-area two-dimensional MoS$_2$ films. *Nanoscale* **2015**, *7*, 1688–1695.

30. Bilgin, I.; Liu, F.; Vargas, A.; Winchester, A.; Man, M.K.L.; Upmanyu, M.; Dani, K.M.; Gupta, G.; Talapatra, S.; Mohite, A.D.; *et al.* Chemical Vapor Deposition Synthesized Atomically Thin Molybdenum Disulfide with Optoelectronic-Grade Crystalline Quality. *ACS Nano* **2015**, *9*, 8822–8832.

31. Yu, Y.; Li, C.; Liu, Y.; Su, L.; Zhang, Y.; Cao, L. Controlled scalable synthesis of uniform, high-quality monolayer and few-layer MoS$_2$ films. *Sci. Rep.* **2013**, *3*, 1866.

32. Grønborg, S.S.; Ulstrup, S.; Bianchi, M.; Dendzik, M.; Sanders, C.E.; Lauritsen, J.V.; Hofmann, P.; Miwa, J.A. Synthesis of Epitaxial Single-Layer MoS$_2$ on Au(111). *Langmuir* **2015**, *31*, 9700–9706.

33. Liu, K.-K.; Zhang, W.; Lee, Y.-H.; Lin, Y.-C.; Chang, M.-T.; Su, C.-Y.; Chang, C.-S.; Li, H.; Shi, Y.; Zhang, H.; *et al.* Growth of large-area and highly crystalline MoS$_2$ thin layers on insulating substrates. *Nano Lett.* **2012**, *12*, 1538–1544.

34. Tao, J.; Chai, J.; Lu, X.; Wong, L.M.; Wong, T.I.; Pan, J.; Xiong, Q.; Chi, D.; Wang, S. Growth of wafer-scale MoS$_2$ monolayer by magnetron sputtering. *Nanoscale* **2015**, *7*, 2497–2503.

31

35. Yue, R.; Barton, A.T.; Zhu, H.; Azcatl, A.; Pena, L.F.; Wang, J.; Peng, X.; Lu, N.; Cheng, L.; Addou, R.; *et al.* HfSe$_2$ Thin Films: 2D Transition Metal Dichalcogenides Grown by Molecular Beam Epitaxy. *ACS Nano* **2015**, *9*, 474–480.

36. Zhang, W.; Li, X.; Jiang, T.; Song, J.; Lin, Y.; Zhu, L.; Xu, X. CVD synthesis of Mo$_{(1-x)}$W$_x$S$_2$ and MoS$_{2(1-x)}$Se$_{2x}$ alloy monolayers aimed at tuning the bandgap of molybdenum disulfide. *Nanoscale* **2015**, *7*, 13554–13560.

37. Feng, Q.; Zhu, Y.; Hong, J.; Zhang, M.; Duan, W.; Mao, N.; Wu, J.; Xu, H.; Dong, F.; Lin, F.; *et al.* Growth of large-area 2D MoS$_{2(1-x)}$Se$_{2x}$ semiconductor alloys. *Adv. Mater.* **2014**, *26*, 2648–2653.

38. Feng, Q.; Mao, N.; Wu, J.; Xu, H.; Wang, C.; Zhang, J.; Xie, L. Growth of MoS$_{2(1-x)}$Se$_{2x}$ (x = 0.41–1.00) Monolayer Alloys with Controlled Morphology by Physical Vapor Deposition. *ACS Nano* **2015**, *9*, 7450–7455.

39. Gong, Y.; Liu, Z.; Lupini, A.R.; Shi, G.; Lin, J.; Najmaei, S.; Lin, Z.; Elías, A.L.; Berkdemir, A.; You, G.; *et al.* Band Gap Engineering and Layer-by-Layer Mapping of Selenium-doped Molybdenum Disulfide. *Nano Lett.* **2014**, *14*, 442–449.

40. Song, J.-G.; Ryu, G.H.; Lee, S.J.; Sim, S.; Lee, C.W.; Choi, T.; Jung, H.; Kim, Y.; Lee, Z.; Myoung, J.-M.; *et al.* Controllable synthesis of molybdenum tungsten disulfide alloy for vertically composition-controlled multilayer. *Nat. Commun.* **2015**, *6*, 7817.

41. Shi, Y.; Zhou, W.; Lu, A.-Y.; Fang, W.; Lee, Y.-H.; Hsu, A.L.; Kim, S.M.; Kim, K.K.; Yang, H.Y.; Li, L.-J.; *et al.* Van der Waals Epitaxy of MoS$_2$ Layers Using Graphene as Growth Templates. *Nano Lett.* **2012**, *12*, 2784–2791.

42. Miwa, J.A.; Dendzik, M.; Grønborg, S.S.; Bianchi, M.; Lauritsen, J.V.; Hofmann, P.; Ulstrup, S. Van der Waals Epitaxy of Two-Dimensional MoS$_2$-Graphene Heterostructures in Ultrahigh Vacuum. *ACS Nano* **2015**, *9*, 6502–6510.

43. Wang, S.; Wang, X.; Warner, J.H. All Chemical Vapor Deposition Growth of MoS$_2$:h-BN Vertical van der Waals Heterostructures. *ACS Nano* **2015**, *9*, 5246–5254.

44. Gong, Y.; Lin, J.; Wang, X.; Shi, G.; Lei, S.; Lin, Z.; Zou, X.; Ye, G.; Vajtai, R.; Yakobson, B.I.; *et al.* Vertical and in-plane heterostructures from WS$_2$/MoS$_2$ monolayers. *Nat. Mater.* **2014**, *13*, 1135–1142.

45. Huang, C.; Wu, S.; Sanchez, A.M.; Peters, J.J.P.; Beanland, R.; Ross, J.S.; Rivera, P.; Yao, W.; Cobden, D.H.; Xu, X. Lateral heterojunctions within monolayer MoSe$_2$-WSe$_2$ semiconductors. *Nat. Mater.* **2014**, *13*, 1096–1101.

46. Han, G.H.; Kybert, N.J.; Naylor, C.H.; Lee, B.S.; Ping, J.; Park, J.H.; Kang, J.; Lee, S.Y.; Lee, Y.H.; Agarwal, R.; *et al.* Seeded growth of highly crystalline molybdenum disulphide monolayers at controlled locations. *Nat. Commun.* **2015**, *6*, 6128.

47. Garidis, K.; Jayakumar, G.; Asadollahi, A.; Litta, E.D.; Hellström, P.-E.; Östling, M. Characterization of bonding surface and electrical insulation properties of inter layer dielectrics for 3D monolithic integration. In Proceedings of the IEEE Ultimate Integration on Silicon (EUROSOI-ULIS), Bologna, Italy, 26–28 January 2015; pp. 165–168.

48. Kim, S.; Konar, A.; Hwang, W.-S.; Lee, J.H.; Lee, J.; Yang, J.; Jung, C.; Kim, H.; Yoo, J.-B.; Choi, J.-Y.; *et al.* High-mobility and low-power thin-film transistors based on multilayer MoS$_2$ crystals. *Nat. Commun.* **2012**, *3*, 1011.

49. Smith, A.D.; Vaziri, S.; Rodriguez, S.; Östling, M.; Lemme, M.C. Wafer scale graphene transfer for back end of the line device integration. In Proceedings of the ULIS 2014—2014 15th International Conference on Ultimate Integration on Silicon, Stockholm, Sweden, 7–9 April 2014; pp. 29–32.

50. Vaziri, S.; Smith, A.D.; Lupina, G.; Lemme, M.C.; Östling, M. PDMS-supported Graphene Transfer Using Intermediary Polymer Layers. In Proceedings of the 2014 44th European Solid State Device Research Conference (ESSDERC), Venice, Italy, 22–26 September 2014; pp. 309–312.

51. Kataria, S.; Wagner, S.; Ruhkopf, J.; Gahoi, A.; Pandey, H.; Bornemann, R.; Vaziri, S.; Smith, A.D.; Ostling, M.; Lemme, M.C. Chemical vapor deposited graphene: From synthesis to applications. *Phys. Status Solidi A* **2014**, *211*, 2439–2449.

52. Lupina, G.; Kitzmann, J.; Costina, I.; Lukosius, M.; Wenger, C.; Wolff, A.; Pasternak, I.; Krajewska, A.; Strupinski, W.; Vaziri, S.; *et al.* Residual Metallic Contamination of Transferred Chemical Vapor Deposited Graphene. *ACS Nano* **2015**, *9*, 4776–4785.

53. Gao, J.; Zhao, J.; Ding, F. Transition metal surface passivation induced graphene edge reconstruction. *J. Am. Chem. Soc.* **2012**, *134*, 6204–6209.

54. Gao, J.; Zhang, J.; Liu, H.; Zhang, Q.; Zhao, J. Structures, mobilities, electronic and magnetic properties of point defects in silicene. *Nanoscale* **2013**, *5*, 9785–9792.

55. Penev, E.S.; Bhowmick, S.; Sadrzadeh, A.; Yakobson, B.I. Polymorphism of Two-Dimensional Boron. *Nano Lett.* **2012**, *12*, 2441–2445.

56. Gao, J.; Zhao, J. Initial geometries, interaction mechanism and high stability of silicene on Ag(111) surface. *Sci. Rep.* **2012**, *2*, 861.

57. Gao, J.; Yip, J.; Zhao, J.; Yakobson, B.I.; Ding, F. Graphene nucleation on transition metal surface: Structure transformation and role of the metal step edge. *J. Am. Chem. Soc.* **2011**, *133*, 5009–5015.

58. Artyukhov, V.I.; Liu, Y.; Yakobson, B.I. Equilibrium at the edge and atomistic mechanisms of graphene growth. *Proc. Natl. Acad. Sci. USA* **2012**, *109*, 15136–15140.

59. Gao, J.; Ding, F. The structure and stability of magic carbon clusters observed in graphene chemical vapor deposition growth on Ru(0001) and Rh(111) surfaces. *Angew. Chem. Int. Ed. Engl.* **2014**, *53*, 14031–14035.

Towards the Realization of Graphene Based Flexible Radio Frequency Receiver

Maruthi N. Yogeesh, Kristen N. Parrish, Jongho Lee, Saungeun Park, Li Tao and Deji Akinwande

Abstract: We report on our progress and development of high speed flexible graphene field effect transistors (GFETs) with high electron and hole mobilities (~3000 $cm^2/V \cdot s$), and intrinsic transit frequency in the microwave GHz regime. We also describe the design and fabrication of flexible graphene based radio frequency system. This RF communication system consists of graphite patch antenna at 2.4 GHz, graphene based frequency translation block (frequency doubler and AM demodulator) and graphene speaker. The communication blocks are utilized to demonstrate graphene based amplitude modulated (AM) radio receiver operating at 2.4 GHz.

Reprinted from *Electronics*. Cite as: Yogeesh, M.N.; Parrish, K.N.; Lee, J.; Park, S.; Tao, L.; Akinwande, D. Towards the Realization of Graphene Based Flexible Radio Frequency Receiver. *Electronics* **2015**, *4*, 933–945.

1. Introduction

Research efforts in the field of flexible GHz nanoelectronics have been gaining significant momentum recently. Research and commercial institutions are investing substantial resources in the design of flexible systems and sensors [1–7], including the development of smart conformal devices and tags, wearable electronics, internet of things (IoTS) and other electronic gadgets that are low cost, scalable to large areas, and mechanically robust, which is beyond the capability of conventional bulk semiconductor technologies. Moreover continuous advancements in the material synthesis of intrinsically flexible nanomaterials such as graphene, h-BN, carbon nano-tubes (CNTs) and transition metal dichalcogenides (TMDs, e.g., molybdenum disulfide—MoS_2) [8,9] have encouraged their prospects for practical electronic application. These materials are naturally more suitable for integration with flexible, soft or glass substrates owing to their two dimensional nature and can potentially offer the electronic performance needed for low-power GHz systems.

Graphene is a single sheet of carbon with very high room temperature mobility (>10,000 $cm^2/V \cdot s$), high cutoff frequency (>300 GHz), and high electrical and thermal conductivity [2,3,10]. These properties of graphene make it a suitable candidate for the design of high frequency analog communication circuits, such as amplifiers, mixers, demodulators, and speakers [11]. Recent work on graphene circuits for RF applications includes: demonstration of graphene three stage integrated amplifier on

Si wafer [12] and demonstration of flexible graphene circuit for a variety of digital modulation schemes [13].

In this paper we present the state of the art flexible graphene field effect transistors (GFET) with high intrinsic f_T of 18 GHz that has recently been improved to 100 GHz [14]. We also present the design and realization of the first all graphene based flexible radio frequency system. The following section discusses the design and fabrication of GFETs. Results and discussion section discuss in more detail the small signal modeling, electrical measurements (DC and RF), and mechanical flexibility and reliability evaluation of GFETs. We also discuss the design of individual graphene based radio frequency system blocks—flexible graphite patch antenna, graphene based frequency translation block (GFET doubler and demodulator), graphene speaker, and the demonstration of the first graphene based AM radio receiver.

2. Flexible Graphene RF Transistor Fabrication

Figure 1 shows the typical device structure of our fabricated flexible graphene field-effect transistors. GFETs are typically prepared on smoothened commercial 125 μm-thick polyimide (Kapton) films. Gate fingers (evaporated 2 nm Ti/40 nm Pd) and gate dielectric films (prepared by atomic layer deposition of 10 nm-thick Al_2O_3 or hexagonal boron nitride films) are prepared by electron-beam lithography using JEOL-6000 FSE (JEOL USA Inc., Peabody, MA, 01960, USA). Chemical vapor deposited high-quality graphene films grown on commercial Cu-foils or films are transferred via the conventional wet-transfer process using ammonia persulfate as the Cu-etchant [10]. Active channel area of graphene and metallic electrodes for source and drain contacts are subsequently patterned via additional electron-beam lithography processes. Low power oxygen plasma is used to etch redundant graphene areas and electron-beam evaporation of 2 nm Ti/50 nm Au is used as the contact metals. Figure 1a is 3D view of freestanding polyimide film with multifingered GFET. Figure 1b is the enlarged optical image of a representative GFET fabricated on polyimide substrate. Devices with channel length of 0.25 μm and an effective width of 60 μm (six fingers with 10 μm width) are prepared for electrical measurements, which can afford GHz high frequency performance.

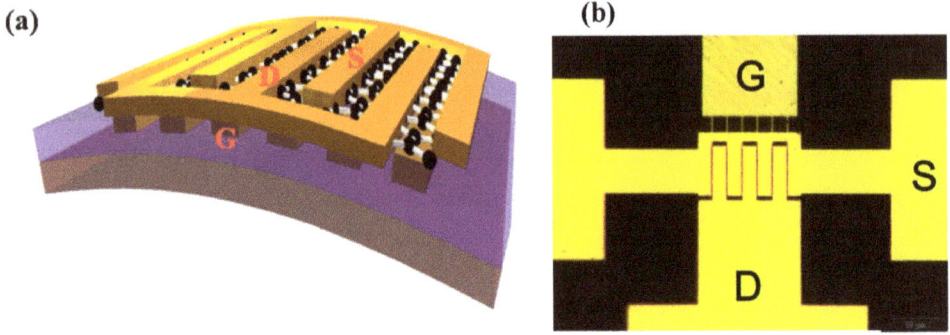

Figure 1. (a) 3D view of multi-fingered GFET on flexible polyimide film; (b) Enlarged optical image of a representative GFET transistor.

3. Results and Discussion

3.1. Compact Model of GFET

We derived a semi-empirical compact model useful for GFET design and analysis. This model shown in Figure 2 has good agreement with DC and RF measurements [15]. In this model we have taken into account the quantum capacitance (C_q), velocity saturation (v_{sat}), contact resistance (R_c) and access resistance (R_a). The derived drain current I_{DS} (see Equation (1)) depends on the graphene channel width (W) and length (L), carrier density (n_i), effective mobility(u_{eff}), drain voltage (V_{DS}), gate voltage (V_{GS}), oxide capacitance (C_{OX}), quantum capacitance (C_q) and temperature (T). This model is described in more detail in our earlier work [15,16]. This model is implemented in Agilent ADS for linear and nonlinear RF circuit simulation.

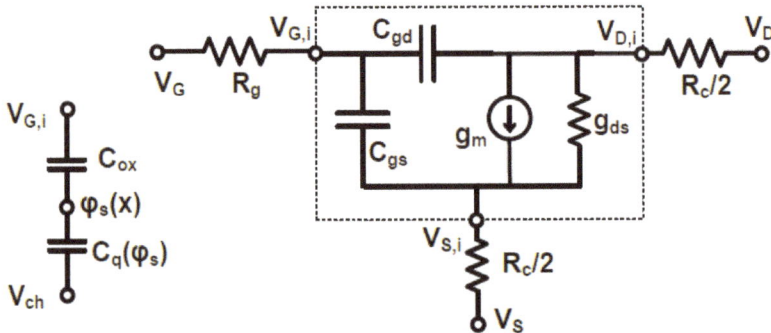

Figure 2. Compact model of GFET (adapted from [17]. Copyright 2013, permission granted by Kristen Parish).

$$I_{DS} = q\frac{W}{L}\mu_{eff}\left(n_i V_{DS} + \left(\frac{(\alpha+1)^2(k_B T \ln 4)^2}{4\pi h^2 v_f^2(\alpha \ln 2 + 1)}\right)\left(-2V_{DS} + V_{DG}\sqrt{1+\delta^2 V_{DG}^2} + V_{GS}\sqrt{1+\delta^2 V_{GS}^2} + \frac{\text{asinh}\delta V_{DG} + \text{asinh}\delta V_{GS}}{\delta}\right)\right) \quad (1)$$

where $\mu_{eff} = \dfrac{\mu}{\sqrt{\left(1+\left(\frac{\mu V_{DS}}{L v_{sat}}\right)^2\right)}}$; $\beta = \dfrac{q^2}{\pi h^2 v_f^2}$; $\delta = \dfrac{2\alpha\sqrt{(\alpha \ln 2+1)}}{(\alpha+1)^2}\dfrac{q}{k_B T \ln 4}$.

3.2. GFET DC Measurement

Electrical measurements of fabricated GFET are routinely done at room temperature. Figure 3 shows the input and output characteristics of a flexible GFET comparing both model and measured data. The extracted carrier mobility of electrons is 9050 cm^2/V·s and that of holes is 3000 cm^2/V·s. Graphene FETs show velocity saturation. We attribute this to the smooth surface (surface roughness <1 nm) on polyimide substrate (with liquid polyimide curing) and high K gate dielectric (Al$_2$O$_3$) which will give rise to well-behaved linear current profile followed by saturation [16].

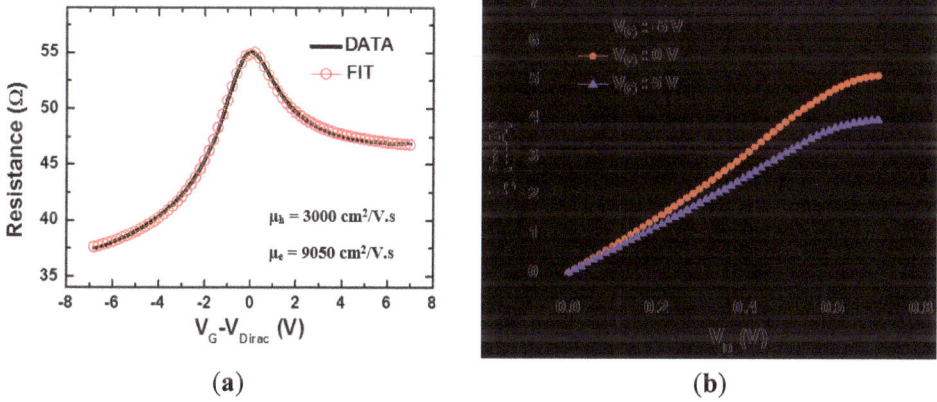

(a) (b)

Figure 3. Electrical properties of graphene transistor (a) Gate modulation of graphene resistance. The drain voltage is set to 10 mV; (b) Drain current (I$_D$) vs. Drain voltage (V$_D$).

3.3. GFET RF Measurement

S-parameter measurements were carried out on another flexible GFET device with ground-signal-ground (GSG) pads at the gate and drain terminals with the source serving as the common ground. The measurement was done using an Agilent VNA (E8361C) and cascade RF probe station. The intrinsic device data is obtained by employing a two-step de-embedding process. The device under test (GFET) can be modeled as shown in Figure 4a. The Y-parameters of the device under test were obtained from extrinsic data, and afterwards the parasitics (obtained from open and short test structures) were de-embedded to determine the intrinsic device frequency response (see Equation (2), adapted from [18]). The open structure will help to determine parallel parasitics (Y_{p1}, Y_{p2}, Y_{p3}) and the short structure will help to determine the series parasitics (Z_{l1}, Z_{l2} and Z_{l3}). These series parasitics are very critical for modern small channel FETs.

$$Y_{GFET} = \left(\left(Y_{dut} - Y_{open} \right)^{-1} - \left(Y_{short} - Y_{open} \right)^{-1} \right)^{-1} \qquad (2)$$

where Y_{GFET} is the Y-parameters of the intrinsic GFET, Y_{dut} is the Y-parameter of DUT (Extrinsic GFET), Y_{short} is the Y-parameter of the DUT with gate, drain and source shorted, Y_{open} is the DUT with no graphene channel. Here, the graphene channel is etched by using reactive ion plasma etch tool. Y_{short} and Y_{open} are deembeded from Y_{dut} to obtain the intrinsic GFET performance (Y_{GFET}).

Figure 4b shows the current gain h_{21} of the intrinsic GFET, from which the 18 GHz f_T is extracted. The extrinsic f_T of the GFET is 1.8 GHz. Figure 4c shows the power gain U (dB) of the intrinsic GFET. We obtained f_{max} of ~3 GHz. f_T and f_{max} are given by Equations (3) and (4). They depend upon the transconductance g_m, gate capacitance C_{gs}, parasitic gate-drain capacitance $C_{p,gs}$, parasitic gate-source capacitance $C_{p\text{-}gd}$ and gate resistance R_g. f_T can be improved by reducing channel length and design of better layouts (reduced parasitics) [19]. f_T and f_{max} are lower for flexible devices compared to devices fabricated on rigid substrates mainly due to poor thermal dissipation of flexible polymer substrate at high fields [19]. Thermo-mechanical breakdown occurs when the local hot spot temperature exceeds the glass transition temperature of the soft substrate [2]. One solution to overcome this challenge is to employ flexible glass substrates, such as Corning "willow" glass, which offers a higher thermal conductivity than polymers or elastomers. Indeed, improved high-frequency graphene transistors have been achieved on bendable willow glass with intrinsic f_T ~100 GHz and record saturation velocity for flexible graphene [14].

$$f_T = \frac{g_m}{2\beta \left(C_{gs} + C_{p,gs} + C_{p,gd} \right) \left(\left(R_{p,s} + R_{p,d} \right) g_d + 1 \right) + C_{p,gd} g_m \left(R_{p,s} + R_{p,d} \right)} \qquad (3)$$

$$f_{max} = \frac{f_T}{2 g_d \left(R_{p,s} + R_{gate} \right) + 2\beta f_T C_{p,gd} R_{gate}} \qquad (4)$$

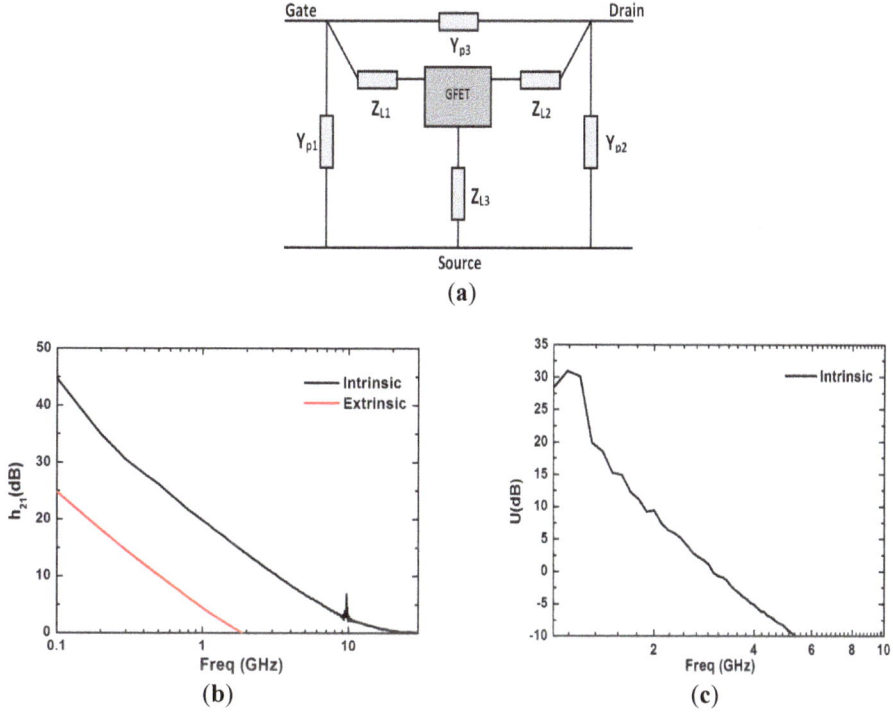

Figure 4. (a) RF model of GFET under test, where Y_{p1}, Y_{p2} and Y_{p3} are parallel parasitics and Z_{L1}, Z_{L2} and Z_{L3} are series parasitics (They represent GFET contact pads and interconnects). (b) Current Gain (h_{21}) *vs.* Frequency. The intrinsic f_T ~18 GHz; (c) Power gain (U) in dB *vs.* Frequency. f_{max} ~3 GHz. These devices are measured at the following voltage conditions (V_{drain} ~0.4 V, V_{gate} ~0.2 V).

3.4. GFET Flexibility and Reliability

Mechanical flexibility of GFET is typically evaluated using custom-made manual bending test fixtures. DC probes were landed on source, drain and gate contacts. The bending of the device was increased and simultaneously DC measurements can be conducted for *in situ* studies. This process can be repeated until device breakdown. Our contemporary graphene transistors with patterned dielectrics can sustain bending strain of around 8.6%. At this tensile strain, the channel-resistance and mobility remained within 80% of its initial value. Strains greater than 8.6% resulted in gate dielectric breakdown and increased gate leakage [8,19]. To improve device reliability and robustness, GFETs were encapsulated with Si_3N_4 and a fluropolymer (e.g., Cytop). The devices were exposed to DI-water and other wet environments to evaluate its liquid resilience. The device performance was within 80% of its initial properties even after two days of immersion in water [19].

39

3.5. GFET Based Flexible Receiver

Figure 5 shows the block diagram of a relatively simple graphene based radio frequency communication system. It consists of a graphite antenna which receives the RF signal. This is followed by an input match network for impedance matching between the antenna and graphene based frequency translation block (mixer or demodulator) that function as an RF signal down-converter. The low frequency output of the frequency translation block or mixer is connected to a graphene speaker through an output matching network. In the following subsections we describe the realization of graphene RF communication circuits.

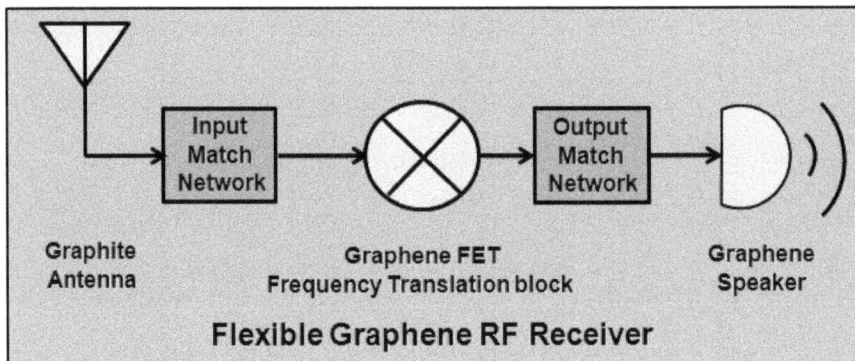

Figure 5. Schematic of a Flexible graphene based RF system.

3.5.1. Graphite Flexible Antenna

We designed and fabricated a graphite based microstrip patch antenna at 2.4 GHz on flexible polyimide substrate (Figure 6a). The performance of this antenna is similar but relatively more lossy than copper based patch antenna. This antenna is designed towards the goal of achieving an all graphene/carbon flexible RF receiver system. The antenna was designed using a graphite sheet since it is more conductive than monolayer graphene at microwave frequencies [20]. Graphite sheets offer superior flexibility and thermal conductivity compared to copper, albeit lower electrical conductivity. It is expected that flexible planar antennas based on graphite sheets are compatible with roll to roll manufacturing for large-area or high volume applications. In addition, the DC conductivity of intercalated graphite antennas can be tuned for the design of sensors and reconfigurable RF systems [21]. The antenna simulation results from a 3D electromagnetic solver, (CST) and experimental measurements are shown in Figure 6b. The antenna dimensions are width $W = 39.11$ mm and length $L = 30.95$ mm. The far field radiation pattern is shown in Figure 6c.

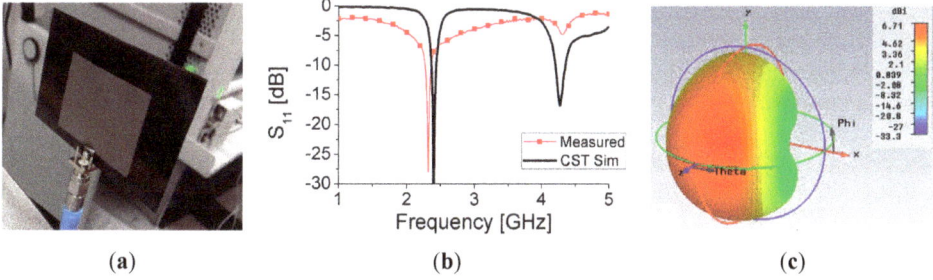

Figure 6. (a) Graphite Patch Antenna; (b) Measured and simulated response; (c) Simulated radiation pattern (adapted from [17], Copyright 2013, permission granted by Kristen Parrish).

To investigate the flexibility of the antenna, convex bending tests were carried out as shown in Figure 7a. The antenna was placed in the middle of a test vice fixture and mechanically bent (see Figure 7b). The S_{11} was measured *in situ* for each bending radius (see Figure 7c). The resonant frequency changes are relatively small down to ~30 mm bending radius. This result offers new opportunity for carbon/graphene RF electronics that can be integrated with flexible substrates, including bendable GHz electronics and antennas.

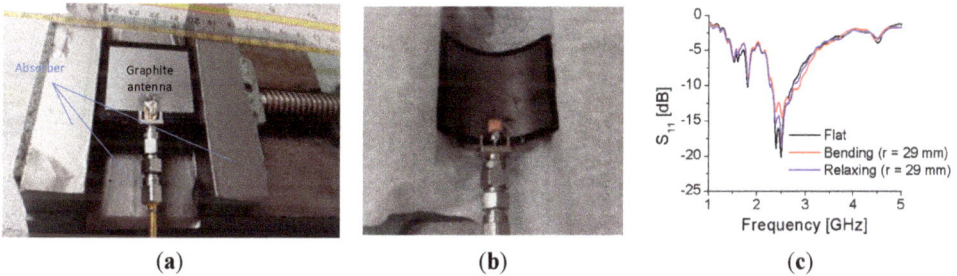

Figure 7. (a) Convex bending of graphite antenna; (b) Antenna is placed in the middle of test vice fixture; (c) Antenna response with convex bending of r = 29 mm (adapted from [17], Copyright 2013, permission granted by Kristen Parrish).

3.5.2. Frequency Translation Block

We describe in the following subsections two types of frequency translation blocks, GFET doublers and demodulators. GFET doubler is a frequency multiplier which can enable the realization of very high frequency communication circuits on soft substrates. On the other hand, a GFET demodulator described here does AM demodulation from carrier frequency to baseband. In general, graphene transistors can also be modified to serve as a demodulator for a variety of communication

41

schemes, such as amplitude shift keying (ASK), frequency shift keying (FSK) and phase shift keying (PSK) modulations [13].

(1) GFET Doubler

We designed and fabricated a graphene doubler on flexible PI substrate. We will first discuss the operating principle of a GFET doubler. The input and output characteristics of an ideal square law transistor doubler are depicted in Figure 8.

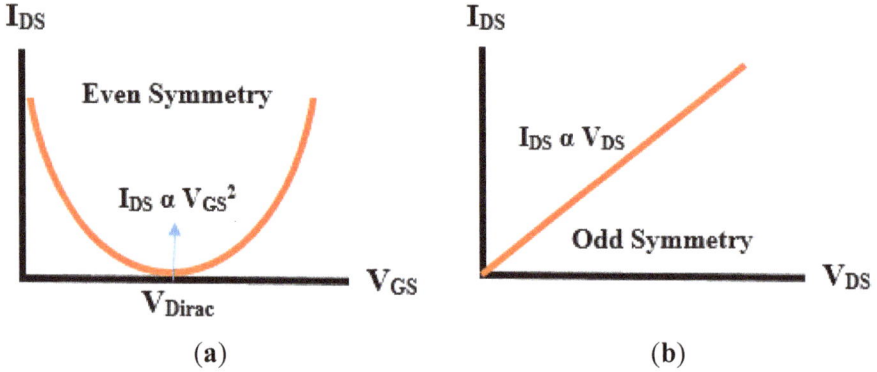

Figure 8. (a) Transfer and (b) Output Characteristics of an ideal square law transistor doubler (adapted from [22], Copyright 2011, AIP Publishing LLC).

It has ambipolar input characteristics (even symmetry) due to current conduction by both holes and electrons. For positive gate bias ($V_{GS} > V_{dirac}$) electrons conduct and for negative gate bias ($V_{GS} < V_{dirac}$) holes conduct. Ideally, the Dirac voltage should be around 0 V. The drain current I_{DS} is proportional to V_{GS}^2. On the other hand, the output current characteristic is linear (odd symmetry) with respect to the drain bias, V_{DS}. Using these two properties of an ideal square law device in our transistor model (Figure 2) we can derive an approximate expression for the output voltage:

$$|v_{ds}| = \frac{g_0 R_L \gamma V_{DS} v_{in}^2 \left(\cos 2\omega_{in} t \right)}{2n_0 + g_0 R_L \left(2n_0 + \gamma v_{in}^2 \right)} \tag{5}$$

where g_0 and γ are scaling factors, and R_L is the load resistor. These parameters and coefficients are described in detail in our previous works [17,22].

In Equation (5) we have neglected higher order terms in the simplified analysis. This working description clearly shows that GFET biased at Dirac point can be used as a frequency doubler [17]. In practice, there are several sources of non-idealities that can degrade doubler performance including charged impurities and contact resistance with consequent reduction of electron-hole symmetry and gate modulation.

We fabricated a prototype GFET doubler biased at the Dirac point using the above concept. Figure 9a shows the schematic of the frequency doubler. A function generator is used to apply the RF input. Bias-Tees are used for input and output routing for DC bias, and an oscilloscope as the load to the doubler. Figure 9b shows the power spectrum. The conversion gain of our doubler is −29.5 dB [16]. This is relevant in the context of (flexible) RF transmitters which typically requires frequency upconversion.

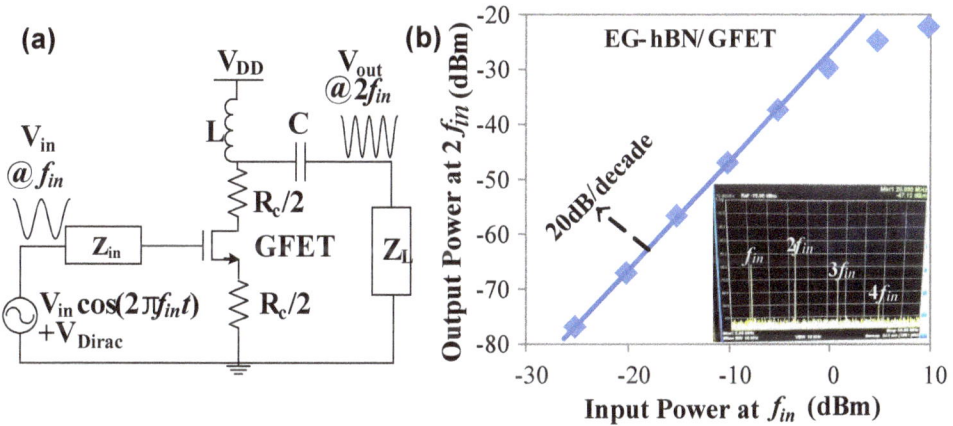

Figure 9. (**a**) GFET Doubler; and (**b**) Power Spectrum (adapted from [16], Copyright 2013, permission granted by IEEE).

(2) GFET AM Demodulator

In our goal towards the demonstration of a flexible graphene AM radio, we designed and fabricated a flexible GFET demodulator. The demodulator circuit is as shown in Figure 10a. The input is an AM modulated signal (2 KHz modulation frequency, 400 MHz carrier frequency). The output is connected to an oscilloscope through a low pass filter. Smoothening capacitors are used at the output to limit the spectra to the audio range. The GFET is biased close to the Dirac point. The ambipolar characteristic of GFET at the Dirac point (see Equation (5)) makes it to act as a square law AM demodulator. Low pass filtering is necessary at output to filter out higher order harmonics that are also produced. The gain of the demodulator is ~−34 dB (Figure 10b,c). The gain performance is on the low end for the current devices due to parasitics from the source and drain contacts and interconnects. The GFET demodulator, resistors and capacitors (C_{c1} and C_{c2} surface mount device components) are mounted on the flexible polyimide substrate.

43

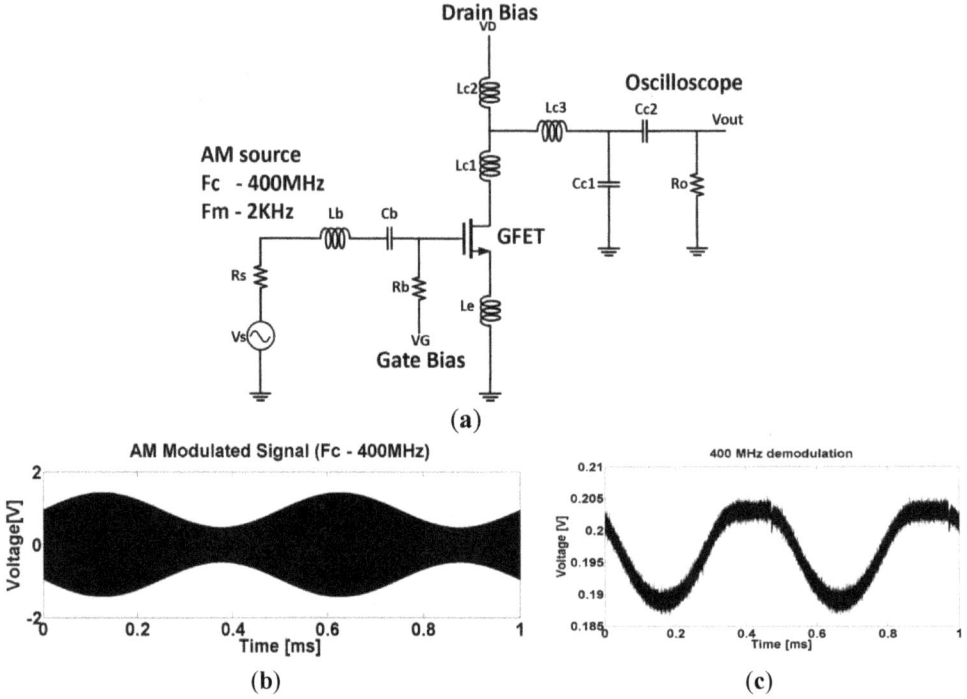

Figure 10. (a) GFET Demodulator (b) AM input and (c) Demod output (adapted from [17], Copyright 2013, permission granted by Kristen Parrish).

3.5.3. Flexible Graphene Speaker

Graphene along with its outstanding electrical and mechanical properties has very good thermal-acoustic characteristics. It offers very small heat capacity per unit area making it suitable as a thermo-acoustic speaker [23,24]. These graphene speakers are becoming quite attractive owing to its potentially low manufacturing cost, light weight, low power consumption and also high flexibility. Recently, Tian *et al.* demonstrated flexible graphene based speaker on a paper substrate [23,24]. In this light, we have designed and fabricated a flexible mono layer graphene speaker on PI substrate as shown in Figure 11. CVD grown monolayer graphene (1 cm^2) is transferred onto a smoothened PI substrate. Gold/Silver metal contacts are deposited at the two edges of graphene. Baseband demodulated signal from receiver is applied to the two contacts. These signals create thermal oscillations at the graphene-air interface, which are then converted to sound waves.

44

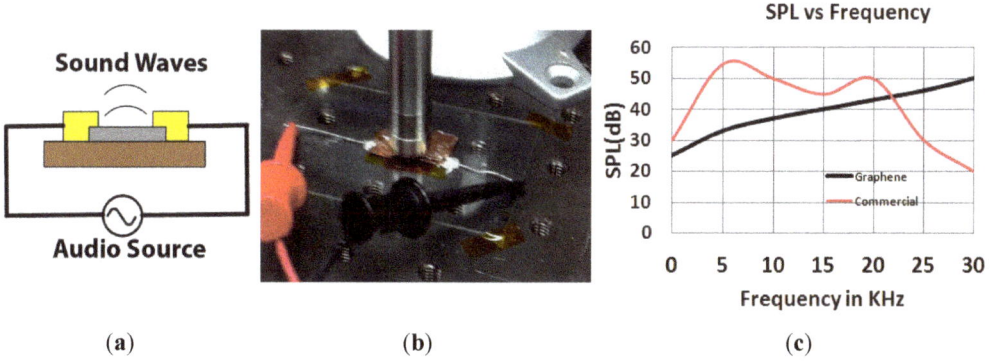

Figure 11. (a) Side view of graphene speaker (grey—monolayer graphene, yellow—gold contacts, brown—PI substrate); (b) Fabricated graphene speaker and microphone test setup; (c) SPL *vs.* frequency plot comparing flexible graphene speaker with a commercial speaker. (adapted from [25], Copyright 2014, IEEE)

The measurement of the sound waves from this speaker is performed using a microphone and signal analyzer (See Figure 11b). The distance between the microphone and speaker is kept at 3 cm. This speaker performance is as shown in Figure 11c. It also shows the comparison with commercial speaker. These speakers provide a sound pressure level (SPL) of 25–50 dB in the audio frequency range. The speaker power at high frequencies is given by Equation (6) [23]. It mainly depends on frequency of sound (f), heat capacity of air (γ), velocity of sound in air (v_g) thermal effusivity of PI substrate (e_s) and air (e_a) [23].

$$P_{speaker} = \frac{Ro}{\sqrt{2}r_o} \frac{\gamma - 1}{v_g} \frac{e_g}{e_s + a_c + e_g} q_o \tag{6}$$

Future research towards monolithic graphene nanosystems requires integration of the speaker with the AM radio receiver. The performance of this speaker at low frequency can be improved by high quality graphene film, waveguide design and fabrication of suspended graphene speakers.

3.5.4. Graphene AM Radio Receiver

A graphene based flexible AM radio receiver at 2.4 GHz was designed and fabricated based on the block diagram of Figure 9. The receiver consists of a graphite antenna, GFET demodulator and speaker. The receiver is fabricated on a polyimide flexible substrate. The AM input signal is obtained from a signal generator (2.45 GHz carrier signal) and transmitted by a horn antenna. The signal is received by the graphite antenna and demodulated using the GFET demodulator based on the

45

ambipolar graphene characteristics at the Dirac point. The output of the demodulator is connected to the speaker [25].

The schematic of the AM radio receiver is shown in Figure 12a. The input matching network is used for impedance matching between the antenna (L_a, R_a) and GFET input impedance. The output matching network has two functions: (i) impedance match between output of GFET and graphene speaker load (R_o) and (ii) low pass filtering of the demodulated signal. Figure 12b shows the fabricated receiver. The passive components—inductors, capacitors and resistors are currently realized using surface mount device (SMD) components. The receiver measurements were carried out in a custom built anechoic chamber. Bending tests (~30 mm radius) were successfully carried out. The performance changes are small due to the flexibility of substrate, antenna, graphene FET and transmission lines.

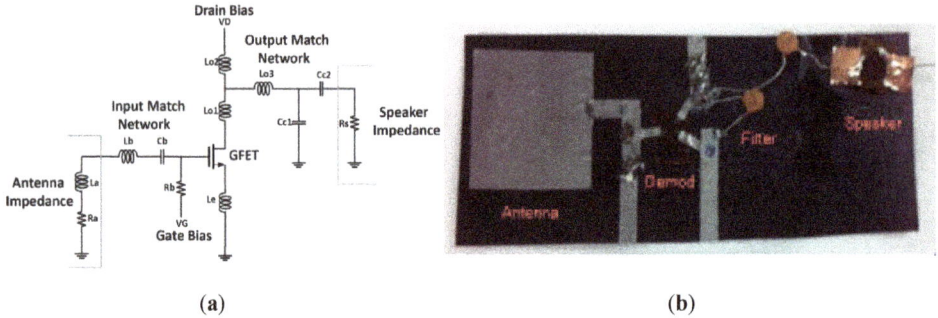

(a)   (b)

Figure 12. (**a**) Schematic of a graphene-based AM radio receiver; (**b**) Fabricated AM radio receiver on a flexible PI (Kapton) substrate (adapted from [25], Copyright 2014, IEEE)

The link budget of the flexible graphene receiver is estimated to be −46dB. In this estimation we have considered path loss to be −1dB/cm for 6cm signal propagation through air, graphite antenna loss to be 5dB, and GFET demodulator circuit gain to be -35dB. The current AM receiver suffers from very high noise output due to lossy (un-optimized) demodulator. To obtain larger signal to noise ratio and amplitude, a GFET low noise amplifier could be incorporated at the input. The performance could be further improved with optimized design of the demodulator and interconnects.

4. Conclusions

In this article on the research progress towards flexible integrated graphene nanosystems, we demonstrate the feasibility of an all graphene/carbon flexible RF communication receiver. We have shown state of the art flexible GFET with high cutoff frequency in the microwave GHz regime. We have also investigated the realization and performance of GFET doubler and demodulator, flexible graphene speaker, flexible graphite antenna and carbon based AM receiver. Improved receiver performance can be achieved with the use of semiconducting layered atomic materials such as MoS_2 and black phosphorous for high-gain baseband amplifiers and mixed signal circuits.

Acknowledgments: This work is supported in part by the Office of Naval Research (ONR) and the National Science Foundation (NSF) Nascent Engineering research center (ERC) under Cooperative Agreement No. EEC-1160494. The research was carried out at the Microelectronics Research Center at The University of Texas at Austin. We acknowledge the help of He Tian and Tian-Ling Ren of Tsinghua University with graphene speaker measurements.

Author Contributions: All authors contributed to the research described in this work. M.N.Y., K.N.P., J.L., D.A conceived and designed the experiments; M.N.Y., K.N.P. S.P. performed the experiments; M.N.Y. and K.N.P. analyzed the data; M.N.Y., D.A. wrote the paper.

Conflicts of Interest: The authors declare no conflict of interest.

References

1. Russer, P.; Fitchner, N. Nanoelectronoics in Radio Frequency technology. *IEEE Microw. Mag.* **2010**, *11*, 119–135.
2. Akinwande, D.; Petrone, N.; Hone, J. Two-dimensional flexible nanoelectronics. *Nat. Commun.* **2014**, *5*.
3. Schwierz, F. Graphene Transistors: Status, Prospects, and Problems. *Proc. IEEE* **2013**, *101*, 1567–1584.
4. Akinwande, D.; Tao, L.; Yu, Q.; Lou, X.; Peng, P.; Kuzum, D. Large-Area Graphene Electrodes: Using CVD to facilitate applications in commercial touchscreens, flexible nanoelectronics, and neural interfaces. *Nanotechnol. Mag. IEEE* **2015**, *9*, 6–14.
5. Yeh, C.-H.; Lain, Y.-W.; Chiu, Y.-C.; Liao, C.-H.; Moyano, D.R.; Hsu, S.S.H.; Chiu, P.-W. Gigahertz Flexible Graphene Transistors for Microwave Integrated Circuits. *ACS Nano* **2014**, *8*, 7663–7670.
6. Nathan, A.; Ahnood, A.; Cole, M.T.; Sungsik, L.; Suzuki, Y.; Hiralal, P.; Bonaccorso, F.; Hasan, T.; Garcia-Gancedo, L.; Dyadyusha, A.; *et al.* Flexible Electronics: The Next Ubiquitous Platform. *Proc. IEEE* **2012**, *100*, 1486–1517.
7. Kim, D.H.; Lu, N.; Ma, R.; Kim, Y.S.; Kim, R.-H.; Wang, S.; Wu, J.; Won, S.M.; Tao, H.; Islam, A.; *et al.* Epidermal Electronics. *Science* **2011**, *333*, 838–843.
8. Lee, J.; Ha, T.; Li, H.; Parrish, K.N.; Holt, M.; Dodabalapur, A.; Ruoff, R.S.; Akinwande, D. 25 GHz Embedded-Gate Graphene Transistors with High-K Dielectrics on Extremely Flexible Plastic Sheets. *ACS Nano* **2012**, *7*.
9. Chang, H.; Yang, S.; Lee, J.; Tao, L.; Hwang, W.; Jena, D.; Lu, N.; Akinwande, D. High-Performance, Highly Bendable MoS2 Transistors with High-K Dielectrics for Flexible Low-Power Systems. *ACS Nano* **2013**, *7*, 5446–5452.
10. Li, X.; Cai, W.; An, J.; Kim, S.; Nah, J.; Yang, D.; Piner, R.; Velamakanni, A.; Jung, I.; Tutuc, E.; *et al.* Large-area synthesis of high-quality and uniform graphene films on copper foils. *Science* **2009**, *324*, 1312–1314.

11. Wang, H.; Hsu, A.L.; Palacios, T. Graphene electronics for RF applications. *Microw. Mag. IEEE* **2012**, *13*, 114–125.

12. Han, S.; Garcia, A.V.; Oida, S.; Jenkins, K.A.; Haensch, W. Graphene radio receiver integrated circuit. *Nat. Commun.* **2014**, *5*.

13. Lee, S.; Lee, K.; Liu, C.; Kulkarni, G.S.; Zhong, Z. Flexible and transparent all-graphene circuits for digital quaternary modulations. *Nat. Commun.* **2012**, *3*.

14. Park, S.; Zhu, W.; Chang, H.-Y.; Yogeesh, M.N.; Ghosh, R.; Banerjee, S.; Akinwande, D. High-frequency prospects of 2D nanomaterials for flexible nanoelectronics from baseband to sub-THz devices. In Proceedings of the 2015 IEEE International Electron Devices Meeting (IEDM) Technical Digest, Washington, DC, USA, 7–9 December 2015; pp. 32.1.1–32.1.4.

15. Parrish, K.N.; Ramón, M.E.; Banerjee, S.K.; Akinwande, D. A Compact Model for Graphene FETs for Linear and Non-linear Circuits. In Proceedings of the SISPAD 2012, Denver, CO, USA, 5–7 September 2012.

16. Lee, J.; Ha, T.J.; Parrish, K.N.; Chowdhury, SF.; Tao, L.; Dodabalapur, A.; Akinwande, D. High performance current saturating graphene field-effect transistor with hexagonal boron nitride dielectric on flexible polymeric substrates. *IEEE Electron Device Lett.* **2013**, *34*, 172–174.

17. Parish, K. Nanoscale Graphene for RF Circuits and System. Ph.D. Dissertation, University of Texas at Austin, Austin, TX, USA, August 2013.

18. Koolen, M.C.A.M.; Geelen, J.A.M.; Versleijen, M.P.J.G. An Improved De-Embedding Technique for on-Wafer High-Frequency Characterization. In Proceedings of the IEEE 1991 Bipolar Circuits and Tech Meeting, Minneapolis, MN, USA, 9–10 September 1991.

19. Lee, J. Graphene Field Effect Transistors for Flexible High Frequency Electronics. Ph.D. Dissertation, University of Texas at Austin, Austin, TX, USA, May 2013.

20. Panasonic PGS Graphite Sheets. Available online: http://industrial.panasonic.com/lecs/www-data/pdf/AYA0000/AYA0000CE2.pdf (accessed on 30 June 2014).

21. Ferendeci, A.; Devlin, C.L.H. Microwave Charecterization of Graphite and Intercalated Graphite as an Antenna Element. In Proceedings of the 2012 IEEE National Aerospace and Electronics Conference (NAECON), Dayton, OH, USA, 25–27 July 2012.

22. Parrish, K.N.; Akinwande, D. Even-odd symmetry and the conversion efficiency of ideal and practical graphene transistor frequency multipliers. *Appl. Phys. Lett.* **2011**, *99*, 223512.

23. Tian, H.; Ren, T.; Xie, D.; Wang, Y.; Zhou, C.; Feng, T.; Fu, D.; Yang, Y.; Peng, P.; Wang, L.; *et al.* Graphene-on-Paper sound source devices. *ACS Nano* **2011**, *5*, 4878–4885.

24. Suk, J.W.; Kirk, K.; Hao, Y.; Hall, N.A.; Ruoff, R.S. Thermoacoustic sound generation from monolayer graphene for transparent and flexible sound sources. *Adv. Mater.* **2012**, *24*, 6342–6347.

25. Yogeesh, M.N.; Parrish, K.N.; Lee, J.; Tao, L.; Akinwande, D. Towards the design and fabrication of graphene based flexible GHz radio receiver systems. In Proceedings of the 2014 IEEE MTT-S International Microwave Symposium (IMS), Tampa, FL, USA, 1–6 June 2014.

Graphene and Two-Dimensional Materials for Optoelectronic Applications

Andreas Bablich, Satender Kataria and Max C. Lemme

Abstract: This article reviews optoelectronic devices based on graphene and related two-dimensional (2D) materials. The review includes basic considerations of process technology, including demonstrations of 2D heterostructure growth, and comments on the scalability and manufacturability of the growth methods. We then assess the potential of graphene-based transparent conducting electrodes. A major part of the review describes photodetectors based on lateral graphene p-n junctions and Schottky diodes. Finally, the progress in vertical devices made from 2D/3D heterojunctions, as well as all-2D heterostructures is discussed.

Reprinted from *Electronics*. Cite as: Bablich, A.; Kataria, S.; Lemme, M.C. Graphene and Two-Dimensional Materials for Optoelectronic Applications. *Electronics* **2016**, *5*, 13.

1. Introduction

Two-dimensional (2D) materials are very promising with respect to their integration into optoelectronic devices. Even though the technology readiness levels are still low and device manufacturability and reproducibility remain a challenge, graphene technology can be found in research labs around the globe. The same challenges apply to other 2D materials, like silicene, germanene, stanene and phosphorene, or for transition metal dichalcogenides (TMDs) to a much greater extent. Nevertheless, the scientific community has been able to demonstrate the enormous potential of 2D materials in numerous prototype devices and systems. This article discusses the state-of-the-art of developments in optoelectronic devices based on 2D materials and their applications.

Optoelectronic devices are electronic devices that use either electric charge to generate light, like light emitting diodes and lasers, or use light to generate electric current, like photovoltaic devices and photodetectors (PDs). The field can be subdivided into different areas depending on the physical mechanisms responsible for photon or charge generation that are exploited for device operation. Photoemission, radiative recombination, stimulated emission, photoconductivity and the photoelectric effect are examples for the mechanisms that optoelectronic devices might exploit. This review focuses on devices utilizing photoconductivity and the photoelectric effect.

To maximize the amount of photons to be converted into electron/hole pairs, a high material absorbance is required. Consequently, the absorbance (A) and the transmittance (T) are the key parameters for optoelectronic materials. The integration of 2D materials into photonic devices is promising, as they tend to absorb a significant amount of photons per layer. For example, graphene has a broadband light absorption of 2.3% below 3 eV [1] that linearly scales with the number of layers [2], despite being only one atom thick. The absorbance of graphene depends on the fine structure constant $\alpha = \dfrac{e^2}{\hbar c}$, where c is the speed of light, e is the charge of an electron and $\hbar$ is Planck's constant divided by 2π (A = $\pi\alpha$ = 2.3%) [2]. Taking into account the frequency independent optical ac conductance G(ω):

$$G\left(\omega\right) = G_0 = \frac{e^2}{4\hbar} \tag{1}$$

of a graphene monolayer [3], its T can be calculated by:

$$T = (1 + 2\pi G_0/c)^{-2} = (1 + 0.5\alpha\pi)^{-2} \approx 0.977 \tag{2}$$

The reflectance of graphene is less than 0.1% in the visible region [4]. Even though the absorbance per graphene layer is quite high, thin graphene films have an overall high broadband transmission level, which is of high significance for transparent conductive electrodes. In addition, single- and bi-layer graphene sheets may become fully transparent as a result of Pauli blocking [5].

Compared to graphene, TMDs, like molybdenum disulfide (MoS$_2$), tungsten disulfide (WS$_2$) and molybdenum diselenide (MoSe$_2$), exhibit even higher absorption in the visible and the near infrared (NIR) range, *i.e.*, above their respective energy band gaps, which makes this 2D material class an ideal candidate to act as the thinnest photo-active materials [6]. The high optical absorption in TMDs can be explained by dipole transitions between localized d states and excitonic coupling of such transitions [6,7]. Due to its high bandgap and absorption coefficient in the deep ultraviolet (UV) region, hexagonal boron nitride (h-BN) has been proposed as a potential UV photodetector [8].

2. Technology of 2D Materials

The key to future commercial uptake of 2D technology is the availability of large-scale material growth or synthesis methods. Chemical vapor deposition (CVD) is a well-established technique in the semiconductor industry for growing a variety of conventional materials. Today, CVD can in principle fulfill the requirement of the large-scale growth of graphene and many 2D materials of the TMD family [9,10]. However, the reproducible synthesis of TMDs on a large scale is still challenging, and the process technology is far from optimized. For example, it is not possible today to control exactly the number of atomic layers across a wafer or to grow monocrystalline films without grain boundaries.

The graphene CVD process is based on the decomposition of hydrocarbons on catalytic/metallic surfaces, such as copper, at temperatures above 800 °C [9]. Growth temperatures as low as 300 °C have also been achieved using a microwave plasma-assisted CVD method [11]. A second bottleneck for large-scale production of graphene devices with a low defect density is the process required to transfer the graphene films from the catalytic surface to the destined substrate. A majority of researchers use a polymer-assisted graphene transfer (e.g., polymethyl methacrylate (PMMA)), as this method is capable of handling chip or wafer-sized graphene films [12]. The method typically requires that liquids are present during the transfer step and is therefore also called wet transfer. Another approach is to use polydimethylsiloxane (PDMS) stamps, and this requires defined and controllable conditions for pressing and releasing stamps on/from the 2D surface [13]. The PDMS stamp method is classified as dry transfer. A general issue with both wet and dry transfer methods is that they often result in polymer residues on the graphene surface. In addition, transfer can induce damage or folding of the graphene films. The combined process of catalytic growth and transfer includes a step where the graphene is removed from its growth catalyst, *i.e.*, most often copper. This can be achieved through wet etching of copper with a suitable acid or through (electrochemical) delamination. There is a high risk that this step leaves copper contaminants on the graphene that greatly reduce device performance and reliability [14]. A thorough comparison and evaluation of graphene transfer methods can be found in [15].

In 2010, Bae *et al.* demonstrated a concept for roll-to-roll production of graphene. They synthesized 30 inch-wide graphene films using a thermal-release tape-based transfer with good electrical and physical properties [16]. The synthesis of polycrystalline CVD graphene on 300 mm wafers has been demonstrated, showing more than 95% monolayer uniformity [17].

Charge carrier mobility is a good measure for the quality of graphene. So far, the highest mobility values have been reported for mechanically-exfoliated graphene, a technique not suitable for large-scale production. Recently, CVD graphene has been reported with Hall mobility values of 350,000 cm^2/Vs at 1.6 K and 50,000 cm^2/Vs at room temperature in devices made from single crystallites on hexagonal boron nitride [13]. These examples serve to demonstrate that chemical vapor deposition (CVD) may become a manufacturable method for graphene production in the future.

3. Transparent Conductive Electrodes

Photovoltaic (PV) cells, photo detectors (PDs), light emitting devices (LED), liquid crystal displays (LCD), flexible organic LEDs, flexible smart windows, bistable displays or touch screens are devices that altogether rely on electrodes with $T > 80\%$ in combination with a low sheet resistance (R_S).

The R_S of touch displays for example should not exceed 500 Ω/ð [18]. Applications like smart windows require R_S values of about 400 Ω/ð, flexible LCDs 300 Ω/ð or less, flexible OLEDs about 100 Ω/ð and high-efficiency solar cells 50 Ω/ð or less [18].

State-of-the-art transparent conductive electrodes (TCEs) predominantly consist of semiconductor-based transparent conductive oxides (TCOs) that can fulfill these requirements. While indium (In) tin oxide (ITO) is the most popular representative, there are several other TCOs available, namely intrinsic indium oxide (In_2O_3), zinc oxide (ZnO) [19] or tin oxide (SnO_2) [20]. ITO generally obtains $T \approx 80\%$ (at 550 nm) and 10 Ω/ð on silicon dioxide (SiO_2) [21] or less than 300 Ω/ð on polyethylene terephthalate (PET), respectively [22].

ITO is comparatively expensive, and its electro-/optical properties change with the indium content and deposition technologies [17]. Although it is the most popular TCE, ITO is brittle and therefore not suitable for flexible electronic or photonic applications [20], and may be ultimately limited by the amount of indium available on the planet. Nevertheless, the first flexible organic displays have been demonstrated by Samsung at the International Consumer Electronics Show 2013 [23]. Gadgets like flexible organic thin film transistor displays for smart watches, wearable devices or consumer electronics are on the market [24]. However, future TCEs for reliable and flexible optoelectronic applications will require smart and eco-friendly high-performance TCO alternatives. Promising candidates to compete with established TCEs are thin metal (wire) films, carbon nanotubes (CNT) and graphene electrodes, which offer flexibility and efficient low- and high-frequency operation in combination with a broad wavelength transparency [25].

Graphene in particular is attracting considerable attention with its broadband transparency exceeding the performance of single-walled carbon nanotubes [26,27], metallic films [28] and ITO [21]. The fundamental limit in R_S and T in intrinsic single-layer graphene (SLG) is 6 k Ω/ð [29] and 97.7% [2], respectively. While intrinsic SLG is highly suitable to be used as a TCE in photonic devices with regards to T, R_S remains far below the specifications for a competitive TCE. At the expense of T, R_S can significantly be reduced by stacking SLG to reach $T > 90\%$ and 30 Ω/ð for four-layer graphene [16]. These properties beat the "minimum industrial standard" in terms of R_S and T reached by ITO films for the first time [27]. The growth of SLG with $T = 95\%$ and 7×10^5 Ω/ð at 300 °C using a microwave plasma has also been successfully shown by Yamada $et\ al.$, which potentially makes the graphene deposition technology compatible with low temperature processing [11]. The high mobility reported in CVD-graphene can result in R_S values comparable to ZnO/Ag/ZnO [30], TiO_2/Ag/TiO_2, CNTs and ITO combined with an identical or even higher transparency [29]. Nevertheless, alternative nanotechnologies, such as silver nanowire meshes [31], show promising figures of merit ($T = 85\%$ and 13 Ω/ð).

These technologies, possibly in combination with graphene as the base material, may lead to superior performance of pure graphene-based TCEs.

4. Photodetectors

Photodetectors (PDs) typically consist of p-n junctions, *i.e.*, one n-doped and a complementarily p-doped area that form a space-charge region at the interface. Crystalline silicon (x-Si) p-n junctions cover a spectral bandwidth from the UV (λ > 190 nm) to the NIR (λ < 1100 nm) spectrum and are the most conventional PDs available. Typical sensor specifications are: dark currents of tens of pA, rise times below 1 μs, a maximum spectral response (SR) in the range of 600 mAW^{-1} and cutoff frequencies in between 20 and 30 MHz. The SR is an area independent relative measure of the wavelength-dependent photocurrent. The position and amplitude of the peak SR can vary by orders of magnitude, depending on technological parameters like junction depth or doping concentrations. The main applications of x-Si sensors are in the fields of digital imaging systems and optical switches. Si avalanche photodiodes (APDs) are a special type of PD, which are optimized for low light level detection. Their bandwidth commonly ranges from 320 nm–1150 nm. InGaAs pin-diodes are detectors used in optical communication systems or power meters, covering in particular NIR bands in between 900 nm and 1700 nm. Sensitivities of ~1.1 A/W or less are typical in such diodes. In the past, germanium (Ge)-based detectors have been used as alternatives to low-cost InGaAs PDs. These detectors celebrated a rebirth in integrated optoelectronics, as it became possible to deposit Ge on Si directly [32]. Recently, tin-doped Ge films (GeSn) are being investigated as alternative semiconductors with reduced direct band gaps [33]. High-stability GaAsP diodes with sensitivities ranging from the near-UV (280 nm) to 680 nm and GaP Schottky diodes with high UV sensitivities (190 nm–550 nm) are exotic material examples that expand the wide range of active absorbers available for PDs. However, every single absorber is optimized only for a specific wavelength regime. Up to now, there is no "all-in-one" PD commercially available that covers a bandwidth including the UV-regime below 190 nm, the visible and the IR spectrum above 1150 nm at the same time. The synthesis of TMDs and novel 2D materials might become one of the key enabling technologies in the near future to extend the spectral bandwidth of existing photonic devices. Graphene in particular plays a significant role in research towards extending the spectral bandwidth of PDs due to its almost linear absorption of photons over the complete electromagnetic spectrum. In the following section, the performance, especially the SR of PDs and optical waveguides, of photonic devices based on 2D materials is reviewed.

4.1. Graphene Photodetectors

Graphene offers highly efficient tuning of the carrier type and density, which can be utilized to induce p–n junctions, e.g., by electrostatic gating [34]. Other possibilities to create p–n junctions in graphene or related 2D materials are chemical doping [35] or by exploiting the work function difference between the intrinsic material (e.g., $\Delta W = 4.45$ eV for graphene [36]) and metal contacts [37–39].

Examples of mono- and bi-layer graphene photodetectors with electrostatically-tunable p–n junctions were reported by Lemme *et al.* [34]. Here, the graphene p-n junctions were formed with a dual gate configuration, *i.e.,* a local top gate electrode and a global back gate (Figure 1a,b). The device thus allowed tuning the carrier type and density under the top gate independently from the graphene outside the top gate region (Figure 1c). Thus, operation in four different regimes is possible, *i.e.,* with p-n, p-p$^+$, n-p and n-n$^+$ junctions at the edges of the top gate (Figure 1d). A photo responsivity of ~1.4 mAW^{-1} was demonstrated with scanning photocurrent measurements ($\lambda = 532$ nm, $P = 30$ μW). It was further shown that the photocurrent was highest when the devices were operated as p-n or p junctions. This was consistent with theory and could be described by the photo-thermoelectric effect (PTE). Similar investigations on dual-gated monolayer and bilayer graphene p-n junctions were reported by Gabor *et al.* [40]. They concluded that nonlocal hot carrier transport with a long-lived and spatially-distributed hot carrier population dominates the intrinsic photoresponse in graphene over the photovoltaic effect [40]. Subsequent experimental and theoretical work supports that the efficiency of lattice cooling in graphene is quite poor, so that hot carriers are responsible for nonlocal transport [41–43]. The highest SR for a graphene p–n junction reported so far is 10 mAW^{-1} at 514.5 nm [44]. This device is similar to those reported in [34] and [40], but optimized, as the edge contacted graphene additionally contains a suspended area in the device center.

(a)

(b)

(c)

(d)

(e)

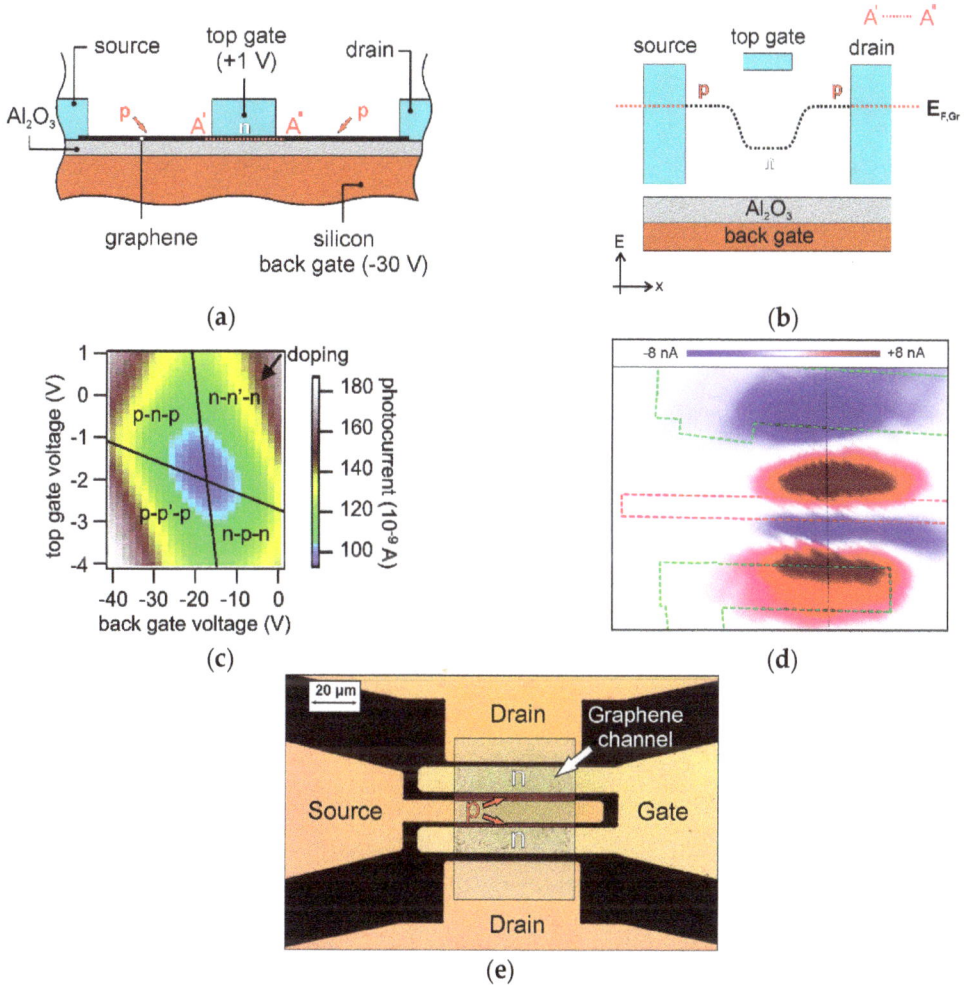

Figure 1. (a) Schematic of a top- and back-gate tunable graphene phototransistor; (b) Schematic of the band structure of a symmetric phototransistor with a top and a back gate; (c) Drain current (color scale) as a function of the back and top gate voltages. The black lines indicate the location of the charge neutrality (Dirac) point of the global (back gated) and local (top gated) device. The phototransistor can be operated in four distinct regions (*i.e.*, p-n-p *etc.*); (d) Photocurrent map of a lateral phototransistor. The photocurrent is highest at the location of p-n junctions, *i.e.*, next to the metal contacts and the top gate region (modified from [34]); (e) Optical micrograph of a graphene phototransistor. Examples for p and n regions are indicated.

Fixed p-n junctions can be readily obtained in metal–graphene–metal assemblies, which have been used with back gate electrodes to demonstrate photocurrents [45–48]. Recently, Withers *et al.* substituted metal electrodes with $FeCl_3$-intercalated multilayer graphene to assemble an "all-graphene"-based PD that generates a PTE-induced photovoltage of ~0.1 VW^{-1} [49]. Mueller *et al.* demonstrated a graphene-metal PD for visible and NIR wavelengths with a bandwidth larger than 40 GHz, an internal quantum efficiency (IQE) of 10% and an external quantum efficiency (EQE) of 0.5% [37]. This device performed well as an optical data link up to 10 Gbit/s. An asymmetric metal–graphene–metal configuration (*i.e.,* two different metals) was used to break the mirror symmetry of the internal electric-field profile to induce the photovoltaic effect (PVE). The two metals effectively doped the graphene differently by charge transfer as a result of work function differences at the metal-graphene interfaces (Figure 2). We note that the discussion about the underlying mechanisms of the photocurrent in metal-graphene junctions is still on-going [50].

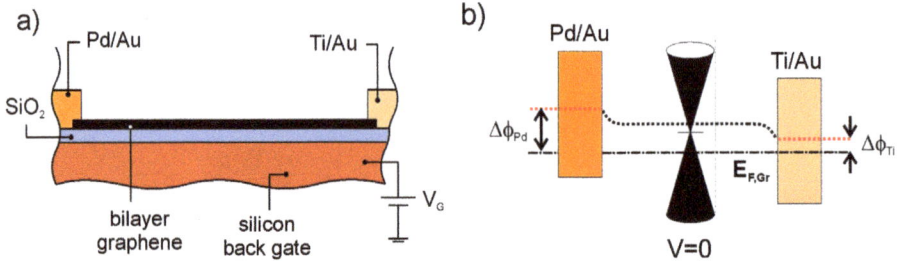

Figure 2. (**a**) Schematic of a bilayer graphene photodetector (PD) with asymmetric metal contacts; (**b**) schematic of the band structure of the device in (a) (after [37]).

Graphene PDs have been integrated into silicon waveguides in a process compatible with Si complementary metal oxide semiconductor (CMOS) technology for potential applications as on-chip optoelectronic couplers [51,52] or modulators [53]. This concept was further explored by Schall *et al.,* who successfully demonstrated waveguide-integrated CVD graphene PDs with a bandwidth of 41 GHz and an extrinsic response of 16 mAW^{-1} [54]. The detection of data signals up to 50 Gbit·s^{-1} is comparable to the performance of state-of-the-art germanium-based Si waveguide-coupled PDs [32,55]. A similar graphene/Si waveguide-integrated PD by Wang *et al.* exhibits a sensitivity of 0.13 mAW^{-1} at mid-IR wavelengths (λ = 2.75 μm). Here, the SR was described by a combination of the PTE and the PVE [25].

Schottky diodes represent an alternative to p-n junctions as photodetectors. Schottky diodes consist of metal-semiconductor junctions that can typically be operated at higher frequencies than semiconductor p-n junctions. PDs require a depletion region where electrical charge can be generated and extracted under reverse bias conditions that separate electrons and holes. Simple vertical Schottky junctions can be obtained by placing graphene on suitable semiconductors (Figure 3a,b). Chen *et al.* presented electrical measurements on graphene/Si Schottky diodes with a junction area of 92 μm^2 [56]. They achieved an ideality factor of $n = 4.89$ at room temperature, which is a quality indicator, where $n = 1$ implies that the device can be modeled by the ideal Shockley diode equation. State-of-the-art bulk semiconductor PDs achieve values in the range of $n = 1$–2. We reported large area graphene/Si Schottky diodes by transferring large-area (2×4 cm^2) single-layer CVD-grown graphene onto n-type Si (Figure 1c, black solid line) [57], demonstrating the integration of 2D materials into the existing Si platform. The simple, yet scalable fabrication process yielded devices with peak sensitivity of 270 mAW^{-1} at 992 nm and an ideality factor of approximately 1.5. When fabricated on p-type Si (Figure 3c, red dashed line), the devices do not exhibit rectification, because the graphene is also p-type when measured in ambient air due to humidity [58] and/or other adsorbates. Figure 3d compares the SR of a graphene/n-Si Schottky diode with that of a commercial x-Si p-n PD. The absolute response of the graphene/Si diode reaches approximately 50% of the reference values. This may be further improved by optimizing the contact resistance between the SLG and the x-Si substrate, a general issue with graphene-based devices [59]. An additional feature is worth mentioning: the SR of the commercial PD vanishes for photon wavelengths $\lambda > 1200$ nm due to the electronic band gap of Si. The graphene/Si heterodevice, in contrast, clearly remains sensitive in the NIR region (inset in Figure 3d). In that regime, the absorption in SLG enables detection, albeit at much lower responsivity values due to the monolayer absorption of approximately 2.3%. An optimized graphene/Si Schottky diode was reported by An *et al.* with a junction area of 25 mm^2 on n-type silicon [60]. Here, three-layer graphene was doped by pyrenecarboxylic acid (PCA). The device reached a maximum SR of ~435 mAW^{-1}, which corresponds to an IQE of ~65% at a peak wavelength in the range of 850 nm–900 nm. This performance is comparable to state-of-the-art bulk x-Si PDs. The device remained stable even after 10 days and could withdraw 1000 light-switching cycles with a current variation of $\pm 2.5\%$ (dark current) and $\pm 5\%$ (photocurrent) without noticeable drift phenomena. A similar non-doped SLG device showed a maximum SR of ~320 mAW^{-1}. Kim *et al.* proposed a graphene/Si heterojunction diode as a chemical sensor [61]. They utilized the fact that the graphene work function can be effectively modified through exposure to liquids or gases. This modulates the Schottky barrier height of the diodes and, hence, their resistance. The authors demonstrated considerable long-term stability and repeatability and suggested such devices as a platform for applications in gas, bio- and environmental sensing.

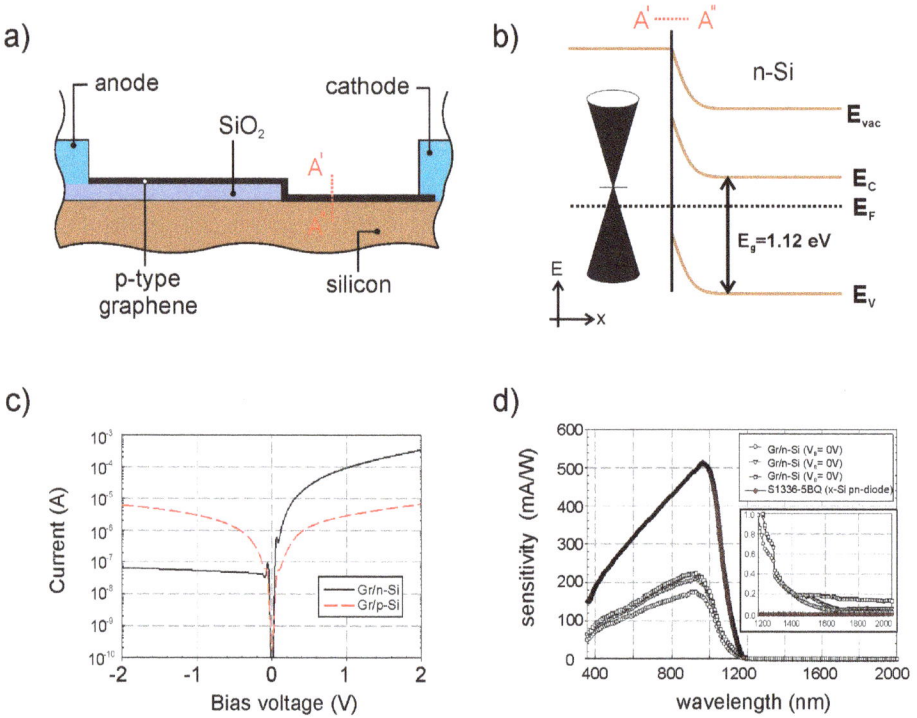

Figure 3. (**a**) Schematic of a graphene/Si heterojunction PD; (**b**) schematic of the band structure of the device in (a); (**c**) I-V curves of graphene-Si diodes on n-type (black solid line) and p-type (red dashed line) Si substrates; (**d**) comparison of the bias-dependent spectral response of a graphene/Si PD with a state-of-the-art bulk p-n junction Si PD. Inset: magnification of the spectral response in the IR region, where the response is entirely due to the graphene [57].

While the extrinsic response of graphene-based devices is generally below ~1 AW^{-1}, which is commonly reached by non-avalanche Ge detectors, the IQE of a variety of reported graphene PDs is quite high, with values reaching 30%–60% [44,45]. This indicates that the response is not limited by the intrinsic performance of graphene, but rather by non-optimized device layouts or by the electrical contacts. In addition, carrier multiplication in graphene may improve the quantum efficiencies in graphene-based devices to over 100%, if specific conditions are fulfilled [43,62,63].

In addition to these efforts, different strategies have been proposed and demonstrated to enhance the light absorption in graphene-based PDs, which include graphene nanostructures [64], chemical doping [65], plasmonic nanostructures in graphene [66] and the integration of graphene into photonic cavities [67,68]. In graphene, enhancement factors of 15 have been achieved using plasmonic nanostructures [69]. Other strategies combine graphene with different nanostructured materials to form graphene-based hybrid materials. Some hybrid materials that have been proposed include polymers [70], plasmonic nanostructures on graphene [71] and one-dimensional (1D) nanorods [72] or zero-dimensional (0D) quantum dots (QDs) [73]. The advantage of graphene hybrids is that the deposited materials do not only act as light harvesters, but they also provide interfaces or hetero-junctions, which can facilitate the separation of excitons, thereby extending the functionality of graphene-based optoelectronic devices. QDs are photosensitive nanostructures with tunable spectral properties, e.g., by controlling their size and shapes. This gives an advantage over other nanostructures, because desired optical properties can be achieved over a broad range of the electromagnetic spectrum. Hybrid graphene—QD phototransistors with ultrahigh gain (*i.e.*, the number of charged carriers generated per incident photon) using colloidal PbS quantum dots have demonstrated photoresponsivity of up to 10^7 AW^{-1} [74,75]. This high value was attributed to efficient carrier generation in the QDs and subsequent charge transfer to graphene. Flexible infra-red PDs based on CVD graphene and PbS QDs have also exhibited high photoresponsivity [76]. Ligand-capped colloidal PbS QDs in hybrid graphene—QD phototransistors reached a photosensitivity of 10^9 AW^{-1} [77]. This was attributed to efficient charge transfer between the graphene and the QDs through the optimized thickness of the capping ligands. Devices with layered and bulk heterojunctions with hybrid graphene—PbSe QD materials have also been studied to investigate the influence of heterojunctions on the photoresponsivity [78].

4.2. TMD Photodetectors

The 2D TMDs, like MoS$_2$, are very promising for photonic and optoelectronic applications. They exhibit even higher absorption coefficients than graphene, even though they are typically limited in their spectral bandwidth by their electronic band gaps. In this section, we discuss TMD-based p-n diodes and heterojunction devices of both TMD-bulk semiconductors and exclusively of 2D materials.

Examples of TMD PDs include lateral symmetric (reverse) Schottky diode configurations, which utilize the built in potentials at the MoS$_2$-metal contacts to drive photocurrents [79,80]. Sanchez-Lopez *et al.* reported impressive responsivity of 880 mAW^{-1} at 561 nm from an exfoliated MoS$_2$ flake [79]. This number was calculated from photocurrents measured under ultra-low illumination intensities of 24 µWcm^{-2}. However, the noise equivalent power (NEP) of 1.8×10^{-15} WHz$^{-1/2}$ in the MoS$_2$ photoconductor is considerably lower compared to state-of-the-art Si avalanche photodetectors (3×10^{-14} WHz$^{-1/2}$) [81]. Zhang *et al.* carried out a similar experiment on CVD-grown MoS$_2$ and showed an even more impressive responsivity of 2200 AW^{-1} in a vacuum and 780 AW^{-1} in ambient air [80]. The authors attributed this difference mainly to enhanced carrier recombination in ambient air. In fact, environmental factors often influence the behavior and performance of TMD PDs. Kufer *et al.* have demonstrated an encapsulated MoS_2 detector that is independent of the ambient air, but becomes highly tunable through an electrostatic gate with responsivity ranging from $10-10^4$ AW^{-1} [82]. A more complex structure was reported by Pospischil *et al.*, who used two independent gate electrodes to induce a lateral, tunable p-n junction in single-layer exfoliated tungsten diselenide (WSe_2) [83]. The WSe_2 flake was electrostatically doped to act either as a solar cell, a PD or an LED with a power conversion efficiency of ~0.5% and an electroluminescence efficiency of ~0.1%. In PD mode, a responsivity of 16 mAW^{-1} was obtained.

Vertical p-n junctions can be fabricated by stacking 2D materials on bulk materials or on other 2D materials, similar to the graphene/Si devices described in Section 4.1 [57]. While Lopez-Sanchez *et al.* investigated exfoliated MoS_2 flakes on Si [84], Yim *et al.* demonstrated the fabrication of large-area CVD MoS_2/Si devices, where a sputtered Mo film was sulfurized in a highly controllable and reproducible manufacturing process [85]. Multispectral measurements showed both the absorption signatures of Si and bulk MoS_2, with the sensitivity bandwidth limited by the Si band gap [85].

Recently, several groups have reported heterostructures based entirely on 2D materials, also referred to as van der Waals (vdW) heterostructures. Furchi *et al.* showed results from an atomically-thin p-n diode consisting of exfoliated flakes of monolayer MoS_2 and monolayer WSe_2 [86]. This device exhibited a responsivity of 11 mAW^{-1} at 650 nm. A very similar device reported by Lee *et al.* exhibited a responsivity of 2 mAW^{-1} at 532 nm [87]. The authors concluded that the photocurrent is dominantly caused by the PV rather than the PTE effect. A similar device layout with identical materials was chosen by Cheng *et al.*; but here, one of the materials (WSe_2) was grown by CVD, and an exfoliated MoS_2 flake was used [88]. This device exhibited an EQE of up to 12%. A gallium telluride (GaTe)—MoS_2 vdW heterodiode by Wang *et al.* displayed a very remarkable responsivity of over $20 AW^{-1}$ [89]. A combination of graphene, h-BN and a TMD by Britnell *et al.* consisted of a h-BN/Gr/WS_2/Gr/h-BN stack [90]. This multilayer vertical device acts as a flexible PD or solar cell and was fabricated on a polyethylene terephthalate (PET) film. A responsivity of 0.1 AW^{-1} was achieved at 633 nm with an EQE of 30%.

In general, the area of research on van der Waals heterostructures is in a very early stage. Given the recent advances in the CVD growth of 2D materials, it seems now possible to scale up the synthesis of 2D material-based heterostructures with good control of the layer thickness and quality [91]. Han *et al.* have developed an atmospheric pressure CVD technique to grow in-plane heterostructures of h-BN and graphene [92]. Figure 4a shows schematics and optical micrographs of the continuous growth of these heterostructures. They mention that by controlling the growth conditions, one can obtain a relatively sharp interface in such a structure. Gao *et al.*

developed a CVD method to controllably grow vertical and lateral heterostructures of h-BN and graphene on Cu foils [93]. They used a novel temperature-triggered reaction process to selectively grow such heterostructures. Figure 4b depicts the growth strategy followed by the authors, where they have used benzoic acid as a carbon source, which decomposes to CO_2 and other hydrocarbons at temperatures higher than 500 °C. The released CO_2 was found to etch h-BN at temperatures above 900 °C, thus exposing the Cu surface for lateral growth of graphene. At lower temperatures, h-BN was not etched by CO_2, and graphene was grown directly on its surface. Their work is quite promising for the fabrication of high performance graphene-based devices. Successful large-area growth of TMDs, like MoS_2, WS_2, $MoSe_2$ and WSe_2, using CVD has motivated researchers to grow these materials simultaneously or on other 2D materials for heterostructures. Shi *et al.* have synthesized MoS_2/graphene heterostructures using ammonium thiomolybdate as a precursor and CVD graphene on Cu foil as a substrate [94]. They were able to achieve single crystalline hexagonal flakes of MoS_2 with a lateral size ranging from several hundred nanometers to several micrometers. Figure 4c shows a schematic of the growth process and a representative AFM phase image revealing the presence of graphene and MoS_2 on Cu foil. Lin *et al.* have demonstrated the direct synthesis of MoS_2, WS_2 and h-BN on epitaxial graphene grown on SiC substrates [95]. Figure 4d shows the atomic arrangement of the heterostructures as revealed by transmission electron microscope (TEM) investigations. By performing detailed structural and morphological studies, they found that the properties of fabricated heterostructures depend significantly on the underlying graphene template. It was observed that wrinkles and defects on the graphene surface act as nucleation sites for the lateral growth of overlayers. The photosensor based on such a structure generated a power-dependent photocurrent ranging from 150–550 nA at $V_{ds} = 1$ V with a laser power ranging from 4–40 µW under a constant excitation wavelength of 488 nm. The device exhibited a photoresponsivity of 40 mAW^{-1} with a 15 µm-long channel at $V_{ds} = 1$ V and $V_{bg} = 0$ V. Further, they demonstrated an improvement of about 10^3 in the photoresponse of MoS_2 grown on graphene as compared to bare MoS_2.

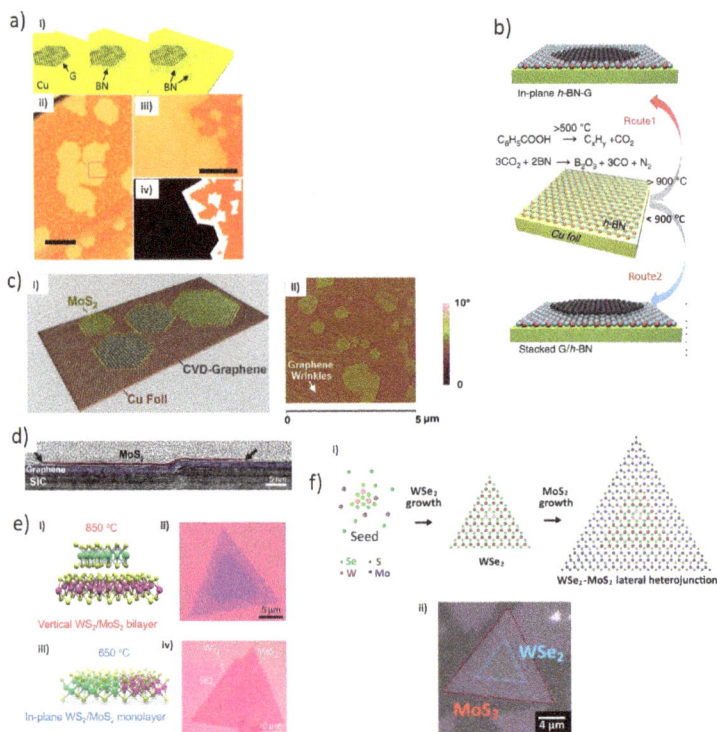

Figure 4. (**a**) Schematic of the continuous growth of graphene/hexagonal boron nitride (h-BN) heterostructures on Cu foil (i). Low (scale bar is 50 μm) (ii) and high magnification (scale bar is 10 μm) (iii) image of graphene/h-BN heterostructures on oxidized Cu substrate; (iv) false color image. Black, white and brown regions indicate graphene, h-BN and bare Cu regions in panel (iii), respectively [92]; (**b**) Schematic illustration of the temperature-triggered switching growth between in-plane and vertical graphene/h-BN heterostructures, depicted by Route 1 and Route 2, respectively [93]; (**c**) Schematic of hexagonal MoS_2 layers grown on a CVD graphene template (i). AFM phase image of MoS_2 grains on graphene (ii) [94]; (**d**) Cross-sectional high resolution TEM of MoS_2 layer grown on epitaxial graphene [95]; (**e**) Schematic and optical image of the vertically-stacked heterostructure synthesized at 850 °C (i, ii) and lateral WS_2/MoS_2 heterostructure synthesized at 650 °C (iii, iv) [96]; (**f**) Schematic illustration of the sequential growth of the monolayer in-plane WSe_2/MoS_2 heterostructures (i) and optical micrograph of the heterostructures depicting the contrast between two layers [97]. Reproduced with permission from: (a) [92] copyright© 2013 American Chemical Society; (b) [93] copyright© 2015 Macmillan Publishers Ltd.; (c) [94] copyright© 2012 American Chemical Society; (d) [95] copyright© 2014 American Chemical Society; (e) [96] copyright© 2014 Macmillan Publishers Ltd.; and (f) [97] copyright© 2015 American Association for the Advancement of Science.

Gong *et al.* reported a one-step growth strategy for fabricating high quality vertically-stacked and in-plane heterostructures based on single-layer MoS_2 and WS_2 by controlling the growth temperature [96]. It was found that a vertically-stacked bilayer consisting of a WS_2 monolayer grown epitaxially on a MoS_2 monolayer is preferred at high temperatures, whereas lateral epitaxy of WS_2 at MoS_2 edges is dominating at low temperatures of around 650 °C. Figure 4e shows the schematic and optical images of vertically-stacked and in-plane heterostructures of single-layered WS_2 and MoS_2. A strong interlayer excitonic transition in vertically-stacked layers and a strong photoluminescence enhancement with the formation of intrinsic p-n junctions in lateral structures attested the formation of atomically-sharp and clean interfaces. The devices based on lateral heterostructures were found to exhibit a photovoltaic effect under illumination with an open loop voltage of 0.12 V and a closed loop current of 5.7 pA without external gating effects. Li *et al.* reported a two-step growth strategy for fabricating epitaxial WSe_2/MoS_2 heterojunctions by sequential CVD of WSe_2 and MoS_2 [97]. First, single crystalline WSe_2 monolayers were prepared, and then, MoS_2 growth was performed on these monolayers separately. Figure 4f shows the schematic of the sequential growth of monolayer WSe_2/MoS_2 in-plane heterostructures along with an optical micrograph showing the distinct optical contrast between the two. They confirmed the intrinsic p-n junction properties of the heterostructures by measuring the depletion width, photoresponse, rectifying behavior and PVE in the devices made out of these structures. A PVE with an open circuit voltage of 0.22 V and a short circuit current of 7.7 pA was measured under white light illumination at a power density of 1 mW/cm^2. The power conversion efficiency of the device was calculated to be around 0.2% with a small fill factor (FF) of 0.39, which might be due to the high equivalent series resistance of the intrinsic TMD layers.

Table 1 summarizes the experimental results from 2D material photonic devices as described in this review. In particular, the number of integrated layers, the device fabrication technology and electro-/optical benchmark parameters EQE, IQE, photo conversion efficiency (PCE) and SR, are listed where reported. The bold SR entries represent competitiveness and improvements of 2D detectors compared to state-of-the art Si PDs with regard to the absolute SR or the frequency bandwidth.

Table 1. Summary of literature data for 2D material-based photonic devices. The table contains the layer thickness, 2D material fabrication technology and electro-/optical benchmark parameters (external quantum efficiency (EQE), internal quantum efficiency (IQE), photo conversion efficiency (PCE) and spectral response (SR)), where reported. SLG, single-layer graphene.

Graphene Devices	Layer No.	Technology	EQE/IQE/PCE (%)	Max. SR (mAW^{-1})	Ref.
Graphene/Si Schottky junction	SLG	CVD	—	270 at 992 nm 0.17 at 2000 nm	[57]
	3LG p-doped		IQE > 65	435 (850–900 nm)	[60]
	SLG		—	320 (850–900 nm)	[60]
Graphene/Si-waveguide	SLG on Si		—	0.13 at 2.75 μm	[98]
Graphene/GaN Schottky diode	SLG		—	0.23 AW^{-1} at 360 nm	[99]
Graphene on SiO$_2$/Si	Bilayer		—	1.3 at 0.292 THz	[100]
Metal-graphene-metal	Bilayer	Exfoliated flake	EQE = 0.5 IQE = 10	6.1 at 1550 nm (VIS-NIR)	[37]
Graphene p–n-junction	SLG (electr. Doping)		—	1.5 at 532 nm (P = 30 μW)	[34]
	Bilayer		—	5 at 850 nm (T = 40 K)	[40]
	Trilayer	Exfoliated, edge contacted	—	10 at 514.5 nm	[44]
Flexible organic PDs	Graphene/PEDOT:PSS ink	Spray coating	—	0.16 AW^{-1} at 500 nm	[101]
TMD Devices	**Layer No.**	**Technology**	**EQE/IQE/PCE (%)**	**Max. SR (mAW^{-1})**	**Ref.**
MoS$_2$ on SiO$_2$ photoconductor	SL	Exfoliated flakes	—	880 A/W at 561 nm	[79]
WSe$_2$	SL (electr. doped)		PCE = 0.5	Electroluminescent at 1.547 eV	[83]
WSe$_2$/MoS$_2$	SL		EQE = 2.1	11 at 650 nm	[86]
h-BN/Gr/WS$_2$/Gr/h-BN	SL		EQE = 30	0.1 at 633 nm	[90]
MoS$_2$/Si Schottky junction	8.26 nm (12 layer)	CVD	—	8.6	[85]
WSe$_2$-/MoS$_2$ pn-junction	SL p-n diode		EQE = 2.1	11 at 650 nm	[86]
MoS$_2$/graphene photoconductor	SLG		IQE ≈ 15	10^7 at 650 nm	[102]

5. Conclusions

In this work, photonic devices that comprise graphene and other two-dimensional materials and their performance have been reviewed and compared to benchmarks of state-of-the-art bulk optoelectronic devices. Moreover, physical mechanisms that enable photodetection have been discussed. We presented challenges of existing 2D material process technologies, namely contaminations, reproducibility and scalability. A variety of 2D photonic devices that show additional functionalities compared to state-of-the-art bulk devices was presented. Some of the intrinsic properties of graphene, such as the absence of an electronic band gap, carrier multiplication and high carrier mobility, suggest applications as broadband and high speed photodetectors, which have been demonstrated in several instances. Even though the monoatomic nature of graphene presents a limitation in the absorbance, many promising solutions are being pursued. A brief overview of heterostructures of bulk and 2D materials, such as hybrid graphene/Si and MoS_2/Si diodes, has been given. Finally, heterojunctions made up entirely of 2D materials have been reviewed, taking into account the recent advances in large-scale fabrication of 2D crystals. The multitude of reports, the excellent performance and the rapid progress in the field are quite encouraging. Nevertheless, there are severe challenges towards a manufacturable, reliable and reproducible large-scale 2D technology for optoelectronics. We remain confident that these can be met with time and substantial efforts.

Acknowledgments: Funding from the European Research Council (ERC, 307311) and the German Research Foundation (DFG LE 2440/1-1, GRK 1564) is gratefully acknowledged.

Author Contributions: A.B., S.K. and M.C.L. wrote the paper.

Conflicts of Interest: The authors declare no conflict of interest.

References

1. Novoselov, K.S.; Fal'ko, V.I.; Colombo, L.; Gellert, P.R.; Schwab, M.G.; Kim, K. A roadmap for graphene. *Nature* **2012**, *490*, 192–200.
2. Nair, R.R.; Blake, P.; Grigorenko, A.N.; Novoselov, K.S.; Booth, T.J.; Stauber, T.; Peres, N.M.R.; Geim, A.K. Fine structure constant defines visual transparency of graphene. *Science* **2008**, *320*, 1308.
3. Kuzmenko, A.B.; van Heumen, E.; Carbone, F.; van der Marel, D. Universal optical conductance of graphite. *Phys. Rev. Lett.* **2008**, *100*, 117401.
4. Pang, S.; Hernandez, Y.; Feng, X.; Müllen, K. Graphene as transparent electrode material for organic electronics. *Adv. Mater.* **2011**, *23*, 2779–2795.
5. Li, Z.Q.; Henriksen, E.A.; Jiang, Z.; Hao, Z.; Martin, M.C.; Kim, P.; Stormer, H.L.; Basov, D.N. Dirac charge dynamics in graphene by infrared spectroscopy. *Nat. Phys.* **2008**, *4*, 532–535.
6. Bernardi, M.; Palummo, M.; Grossman, J.C. Extraordinary sunlight absorption and one nanometer thick photovoltaics using two-dimensional monolayer materials. *Nano Lett.* **2013**, *13*, 3664–3670.
7. Splendiani, A.; Sun, L.; Zhang, Y.; Li, T.; Kim, J.; Chim, C.-Y.; Galli, G.; Wang, F. Emerging photoluminescence in monolayer MoS_2. *Nano Lett.* **2010**, *10*, 1271–1275.
8. Li, J.; Majety, S.; Dahal, R.; Zhao, W.P.; Lin, J.Y.; Jiang, H.X. Dielectric strength, optical absorption, and deep ultraviolet detectors of hexagonal boron nitride epilayers. *Appl. Phys. Lett.* **2012**, *101*, 171112.

9. Kataria, S.; Wagner, S.; Ruhkopf, J.; Gahoi, A.; Pandey, H.; Bornemann, R.; Vaziri, S.; Smith, A.D.; Ostling, M.; Lemme, M.C. Chemical vapor deposited graphene: From synthesis to applications. *Phys. Status Solidi A* **2014**, *211*, 2439–2449.

10. Chhowalla, M.; Shin, H.S.; Eda, G.; Li, L.-J.; Loh, K.P.; Zhang, H. The chemistry of two-dimensional layered transition metal dichalcogenide nanosheets. *Nat. Chem.* **2013**, *5*, 263–275.

11. Yamada, T.; Kim, J.; Ishihara, M.; Hasegawa, M. Low-temperature graphene synthesis using microwave plasma CVD. *J. Phys. Appl. Phys.* **2013**, *46*, 063001.

12. Smith, A.; Vaziri, S.; Rodriguez, S.; Östling, M.; Lemme, M.C. Wafer scale graphene transfer for back end of the line device integration. In Proceedings of the 2014 15th International Conference on Ultimate Integration on Silicon (ULIS), Stockholm, Sweden, 7–9 April 2014.

13. Banszerus, L.; Schmitz, M.; Engels, S.; Dauber, J.; Oellers, M.; Haupt, F.; Watanabe, K.; Taniguchi, T.; Beschoten, B.; Stampfer, C. Ultrahigh-mobility graphene devices from chemical vapor deposition on reusable copper. *Sci. Adv.* **2015**, *1*, e1500222.

14. Lupina, G.; Kitzmann, J.; Costina, I.; Lukosius, M.; Wenger, C.; Wolff, A.; Vaziri, S.; Östling, M.; Pasternak, I.; Krajewska, A.; *et al.* Residual metallic contamination of transferred chemical vapor deposited graphene. *ACS Nano* **2015**.

15. Wagner, S.; Weisenstein, C.; Smith, A.D.; Östling, M.; Kataria, S.; Lemme, M.C. Graphene transfer methods for the fabrication of membrane-based NEMS devices. *Microelectron. Eng.* **2016**, *159*, 108–113.

16. Bae, S.; Kim, H.; Lee, Y.; Xu, X.; Park, J.-S.; Zheng, Y.; Balakrishnan, J.; Lei, T.; Kim, H.R.; Song, Y.I.; *et al.* Roll-to-roll production of 30-inch graphene films for transparent electrodes. *Nat. Nanotechnol.* **2010**, *5*, 574–578.

17. Rahimi, S.; Tao, L.; Chowdhury, S.F.; Park, S.; Jouvray, A.; Buttress, S.; Rupesinghe, N.; Teo, K.; Akinwande, D. Toward 300 mm wafer-scalable high-performance polycrystalline chemical vapor deposited graphene transistors. *ACS Nano* **2014**, *8*, 10471–10479.

18. Bae, S.; Kim, S.J.; Shin, D.; Ahn, J.-H.; Hong, B.H. Towards industrial applications of graphene electrodes. *Phys. Scr.* **2012**, *2012*, 014024.

19. Bablich, A.; Watty, K.; Merfort, C.; Seibel, K.; Boehm, M. A novel high-dynamic A-Si:H multicolor pin-detector with ZnO:Al front and back contacts. In Proceedings of the 2010 MRS Fall Meeting: Symposium AA—Group IV Semiconductor Nanostructures and Applications, Boston, MA, USA, 29 November–3 December 2010; Volume 1305.

20. Hamberg, I.; Granqvist, C.G. Evaporated Sn-doped In_2O_3 films: Basic optical properties and applications to energy-efficient windows. *J. Appl. Phys.* **1986**, *60*, R123–R160.

21. Minami, T. Transparent conducting oxide semiconductors for transparent electrodes. *Semicond. Sci. Technol.* **2005**, *20*, S35–S44.

22. Granqvist, C.G. Transparent conductors as solar energy materials: A panoramic review. *Sol. Energy Mater. Sol. Cells* **2007**, *91*, 1529–1598.

23. Engadget. Samsung Showcases 4.5-Inch Flexible AMOLED, May Actually Mass Produce This One. Available online: http://www.engadget.com/2010/11/04/samsung-showcases-4-5-inch-flexible-amoled-may-actually-mass-pr/ (accessed on 1 October 2015).

24. Pocket-Lint. Plastic Logic Shows Off Colour e-Paper Display Smart Watch Concept: The Future of Wearable Tech? Available online: http://www.pocket-lint.com/news/120209-plastic-logic-colour-e-paper-smart-watch-concept (accessed on 1 October 2015).

25. Koppens, F.H.L.; Mueller, T.; Avouris, P.; Ferrari, A.C.; Vitiello, M.S.; Polini, M. Photodetectors based on graphene, other two-dimensional materials and hybrid systems. *Nat. Nanotechnol.* **2014**, *9*, 780–793.

26. Wu, Z.; Chen, Z.; Du, X.; Logan, J.M.; Sippel, J.; Nikolou, M.; Kamaras, K.; Reynolds, J.R.; Tanner, D.B.; Hebard, A.F.; *et al.* Transparent, conductive carbon nanotube films. *Science* **2004**, *305*, 1273–1276.

27. De, S.; Coleman, J.N. Are there fundamental limitations on the sheet resistance and transmittance of thin graphene films? *ACS Nano* **2010**, *4*, 2713–2720.

28. Lee, J.-Y.; Connor, S.T.; Cui, Y.; Peumans, P. Solution-processed metal nanowire mesh transparent electrodes. *Nano Lett.* **2008**, *8*, 689–692.

29. Bonaccorso, F.; Sun, Z.; Hasan, T.; Ferrari, A.C. Graphene photonics and optoelectronics. *Nat. Photonics* **2010**, *4*, 611–622.

30. Sahu, D.R.; Lin, S.-Y.; Huang, J.-L. ZnO/Ag/ZnO multilayer films for the application of a very low resistance transparent electrode. *Appl. Surf. Sci.* **2006**, *252*, 7509–7514.

31. De, S.; Higgins, T.M.; Lyons, P.E.; Doherty, E.M.; Nirmalraj, P.N.; Blau, W.J.; Boland, J.J.; Coleman, J.N. Silver nanowire networks as flexible, transparent, conducting films: Extremely high dc to optical conductivity ratios. *ACS Nano* **2009**, *3*, 1767–1774.

32. Vivien, L.; Polzer, A.; Marris-Morini, D.; Osmond, J.; Hartmann, J.M.; Crozat, P.; Cassan, E.; Kopp, C.; Zimmermann, H.; Fédéli, J.M. Zero-bias 40 Gbit/s germanium waveguide photodetector on silicon. *Opt. Express* **2012**, *20*, 1096–1101.

33. Wirths, S.; Geiger, R.; von den Driesch, N.; Mussler, G.; Stoica, T.; Mantl, S.; Ikonic, Z.; Luysberg, M.; Chiussi, S.; Hartmann, J.M. Lasing in direct-bandgap GeSn alloy grown on Si. *Nat. Photonics* **2015**, *9*, 88–92.

34. Lemme, M.C.; Koppens, F.H.L.; Falk, A.L.; Rudner, M.S.; Park, H.; Levitov, L.S.; Marcus, C.M. Gate-activated photoresponse in a graphene p–n junction. *Nano Lett.* **2011**, *11*, 4134–4137.

35. Farmer, D.B.; Golizadeh-Mojarad, R.; Perebeinos, V.; Lin, Y.-M.; Tulevski, G.S.; Tsang, J.C.; Avouris, P. Chemical doping and electron-hole conduction asymmetry in graphene devices. *Nano Lett.* **2008**, *9*, 388–392.

36. Sherpa, S.D.; Kunc, J.; Hu, Y.; Levitin, G.; de Heer, W.A.; Berger, C.; Hess, D.W. Local work function measurements of plasma-fluorinated epitaxial graphene. *Appl. Phys. Lett.* **2014**, *104*, 081607.

37. Mueller, T.; Xia, F.; Avouris, P. Graphene photodetectors for high-speed optical communications. *Nat. Photonics* **2010**, *4*, 297–301.

38. Rao, G.; Freitag, M.; Chiu, H.-Y.; Sundaram, R.S.; Avouris, P. Raman and photocurrent imaging of electrical stress-induced p–n junctions in graphene. *ACS Nano* **2011**, *5*, 5848–5854.

39. Freitag, M.; Low, T.; Xia, F.; Avouris, P. Photoconductivity of biased graphene. *Nat. Photonics* **2013**, *7*, 53–59.

40. Gabor, N.M.; Song, J.C.W.; Ma, Q.; Nair, N.L.; Taychatanapat, T.; Watanabe, K.; Taniguchi, T.; Levitov, L.S.; Jarillo-Herrero, P. Hot carrier–assisted intrinsic photoresponse in graphene. *Science* **2011**, *334*, 648–652.

41. Song, J.C.W.; Rudner, M.S.; Marcus, C.M.; Levitov, L.S. Hot carrier transport and photocurrent response in graphene. *Nano Lett.* **2011**, *11*, 4688–4692.

42. Kim, M.-H.; Yan, J.; Suess, R.J.; Murphy, T.E.; Fuhrer, M.S.; Drew, H.D. Photothermal response in dual-gated bilayer graphene. *Phys. Rev. Lett.* **2013**, *110*, 247402.

43. Tielrooij, K.J.; Song, J.C.W.; Jensen, S.A.; Centeno, A.; Pesquera, A.; Zurutuza Elorza, A.; Bonn, M.; Levitov, L.S.; Koppens, F.H.L. Photoexcitation cascade and multiple hot-carrier generation in graphene. *Nat. Phys.* **2013**, *9*, 248–252.

44. Freitag, M.; Low, T.; Avouris, P. Increased responsivity of suspended graphene photodetectors. *Nano Lett.* **2013**, *13*, 1644–1648.

45. Park, J.; Ahn, Y.H.; Ruiz-Vargas, C. Imaging of photocurrent generation and collection in single-layer graphene. *Nano Lett.* **2009**, *9*, 1742–1746.

46. Lee, E.J.H.; Balasubramanian, K.; Weitz, R.T.; Burghard, M.; Kern, K. Contact and edge effects in graphene devices. *Nat. Nanotechnol.* **2008**, *3*, 486–490.

47. Xia, F.; Mueller, T.; Golizadeh-Mojarad, R.; Freitag, M.; Lin, Y.; Tsang, J.; Perebeinos, V.; Avouris, P. Photocurrent imaging and efficient photon detection in a graphene transistor. *Nano Lett.* **2009**, *9*, 1039–1044.

48. Mueller, T.; Xia, F.; Freitag, M.; Tsang, J.; Avouris, P. Role of contacts in graphene transistors: A scanning photocurrent study. *Phys. Rev. B* **2009**, *79*, 245430.

49. Withers, F.; Bointon, T.H.; Craciun, M.F.; Russo, S. All-graphene photodetectors. *ACS Nano* **2013**, *7*, 5052–5057.

50. Echtermeyer, T.J.; Nene, P.S.; Trushin, M.; Gorbachev, R.V.; Eiden, A.L.; Milana, S.; Sun, Z.; Schliemann, J.; Lidorikis, E.; Novoselov, K.S.; *et al.* Photothermoelectric and photoelectric contributions to light detection in metal–graphene–metal photodetectors. *Nano Lett.* **2014**, *14*, 3733–3742.

51. Pospischil, A.; Humer, M.; Furchi, M.M.; Bachmann, D.; Guider, R.; Fromherz, T.; Mueller, T. CMOS-compatible graphene photodetector covering all optical communication bands. *Nat. Photonics* **2013**, *7*, 892–896.

52. Naiini, M.M.; Vaziri, S.; Smith, A.D.; Lemme, M.C.; Ostling, M. Embedded graphene photodetectors for silicon photonics. In Proceedings of the 2014 72nd Annual Device Research Conference (DRC), Santa Barbara, CA, USA, 22–25 June 2014; pp. 43–44.

53. Liu, M.; Yin, X.; Ulin-Avila, E.; Geng, B.; Zentgraf, T.; Ju, L.; Wang, F.; Zhang, X. A graphene-based broadband optical modulator. *Nature* **2011**, *474*, 64–67.

54. Schall, D.; Neumaier, D.; Mohsin, M.; Chmielak, B.; Bolten, J.; Porschatis, C.; Prinzen, A.; Matheisen, C.; Kuebart, W.; Junginger, B.; *et al.* 50 GBit/s photodetectors based on wafer-scale graphene for integrated silicon photonic communication systems. *ACS Photonics* **2014**, *1*, 781–784.

55. Chen, L.; Lipson, M. Ultra-low capacitance and high speed germanium photodetectors on silicon. *Opt. Express* **2009**, *17*, 7901–7906.

56. Chen, C.-C.; Aykol, M.; Chang, C.-C.; Levi, A.F.J.; Cronin, S.B. Graphene-silicon schottky diodes. *Nano Lett.* **2011**, *11*, 1863–1867.

57. Riazimehr, S.; Schneider, D.; Yim, C.; Kataria, S.; Passi, V.; Bablich, A.; Duesberg, G.S.; Lemme, M.C. Spectral sensitivity of a graphene/silicon pn-junction photodetector. In Proceedings of the 2015 Joint International EUROSOI Workshop and International Conference on Ultimate Integration on Silicon (EUROSOI-ULIS), Bologna, Italy, 26–28 January 2015.

58. Smith, A.D.; Elgammal, K.; Niklaus, F.; Delin, A.; Fischer, A.C.; Vaziri, S.; Forsberg, F.; Råsander, M.; Hugosson, H.; Bergqvist, L.; *et al.* Resistive graphene humidity sensors with rapid and direct electrical readout. *Nanoscale* **2015**, *7*, 19099–19109.

59. Cusati, T.; Fiori, G.; Gahoi, A.; Fortunelli, A.; Lemme, M.C.; Iannaccone, G. Electrical properties of graphene-metal contacts.: A theoretical and experimental study. In Proceedings of The IEEE International Electron Device Meeting IEDM, Washington, DC, USA, 7–9 December 2015.

60. An, X.; Liu, F.; Jung, Y.J.; Kar, S. Tunable graphene–silicon heterojunctions for ultrasensitive photodetection. *Nano Lett.* **2013**, *13*, 909–916.

61. Kim, H.-Y.; Lee, K.; McEvoy, N.; Yim, C.; Duesberg, G.S. Chemically modulated graphene diodes. *Nano Lett.* **2013**, *13*, 2182–2188.

62. Winzer, T.; Knorr, A.; Malic, E. Carrier multiplication in graphene. *Nano Lett.* **2010**, *10*, 4839–4843.

63. Plötzing, T.; Winzer, T.; Malic, E.; Neumaier, D.; Knorr, A.; Kurz, H. Experimental verification of carrier multiplication in graphene. *Nano Lett.* **2014**, *14*, 5371–5375.

64. Zhang, J.; Zhu, Z.; Liu, W.; Yuan, X.; Qin, S. *Graphene Plasmonics for Light Trapping and Absorption Engineering in optoelectronic devices*; Frontiers in Optics: San Diego, CA, USA, 2015.

65. Peters, E.C.; Lee, E.J.H.; Burghard, M.; Kern, K. Gate dependent photocurrents at a graphene p–n junction. *Appl. Phys. Lett.* **2010**, *97*, 193102.

66. Gilbertson, A.M.; Francescato, Y.; Roschuk, T.; Shautsova, V.; Chen, Y.; Sidiropoulos, T.P.H.; Hong, M.; Giannini, V.; Maier, S.A.; Cohen, L.F.; *et al.* Plasmon-induced optical anisotropy in hybrid graphene–metal nanoparticle systems. *Nano Lett.* **2015**, *15*, 3458–3464.

67. Engel, M.; Steiner, M.; Lombardo, A.; Ferrari, A.C.; v Löhneysen, H.; Avouris, P.; Krupke, R. Light–matter interaction in a microcavity-controlled graphene transistor. *Nat. Commun.* **2012**, *3*, 906.

68. Furchi, M.; Urich, A.; Pospischil, A.; Lilley, G.; Unterrainer, K.; Detz, H.; Klang, P.; Andrews, A.M.; Schrenk, W.; Strasser, G.; *et al.* Microcavity-integrated graphene photodetector. *Nano Lett.* **2012**, *12*, 2773–2777.

69. Liu, Y.; Cheng, R.; Liao, L.; Zhou, H.; Bai, J.; Liu, G.; Liu, L.; Huang, Y.; Duan, X. Plasmon resonance enhanced multicolour photodetection by graphene. *Nat. Commun.* **2011**, *2*, 579.

70. Huisman, E.H.; Shulga, A.G.; Zomer, P.J.; Tombros, N.; Bartesaghi, D.; Bisri, S.Z.; Loi, M.A.; Koster, L.J.A.; van Wees, B.J. High gain hybrid graphene–organic semiconductor phototransistors. *ACS Appl. Mater. Interfaces* **2015**, *7*, 11083–11088.

71. Echtermeyer, T.J.; Britnell, L.; Jasnos, P.K.; Lombardo, A.; Gorbachev, R.V.; Grigorenko, A.N.; Geim, A.K.; Ferrari, A.C.; Novoselov, K.S. Strong plasmonic enhancement of photovoltage in graphene. *Nat. Commun.* **2011**, *2*, 458.

72. Dang, V.Q.; Trung, T.Q.; Kim, D.-I.; Duy, L.T.; Hwang, B.-U.; Lee, D.-W.; Kim, B.-Y.; Toan, L.D.; Lee, N.-E. Ultrahigh responsivity in graphene–ZnO nanorod hybrid UV photodetector. *Small* **2015**, *11*, 3054–3065.

73. Konstantatos, G.; Sargent, E.H. Solution-processed quantum dot photodetectors. *IEEE Proc.* **2009**, *97*, 1666–1683.

74. Konstantatos, G.; Badioli, M.; Gaudreau, L.; Osmond, J.; Bernechea, M.; de Arquer, F.P.G.; Gatti, F.; Koppens, F.H.L. Hybrid graphene-quantum dot phototransistors with ultrahigh gain. *Nat. Nanotechnol.* **2012**, *7*, 363–368.

75. Zhang, D.; Gan, L.; Cao, Y.; Wang, Q.; Qi, L.; Guo, X. Understanding charge transfer at PbS-decorated graphene surfaces toward a tunable photosensor. *Adv. Mater.* **2012**, *24*, 2715–2720.

76. Sun, Z.; Liu, Z.; Li, J.; Tai, G.; Lau, S.-P.; Yan, F. Infrared photodetectors based on CVD-grown graphene and PbS quantum dots with ultrahigh responsivity. *Adv. Mater.* **2012**, *24*, 5878–5883.

77. Turyanska, L.; Makarovsky, O.; Svatek, S.A.; Beton, P.H.; Mellor, C.J.; Patanè, A.; Eaves, L.N.; Thomas, R.; Fay, M.W.; Marsden, A.J.; *et al.* Ligand-induced control of photoconductive gain and doping in a hybrid graphene–quantum dot transistor. *Adv. Electron. Mater.* **2015**.

78. Zhang, Y.; Song, X.; Wang, R.; Cao, M.; Wang, H.; Che, Y.; Ding, X.; Yao, J. Comparison of photoresponse of transistors based on graphene-quantum dot hybrids with layered and bulk heterojunctions. *Nanotechnology* **2015**, *26*, 335201.

79. Lopez-Sanchez, O.; Lembke, D.; Kayci, M.; Radenovic, A.; Kis, A. Ultrasensitive photodetectors based on monolayer MoS$_2$. *Nat. Nanotechnol.* **2013**, *8*, 497–501.

80. Zhang, W.; Huang, J.-K.; Chen, C.-H.; Chang, Y.-H.; Cheng, Y.-J.; Li, L.-J. High-gain phototransistors based on a CVD MoS$_2$ monolayer. *Adv. Mater.* **2013**, *25*, 3456–3461.

81. Krainak, M.A.; Sun, X.; Yang, G.; Lu, W. Comparison of linear-mode avalanche photodiode lidar receivers for use at one-micron wavelength. In Proceedings of the SPIE 7681: Advanced Photon Counting Techniques IV, Orlando, FL, USA; 2010; Volume 7681, pp. 76810Y–76810Y-13.

82. Kufer, D.; Konstantatos, G. Highly sensitive, encapsulated MoS_2 photodetector with gate controllable gain and speed. *Nano Lett.* **2015**, *15*, 7307–7313.

83. Pospischil, A.; Furchi, M.M.; Mueller, T. Solar-energy conversion and light emission in an atomic monolayer p-n diode. *Nat. Nano.* **2014**, *9*, 257–261.

84. Lopez-Sanchez, O.; Alarcon Llado, E.; Koman, V.; Fontcuberta i Morral, A.; Radenovic, A.; Kis, A. Light generation and harvesting in a van der Waals Heterostructure. *ACS Nano* **2014**, *8*, 3042–3048.

85. Yim, C.; O'Brien, M.; McEvoy, N.; Riazimehr, S.; Schafer-Eberwein, H.; Bablich, A.; Pawar, R.; Iannaccone, G.; Downing, C.; Fiori, G.; *et al.* Heterojunction hybrid devices from vapor phase grown MoS_2. *Sci. Rep.* **2014**, *4*, 5458.

86. Furchi, M.M.; Pospischil, A.; Libisch, F.; Burgdörfer, J.; Mueller, T. Photovoltaic effect in an electrically tunable van der Waals Heterojunction—Nano letters. *Nano Lett.* **2014**, *14*, 4785–4791.

87. Lee, C.-H.; Lee, G.-H.; van der Zande, A.M.; Chen, W.; Li, Y.; Han, M.; Cui, X.; Arefe, G.; Nuckolls, C.; Heinz, T.F.; *et al.* Atomically thin p–n junctions with van der Waals heterointerfaces. *Nat. Nanotechnol.* **2014**, *9*, 676–681.

88. Cheng, R.; Li, D.; Zhou, H.; Wang, C.; Yin, A.; Jiang, S.; Liu, Y.; Chen, Y.; Huang, Y.; Duan, X. Electroluminescence and photocurrent generation from atomically sharp WSe_2/MoS_2 heterojunction p–n diodes. *Nano Lett.* **2014**, *14*, 5590–5597.

89. Wang, Z.; Xu, K.; Li, Y.; Zhan, X.; Safdar, M.; Wang, Q.; Wang, F.; He, J. Role of Ga vacancy on a multilayer GaTe phototransistor. *ACS Nano* **2014**, *8*, 4859–4865.

90. Britnell, L.; Ribeiro, R.M.; Eckmann, A.; Jalil, R.; Belle, B.D.; Mishchenko, A.; Kim, Y.-J.; Gorbachev, R.V.; Georgiou, T.; Morozov, S.V.; *et al.* Strong light-matter interactions in Heterostructures of atomically thin films. *Science* **2013**, *340*, 1311–1314.

91. Shi, Y.; Li, H.; Li, L.-J. Recent advances in controlled synthesis of two-dimensional transition metal dichalcogenides via vapour deposition techniques. *Chem. Soc. Rev.* **2015**, *44*, 2744–2756.

92. Han, G.H.; Rodríguez-Manzo, J.A.; Lee, C.-W.; Kybert, N.J.; Lerner, M.B.; Qi, Z.J.; Dattoli, E.N.; Rappe, A.M.; Drndic, M.; Johnson, A.T.C. Continuous growth of hexagonal graphene and boron nitride in-plane heterostructures by atmospheric pressure chemical vapor deposition. *ACS Nano* **2013**, *7*, 10129–10138.

93. Gao, T.; Song, X.; Du, H.; Nie, Y.; Chen, Y.; Ji, Q.; Sun, J.; Yang, Y.; Zhang, Y.; Liu, Z. Temperature-triggered chemical switching growth of in-plane and vertically stacked graphene-boron nitride heterostructures. *Nat. Commun.* **2015**, *6*, 6835.

94. Shi, Y.; Zhou, W.; Lu, A.-Y.; Fang, W.; Lee, Y.-H.; Hsu, A.L.; Kim, S.M.; Kim, K.K.; Yang, H.Y.; Li, L.-J.; *et al.* van der Waals Epitaxy of MoS_2 Layers Using Graphene as Growth Templates. *Nano Lett.* **2012**, *12*, 2784–2791.

95. Lin, Y.-C.; Lu, N.; Perea-Lopez, N.; Li, J.; Lin, Z.; Peng, X.; Lee, C.H.; Sun, C.; Calderin, L.; Browning, P.N.; *et al.* Direct synthesis of van der Waals solids. *ACS Nano* **2014**, *8*, 3715–3723.

96. Gong, Y.; Lin, J.; Wang, X.; Shi, G.; Lei, S.; Lin, Z.; Zou, X.; Ye, G.; Vajtai, R.; Yakobson, B.I.; *et al.* Vertical and in-plane heterostructures from WS_2/MoS_2 monolayers. *Nat. Mater.* **2014**, *13*, 1135–1142.

97. Li, M.-Y.; Shi, Y.; Cheng, C.-C.; Lu, L.-S.; Lin, Y.-C.; Tang, H.-L.; Tsai, M.-L.; Chu, C.-W.; Wei, K.-H.; He, J.-H.; *et al.* Epitaxial growth of a monolayer WSe_2-MoS_2 lateral p–n junction with an atomically sharp interface. *Science* **2015**, *349*, 524–528.

98. Wang, X.; Cheng, Z.; Xu, K.; Tsang, H.K.; Xu, J.-B. High-responsivity graphene/silicon-heterostructure waveguide photodetectors. *Nat. Photonics* **2013**, *7*, 888–891.

99. Xu, K.; Xu, C.; Xie, Y.; Deng, J.; Zhu, Y.; Guo, W.; Xun, M.; Teo, K.B.K.; Chen, H.; Sun, J. Graphene GaN-based schottky ultraviolet detectors. *IEEE Trans. Electron Devices* **2015**, *62*, 2802–2808.

100. Spirito, D.; Coquillat, D.; Bonis, S.L.D.; Lombardo, A.; Bruna, M.; Ferrari, A.C.; Pellegrini, V.; Tredicucci, A.; Knap, W.; Vitiello, M.S. High performance bilayer-graphene terahertz detectors. *Appl. Phys. Lett.* **2014**, *104*, 061111.

101. Liu, Z.; Parvez, K.; Li, R.; Dong, R.; Feng, X.; Müllen, K. Transparent conductive electrodes from Graphene/PEDOT:PSS hybrid inks for ultrathin organic photodetectors. *Adv. Mater.* **2015**, *27*, 669–675.

102. Zhang, W.; Chu, C.-P.; Huang, J.-K.; Chen, C.-H.; Tsai, M.-L.; Chang, Y.-H.; Liang, C.-T.; Chen, Y.-Z.; Chueh, Y.-L.; He, J.-H.; *et al.* Ultrahigh-gain photodetectors based on atomically thin Graphene-MoS_2 heterostructures. *Sci. Rep.* **2014**, *4*.

Two-Dimensional Materials for Sensing: Graphene and Beyond

Seba Sara Varghese, Saino Hanna Varghese, Sundaram Swaminathan, Krishna Kumar Singh and Vikas Mittal

Abstract: Two-dimensional materials have attracted great scientific attention due to their unusual and fascinating properties for use in electronics, spintronics, photovoltaics, medicine, composites, *etc.* Graphene, transition metal dichalcogenides such as MoS_2, phosphorene, *etc.*, which belong to the family of two-dimensional materials, have shown great promise for gas sensing applications due to their high surface-to-volume ratio, low noise and sensitivity of electronic properties to the changes in the surroundings. Two-dimensional nanostructured semiconducting metal oxide based gas sensors have also been recognized as successful gas detection devices. This review aims to provide the latest advancements in the field of gas sensors based on various two-dimensional materials with the main focus on sensor performance metrics such as sensitivity, specificity, detection limit, response time, and reversibility. Both experimental and theoretical studies on the gas sensing properties of graphene and other two-dimensional materials beyond graphene are also discussed. The article concludes with the current challenges and future prospects for two-dimensional materials in gas sensor applications.

Reprinted from *Electronics*. Cite as: Varghese, S.S.; Varghese, S.H.; Swaminathan, S.; Singh, K.K.; Mittal, V. Two-Dimensional Materials for Sensing: Graphene and Beyond. *Electronics* **2015**, *4*, 651–681.

1. Introduction

Detection of gas molecules is extremely important in environmental monitoring, industrial chemical processing, public safety, agriculture, medicine and indoor air quality control. For a long time, metal oxide semiconductor gas sensors have played an inevitable role in environmental contaminant detection and industrial process control [1]. Metal oxide semiconductors are the most widely used gas sensing materials due to their numerous advantages such as high sensitivity towards various gases with ease of fabrication, high compatibility with other processes, low cost, simplicity in measurements along with minimal power consumption [2–4]. The high operating temperatures (200 °C to 500 °C) [5], long recovery periods, limited maximum sensitivity (in the range of parts-per-million), low specificity, and limited measurement accuracy [6] basically limited their applications in rapidly changing environment.

The increasing demand for highly sensitive, selective, cost-effective, low power consuming, stable and portable sensors has stimulated extensive research on new sensing materials. One of the most important features of a material to be used for gas sensing is its high surface-to-volume ratio. Nanostructures possess high surface-to-volume ratio which provides large active surface area for the interactions of gas molecules. This strongly favors the adsorption of gases on nanostructures and ultimately leads to highly sensitive sensor performance [7]. One-dimensional (1D) nanostructures are particularly suited for gas detection applications because of their large surface-to-volume ratio, good thermal and chemical stabilities, and sensitivity of electrical properties to changes in the surroundings [8]. Gas sensors using 1D nanostructures such as carbon nanotubes (CNTs) [9–12] and nanowires (NWs) [4,13–15] have demonstrated excellent performance for gas sensing [16,17] with high sensitivity down to parts-per-billion (ppb) levels [18], fast response (time scale in the order of seconds) [19], good selectivity along with low power consumption and miniaturization.

Two-dimensional (2D) materials have captured the interest of research community after the first successful isolation of graphene sheets by micromechanical exfoliation of highly-oriented pyrolytic graphite in 2004 [20]. Graphene, a one-atom-thick sheet of carbon atoms with a 2D hexagonal crystal structure, has shown great promise for applications such as electronics and photonics [21–24], energy conversion and storage [25–29], medicine [30–34], chemical and biological sensing [35–38], etc., due to its interesting physical, chemical, electrical, optical, thermal and mechanical properties [39–42]. Many experimental and theoretical reports on gas sensors made from graphene [43–57] and its derivatives such as graphene oxide (GO) [58–61], reduced graphene oxide (rGO) [62–66] proved that graphene and its derivatives could be used as efficient sensing materials for next-generation gas sensing systems. Inspired by the superior performance of the first 2D material, graphene for gas sensing and many other applications, a lot of research effort has been devoted to the isolation of other 2D materials which exist as strongly bonded stacked layers in bulk crystals and their potential applications. So far, several hundreds of different 2D materials with extraordinary properties are known which include allotropes of various elements such as graphyne, silicene, germanene, phosphorene. etc., and compounds such as transition metal dichalcogenides (TMDs) (for example, MoS_2, $MoSe_2$, WTe_2, TaS_2, $TaSe_2$, etc.), germanane, hexagonal boron nitride (BN), etc. [67]. These family of materials have good optical transparency, excellent mechanical flexibility, great mechanical strength and also peculiar electrical properties, which greatly favors their applications in electronics, optoelectronics [68,69], etc. 2D materials provide a promising platform for the development of ultrahigh sensitive and highly selective sensors by tailoring their rich surface chemistry without any deterioration of their unique optical and electrical properties.

So far, many review articles on 2D materials discussing their novel physical, electronic properties, recent developments in synthesis techniques and their applications in electronics, optoelectronics, energy conversion and storage, gas storage, *etc.*, have been published [67,70–75]. During the past few years, there have been some comprehensive reviews on gas sensors based on graphene and graphene related materials with primary focus on the properties for gas sensing, sensing mechanisms, modifications of graphene for enhanced sensing features, *etc.* [76–78]. To the best of our knowledge, no specific review on the role of two-dimensional materials in gas sensing has been reported. This necessitates a summary of the latest developments in the area of gas sensors by employing graphene and other 2D materials beyond graphene as sensing materials, with emphasis on the sensing performance indicators such as sensitivity, selectivity, detection threshold, response and recovery times. The novel properties of 2D materials that make them perfectly suitable as sensing elements for gas sensor systems are described. In addition to experimental verifications of 2D material based gas sensors, theoretical first-principles studies on the adsorption of various gas molecules on 2D materials are also discussed.

2. Two-Dimensional Materials for Gas Sensing

2D materials are basically single layer materials with thickness of few nanometers or even less. The high surface area of 2D materials provides large number of reactive sites, which make these materials efficient in sensing, catalysis and energy storage technologies. In addition to large surface area, the high electrical conductivities and low noise of the 2D materials also contribute to their indispensable role in future gas detection systems. Due to low electrical noise and high electrical conductivity, a small change in carrier concentration induced by gas exposure could lead to significant changes in electrical conductivity. These features along with the charge transfer between gas molecules and these materials associated with the adsorption of gas molecules make these materials one of the best suitable candidates for the fabrication of gas sensors [79].

The discovery of graphene and understanding of their properties encouraged scientists to investigate the potential applications of graphene and graphene based materials as gas sensing materials [44]. The utilization of graphene as an ultrasensitive sensing element had emerged as one of the most significant applications of graphene. The sensing property of graphene could be attributed to its two dimensional structure which provides advantages such as: (i) Maximum surface area per unit volume ratio and (ii) High sensitivity of electron transport through graphene to the adsorption of gas molecules [36]. The astonishing gas sensing performance of graphene with a 2D structure inspired researchers to investigate other 2D materials, which were discovered after graphene for their gas sensing characteristics. All 2D materials which include graphene, MoS_2, phosphorene, *etc.*, rely on the change in electrical conductivity upon their interaction with gaseous species due to the induced charge transfer by the gas molecules present in the surroundings. Besides graphene, research has also centered on gas sensors based on various 2D materials such as TMDs especially MoS_2 [80–87], WS_2 [88–90], $MoSe_2$ [91] and phosphorene [92,93]. 2D nanostructures in the form of nanosheets (NSs), nanowalls, nanoplates, *etc.*, made from ZnO, NiO, CuO, WO_3, SnO_2, *etc.*, have also proven as successful sensing materials, which could be used as building blocks for the fabrication of gas sensors [94–108].

2.1. Graphene

Graphene, the basic building block of other carbon based allotropes such as graphite, CNTs and fullerenes, was first isolated by Geim and Novoselov of the University of Manchester, for which they were awarded the Nobel Prize in Physics in 2010. Since then, the study of this material has become one of the hottest topics for material scientists due to the exotic thermal, optical, electrical, mechanical and physiochemical properties arising from the two-dimensional crystal structure [20]. The interesting properties of graphene that have been explored so far include linear energy-dispersion at the Dirac point, existence of massless relativistic particles at the Dirac point, quantum Hall-effect at room temperature, exceptionally high charge carrier mobility at room temperature, high electrical and thermal conductivity, great mechanical strength, high optical transparency and high specific surface area [21,39,109–113]. These unique characteristics make graphene a suitable material for a vast variety of applications.

Soon after the discovery of graphene and graphene oxide by mechanical and chemical exfoliation of 3D bulk graphite, scientists started analyzing the usefulness of graphene and its derivatives for gas sensing. 2D-graphene possess large surface area compared to 1D-CNTs due to the fact that every carbon atom in graphene acts as an active site for the gas molecules present in the surroundings to interact with. In addition to this, electrical properties of graphene such as high charge carrier mobilities at room temperature [109], metallic conductivity and low Johnson noise [21,79] also contribute to the rise of graphene as one of the best suitable materials for gas sensing [44]. The defect less high quality crystal structure of graphene being a 2D material screens the (1/f) noise caused by thermal switching better than other 1D and 0D structures [114].

Graphene based gas sensors work on the principle that introduction of gas molecules change the local charge carrier concentration in graphene by either increasing or decreasing the concentration of electrons (as shown in Figure 1a) depending on the nature of gas species (electron donor or acceptor) which leads to corresponding decrease or increase in electrical conductivity [43–45]. Graphene possess the capability to induce noticeable changes in the electronic properties even by a small change in the carrier concentration in graphene by the gas adsorbates due to high signal-to-noise ratio contributed by its features [44].

Several research groups used pristine graphene for sensing gas molecules such as CO_2, NH_3, NO_2, NO, N_2O, O_2, SO_2, H_2O, etc., [47–56] since Schedin et al. [44] demonstrated the first micrometer-sized sensor made from graphene that was capable of detecting gases at the ultimate concentration (single molecule level) at room temperature. This sensor responded rapidly to the attachment and the detachment of a single NO_2 gas molecule from graphene's surface. The sensitivity reported for the first graphene sensor was nearly several orders of magnitude greater than that of the previous sensors. The observed changes in the resistivity curve reflected the type of the gas species, either electron donor or acceptor. Out of the considered gases, the adsorption of NO_2 and H_2O led to decrease in resistivity which indicated their electron acceptor nature, whereas the adsorption of NH_3 and CO led to increase in resistivity which indicated their electron donor nature as shown in Figure 1b. Even after stopping the gas flow, the sensor recovery was achieved only after annealing the device at 150 °C (as shown in Section IV of Figure 1b).

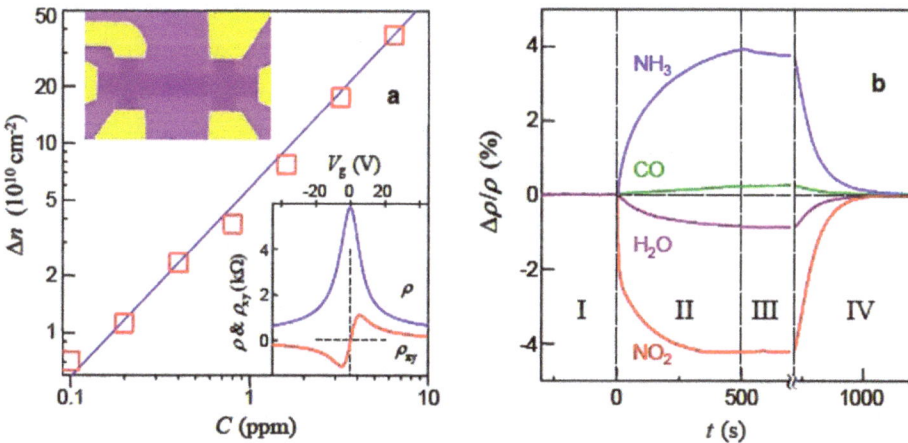

Figure 1. Sensitivity of graphene to chemical doping. (**a**) Concentration Δn of chemically-induced charge carriers in single-layer graphene exposed to different concentrations of NO_2. Upper inset: scanning-electron micrograph of this device. The scale of the micrograph is given by the width of the Hall bar, which is 1 µm. Lower inset: Characterization of the graphene device by using the electric field effect; (**b**) Changes in resistivity ρ caused by graphene's exposure to various gases diluted in concentration 1 ppm. The positive (negative) sign of changes is chosen here to indicate electron (hole) doping. Region I—the device is in vacuum prior to its exposure; II—exposure to a 5 liter volume of a diluted chemical; III—evacuation of the experimental setup; and IV—annealing at 150 °C. Reprinted with permission from Ref. [44]. Copyright, 2007, Nature Publishing Company Ltd.

Chen *et al.* [52] demonstrated the detection of a wide range of gas molecules using pristine graphene synthesized by chemical vapor deposition (CVD) at very low concentrations. Detection limit (DL) as low as 158 parts-per-quadrillion (ppq) to NO gas molecules with a signal-to-noise ratio of 3 was achieved at room temperature under inert atmosphere by continuous *in situ* cleaning of graphene's surface by ultraviolet (UV) light illumination. The sensor also showed fast response, good reproducibility and also 80% of recovery within few minutes of exposure to a mixture of 10 parts-per-trillion (ppt) NO in N_2. The obtained sensitivity was nearly 300% better than similar CNT based sensors and also the DLs were lower than the lowest detection levels reported for other nanosensors [52].

Low-frequency electronic noise spectrum was employed as a sensing parameter to improve the selectivity of graphene to vapors of different chemicals [51]. They observed that the vapors of all considered chemicals (chloroform, methanol, tetrohydrofuran, acetonitrile, ethanol, methylene chloride and toluene) change the electrical resistance of graphene sensors. But only chloroform, methanol, tetrohydrofuran, acetonitrile and ethanol changed the noise spectrum of single layer graphene transistors by inducing Lorentzian components with different characteristic frequencies, whereas toluene and methylene chloride do not modify the noise spectrum. Rumyantsev and his group proved that a single pristine graphene device could achieve good selectivity along with high sensitivity by combining low-frequency noise spectrum measurements along with the change in resistance [51].

Electrochemically exfoliated few layered graphene (FLG, 3- to 10-layer graphene) exhibited good CO_2 and liquid petroleum gas (LPG) sensing performance such as high sensitivity (3.83 and 0.92 for CO_2 and LPG respectively), response time (11 s and 5 s for CO_2 and LPG respectively), and recovery time (14 s and 8 s for CO_2 and LPG respectively) at operating temperatures of 423 K and 398 K respectively [54]. The detection limit of the chemiresistive graphene based sensor was found to be 3 ppm and 4 ppm for CO_2 and LPG respectively with excellent stability at room temperature [54]. The good LPG sensing behavior of the FLG based chemiresistive gas sensor at relatively low temperatures promise their use for practical LPG detection.

The huge cost associated with the fabrication of graphene based devices remains as a great obstacle for widespread gas sensing using graphene. Recently, a simple, low power, low cost resistive gas sensor based on graphene-paper (G-paper) prepared by direct transfer of graphene layers on to paper (as shown in Figure 2a) without any intermediate layers was demonstrated for the first time by Kumar *et al.* [55]. The achieved resolution limit of ~300 ppt was better than other sensors based on graphitic and semiconducting metal oxides using paper as substrate and these sensors were capable of withstanding minor strain. The G-paper strip showed ~65% increase in conductance in 1400 s to the flow of 2.5 ppm NO_2 gas and the conductance decrease by ~15% in 1500 s on stopping the NO_2 flow (as shown in Figure 2b). Cleaning of graphene by UV exposure to remove the adsorbed gas molecules dramatically reduced the recovery time (from hours to 30 s), but the overall response and the characteristic time constant got improved.

Ricciardella *et al.* [56] demonstrated the potential of inkjet printing technique for manufacturing chemi-resistive sensors based on liquid phase exfoliated (LPE) graphene. This technique allows deposition of small ink volumes with a more controlled drying process, which ensures good printed film quality compared to drop casting method. The LPE graphene being a p-type material, showed a decrease in current on exposure to NH_3 (electron donor) and an increase in current on NO_2 exposure (electron acceptor), thus enabling specific detection of these gases. The fabricated sensor exhibited good repeatability upon exposure to both gases at room temperature and atmospheric pressure with relative humidity of 50%, which was found to be independent on the number of printed layers. The perennial issue of graphene based gas sensors operating in environmental conditions such as low reproducibility could be overcome by using inkjet printing technology for sensor fabrication [56].

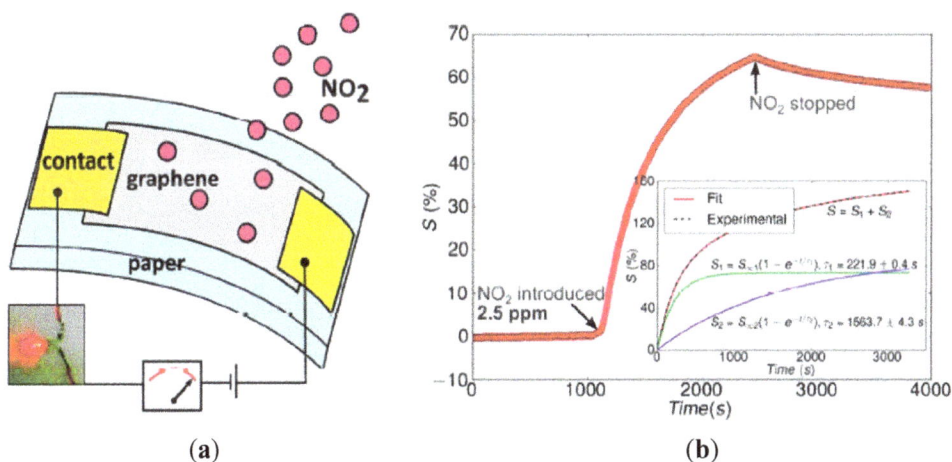

Figure 2. (a) Schematic of a G-paper strip based gas sensor; (b) Response of a G-paper strip to 2.5 ppm of NO_2. Inset shows a fit of double exponential function to the temporal response for 2.5 ppm of NO_2. The two constituent exponentials are also shown along with the estimates of time-constants. Reprinted with permission from Ref. [55]. Copyright, 2015, American Chemical Society.

Apart from experimental studies, theoreticians investigated the adsorption of gas molecules on pristine graphene to understand the interactions of graphene with gas molecules so as to fully exploit the potential of graphene for gas sensing. All theoretical works on pristine graphene using *ab initio* simulations based on quantum mechanics showed weak adsorption of gas molecules on its surface due to the highly stable and strongly bonded carbon atoms in graphene. The capability of graphene to sense gas molecules is usually investigated by inferring the results from the calculations of adsorption energies, charge transfers and the density of states of graphene-gas molecule adsorption systems [43,45–57].

Using first-principles calculations based on density functional theory (DFT), it was found that gas molecules such as H_2O, NH_3, CO, NO_2 and NO are only physically adsorbed on pristine graphene. The electron donating behavior of CO and NH_3 and the electron withdrawing behavior of NO_2 and H_2O demonstrated experimentally was confirmed by simulations done on a graphene model [45]. For the physisorption of hydrogen on graphene flakes, first-principles non-local van der Waals (vdW) density functional (B3LYP-D3) method calculated an adsorption energy of 5.013 (kJ· mol^{-1}), that was found to be in great agreement with the experimental results [57]. All these reports of gas adsorption on pristine graphene showed low adsorption energies due to the inert property of graphene, which makes it unsuitable for practical use.

Graphene oxide, single- or few layered-graphite oxide sheets obtained by chemical oxidation of graphite and subsequent exfoliation in water, are rich in oxygen-containing groups such as epoxies, hydroxyls, carboxyls, *etc.*, on its basal planes and edges which makes it a versatile material for gas detection [58–61]. Prezioso and his co-workers [58] proposed a practical and highly sensitive gas sensor based on large and highly oxidized GO flakes prepared by modified Hummers' method. The sensor device was fabricated by drop casting GO monolayers with an average size of ~30 μm (about 2 orders of magnitude higher than that of commercially available GO) on platinum interdigitated electrodes. They measured a DL of 20 parts-per-billion (ppb) to NO_2, which is the lowest value ever reported with other graphene based gas sensors. Compared to CNTs and reduced GO based sensors, the large number of active sites on GO lead to highly improved sensitivity and hence suited for applications that require high sensitivity but at the cost of slow response. GO has proved to be the best suitable material for humidity sensing applications with ultrahigh sensitivity [59]. Microscale capacitive humidity sensor fabricated using GO films exhibited excellent sensing properties such as a sensitivity of up to 37,800%, which is more than 10 times greater than that of the best capacitive humidity sensors at 15%–95% relative humidity, fast response and fast recovery which are less than 1/4th and 1/2th of that of the conventional sensors respectively [59].

Few theoretical studies of gas molecular adsorption on GO have also been reported [60,61]. The adsorption of ammonia on graphene and graphene oxide investigated using first-principles calculations showed that the surface active sites such as epoxy and hydroxyls on GO surface enhanced the interactions between NH_3 and GO. The adsorption energy of NH_3 on GO was found to be greater than that on pristine graphene due to the presence of hydroxyl and epoxy groups on graphene surface [60].

One of the most popular approach for the synthesis of graphene by thermal [115,116] or chemical reduction [117,118] of GO results in reduced GO (rGO), having almost identical structure of pristine graphene, but with several residual oxygen functional groups. rGO has also found extensive use in gas sensing [62–66] which could be attributed to the presence of some chemically active defect sites even after reduction of GO, high conductivity, capability for surface modification and water dispersibility. Ppb level detection of acetone and other toxic chemicals has been successfully demonstrated by a GO based chemical sensor reduced by hydrazine vapor. They observed that the level of GO reduction dependent on the exposure time to hydrazine vapor influenced the sensing response and the $1/f$ noise [62]. A 360% increase in sensing response to NO_2 gas at room temperature and atmospheric pressure was achieved by GO based field effect transistor (FET) gas sensors upon chemical reduction, on comparison with thermally reduced GO based sensors [63]. Even though rGO had demonstrated as a promising gas sensing material, selective detection of gas molecules remains as a great challenge for enabling practical use, similar to graphene. In this direction, Lipatov et al. fabricated a rGO based gas sensing system that could recognize different alcohols such as ethanol, methanol and isopropanol by making use of the significantly different properties of rGO flakes obtained from the same batch fabrication [64].

Later both experimental and theoretical studies proved that the sensitivity of graphene based gas sensors can be enhanced significantly by introducing dopants such as boron, nitrogen, phosphorus, gallium, chromium, manganese, silicon, sulphur, etc., and defects [119–124]. The introduction of dopants into the crystal structure and defects into the basal plane modifies the physical, chemical and electrical properties of graphene, that could be tailored for improving the sensitivity and selectivity of graphene based gas sensors [125]. Heteroatoms create new active sites on graphene surface that enables strong adsorption of gas molecules which has been proved by theoretical investigations of gas sensing using doped graphene. The strong chemical doping of doped graphene by the gas molecules was evident from the large values of adsorption energies and the charge transfers calculated using ab initio calculations. Chromium and manganese are found to be the best suitable dopants for the sensitive detection of SO_2 [120]. Zhang et al. proved strong chemisorption of H_2S on Fe-doped graphene, suggesting the possibility of employing Fe-doped graphene for H_2S gas detection [121]. It has been found that nitrogen-doped graphene is the best material for selective sensing of CO [122]. First-principles study of the effect of the modification of graphene with Stones-Wales (SW) defect, Al doping and the combination of two on the SO_2 adsorption behavior of graphene showed high chemical reactivity of Al-doped graphene and Al-doped graphene with SW

81

defect towards SO_2 compared to pristine graphene and SW-defected graphene [123]. Vacancy defected graphene along with dopants such as boron, nitrogen and sulphur was found to be more sensitive to formaldehyde than pristine graphene due to the strong chemisorption of formaldehyde on vacancy-defected graphene with dopants [124].

It was also proved experimentally that modification of graphene by dopant atoms enhances the gas sensing properties of graphene [126,127]. Niu *et al.* [127] reported excellent NO_2 sensing ability of N and Si co-doped graphene nanosheets (NSi-GNS) prepared by high-temperature annealing of N and Si-containing graphene oxide-ionic liquid (GO-IL) composites (as shown in Figure 3). In the NSi-GNS, N atoms act as active sites for NO_2 gas adsorption, whereas Si atoms modify the electronic structure significantly. NSi-GNS based gas sensor obtained by annealing of GO-IL at 400 °C (NSi-GNS-400) spread on a glass substrate between the silver electrodes (as shown in Figure 4a), showed high negative sensor response ($-26\% \pm 1\%$) in 21 ppm of NO_2 and good stability even after five sensing cycles (as shown in Figure 4b). The response time and the recovery times observed for NSi-GNS-400 were 68 s and 635 s respectively (as shown in Figure 4c). The response of the NSi-GNS-400 upon NO_2 exposure with concentrations varying from 21 to 1 ppm (as shown in Figure 4d) had a decreasing trend of response with decrease in NO_2 concentration. The high sensitivity of the sensor was evident from the response value of -8.8% at 1 ppm NO_2 [127].

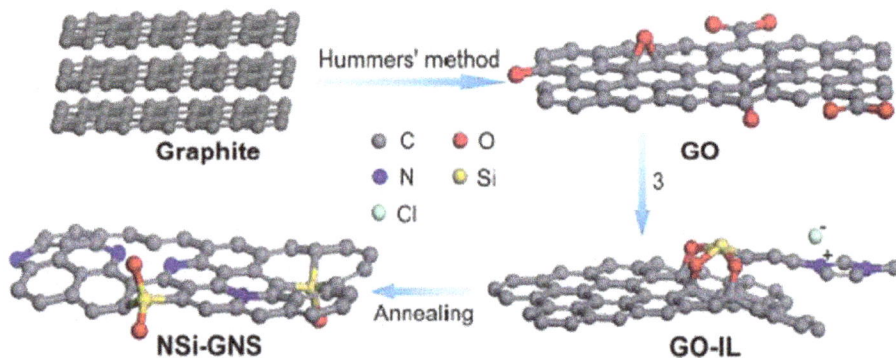

Figure 3. Schematic illustration of the synthesis of N and Si co-doped graphene nanosheets (NSi-GNS) through the high-temperature annealing of N and Si-containing graphene oxide-ionic liquid composite (GO-IL). Reprinted with permission from Ref. [127]. Copyright, 2013, Royal Society of Chemistry.

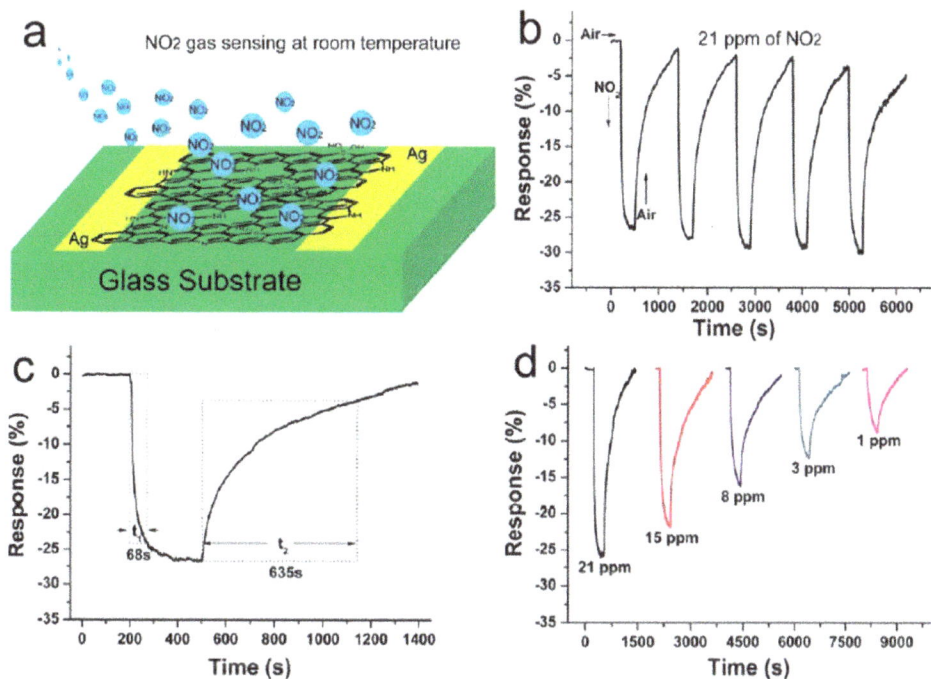

Figure 4. (a) NO$_2$ sensor; (b) the response of NSi-GNS-400 to 21 ppm of NO$_2$; (c) response and recovery time; and (d) response to NO$_2$ with varying concentrations. Reprinted with permission from Ref. [127]. Copyright, 2013, Royal Society of Chemistry.

Apart from dopants and defects, other chemical functionalization methods such as modification with metal and metal oxide nanoparticles (NPs) and polymers have also been studied. Compared to pristine graphene and rGO, sensors based on graphene/rGO modified with nanoparticles of metals or metal oxides have demonstrated highly sensitive and selective sensing behavior [128–135], which could be attributed to large changes in the electronic properties of graphene/rGO upon gas exposure due to the synergistic effects of NPs and graphene/rGO.

Many reports in the literature prove that pristine graphene has poor sensitivity towards hydrogen gas owing to the absence of dangling bonds in the structure. Palladium (Pd) NP decorated CVD grown graphene showed good sensing response of 33% at room temperature to 1000 ppm H$_2$ with a detection limit of 20 ppm, due to the significant increase in resistance of the Pd decorated graphene sensor upon hydrogen injection. The sensor was found to be flexible that no significant degradation in the sensing response upon bending the sensor to a curved geometry [128].

Cho and his co-workers proved that the introduction of aluminium (Al) NPs and Pd NPs on graphene lead to improved sensitivity of graphene to NO_2 and NH_3 gases respectively [129]. Wang *et al.* [130] fabricated gas sensors based on rGO functionalized with platinum (Pt) NPs by mid-temperature thermal annealing and alternating current dielectrophoretic technique, which allowed for efficient sensing of multiple gases. At room temperature, the rGO gas sensors with (without) Pt NPs exhibited sensitivities of 14% (7%), 8% (5%), and 10% (8%), for 1000 ppm H_2, NH_3, and NO gases, respectively. An improvement of 100%, 60% and 25%, to H_2, NH_3, and NO gases was observed for Pt functionalized rGO compared to that of rGO sensors without Pt NP decoration. The recovery/response time for H_2 gas was found to decrease with Pt decoration while for NH_3, and NO, it showed the opposite behavior [130].

Recently, gas sensors based on graphene transistors decorated with SnO_2 NPs exhibited high selectivity, fast response and short recovery (~1 s) to 100 ppm H_2 at 50 °C (as shown in Figure 5a–c) [131]. FETs based on graphene decorated with metal oxide NPs were employed for developing high performance hydrogen gas sensors. The extremely high surface-to-volume ratio and the abundance of dangling bonds by the decoration of graphene with metal oxide NPs, resulted in strong interactions of gas molecules in the surroundings through the grain boundaries and unsaturated bonds of metal oxide NPs on graphene (as shown in Figure 6b), which ultimately resulted in excellent H_2 sensitivity of the graphene-metal oxide NP hybrid, compared to pristine graphene (as shown in Figure 6a). Graphene facilitates quick transfer of electrons from metal oxide NPs due to the insignificant Schottky barrier caused by the matching work functions of NPs and graphene. They observed an increase in output current with increase in H_2 concentration from 1 ppm to 100 ppm with good reproducibility (as shown in Figure 5b). The achieved lowest resolution limit of 1 ppm proves the potential of SnO_2 NP-graphene for high sensitive low level detection of H_2 gas [131].

rGO/ZnO nanocomposites have recently been employed as highly sensitive NO_2 gas sensors at room temperature with fast response and recovery than those based on pristine rGO, which clearly indicate the improvement in the sensing property of rGO by ZnO NP decoration due to the tuning of the semiconducting properties of rGO induced by ZnO [132]. Apart from SnO_2 and ZnO, graphene/rGO decorated with other metal oxides such as Cu_2O [133], WO_3 [134], Fe_2O_3 [135], *etc.*, have also demonstrated to be successful candidates for the detection of various gases due to the higher sensing response as compared to non-decorated graphene/rGO.

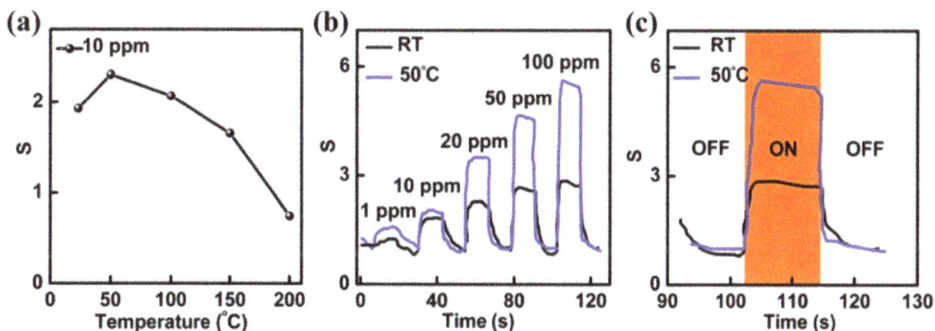

Figure 5. (**a**) The sensitivity of the gas sensor decorated with SnO_2 nanoparticles (NPs) at various temperatures; (**b**) Real-time dynamic response of gas sensors decorated with SnO_2 NPs exposed to different H_2 concentrations at different operation temperatures; (**c**) The response and recovery times of the sensor exposed to the 100 ppm H_2 concentration. Reprinted with permission from Ref. [131]. Copyright, 2015, Royal Society of Chemistry.

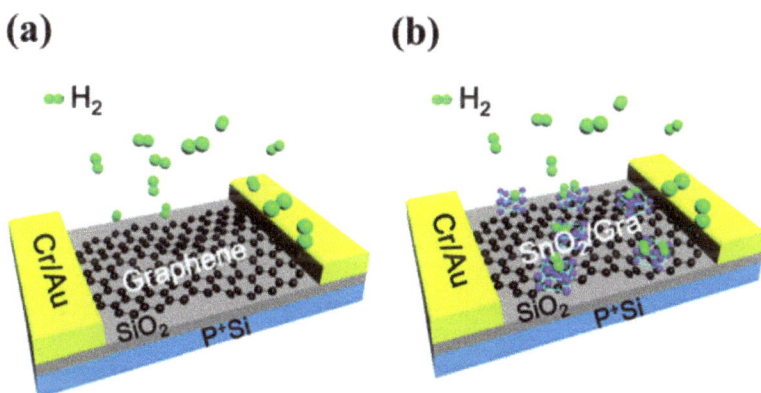

Figure 6. (**a**) The schematic of graphene transistor without obvious sensitivity to hydrogen; (**b**) The schematic of graphene field effect transistor (FET) decorated by metal oxide NPs with obvious sensitivity to hydrogen. Reprinted with permission from Ref. [131]. Copyright, 2015, Royal Society of Chemistry.

Recently, graphene/rGO functionalized with polymers has also emerged as new gas sensing materials due to the improved gas sensing behavior. The investigation of the synergistic behavior of conducting polymers (CPs) such as polyaniline (PANI), polypyrrole (PPy), polypyrene and graphene/rGO for molecular gas sensing resulted in outstanding performance in terms of both sensitivity and selectivity [136]. Graphene-PANI nanocomposites showed much higher sensitivity to the presence of hydrogen gas compared to that of pristine graphene sheets and

PANI nanofibers [137]. Bai *et al.* [138] have fabricated a chemoresistor-type gas sensor using lyophilized GO/PPy composite hydrogel prepared through chemical polymerization of pyrrole in GO aqueous suspension for testing its NH_3 sensing property. They found that the resistance of the GO/PPy composite aerogel increased by about 40% within 600 s, compared with only 7% increase in the case of gas sensor based on electropolymerized PPy film to 800 ppm NH_3 gas (as shown in Figure 7). The superior performance of the composite aerogel along with simple and low-cost fabrication process could be employed for developing highly sensitive and economically feasible gas sensors. In another similar work, the unique electrical properties of rGO and PPy were combined for NH_3 gas sensing application in which the performance comparison of rGO/PPy nanocomposite with intrinsic graphitic materials and nanocomposites of GO, graphene and graphite with PPy have been discussed [139]. The effective electron charge transfer between PPy and NH_3 and the efficient transfer of resistance variation in PPy by the uniformly dispersed rGO in the rGO/PPy nanocomposite enabled rapid and highly sensitive detection of ammonia gas, compared to PPy/GO, PPy/graphene, PPy/graphite and graphite. The excellent reproducibility of the nanocomposite based sensor was due to the ease of the recovery process at a lower temperature of 373 K.

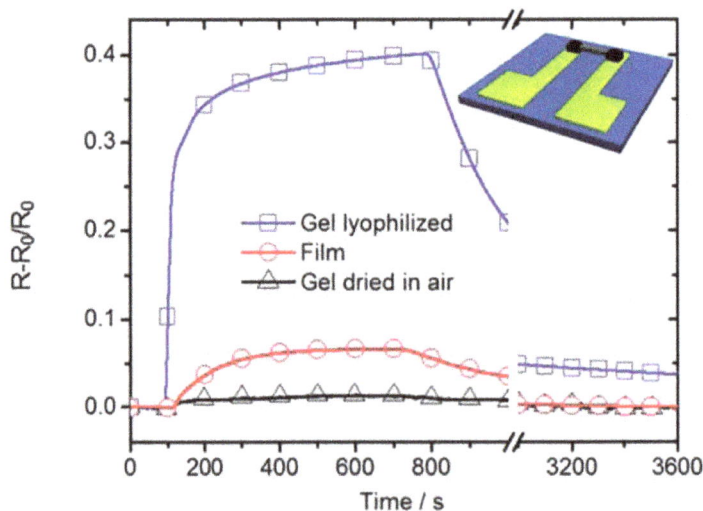

Figure 7. Ammonia gas sensing performance of three devices with sensing elements of lyophilized GO/PPy hydrogel, electrochemically deposited PPy film and GO/PPy hydrogel dried in air. Inset is a sketch of the gas sensor devices. Reprinted with permission from Ref. [138]. Copyright, 2011, Royal Society of Chemistry.

Recently, a chemoresistive gas sensor based on nanocomposites of graphene and polystyrene-sulfonate (rGO/PSS) was reported as a successful trimethylamine

(TMA) gas detector. The rGO/PSS based gas sensor exhibited linear sensing response upon exposure to increasing TMA concentrations (23 to 183 mg/L), along with good repeatability and reproducibility [140]. The decrease in resistance of the GO/(3,4-ethylenedioxythiophene):poly(styrenesulfonate) (PEDOT:PSS) and rGO/PEDOT:PSS composite devices on hydrogen gas exposure suggested that nanocomposites of PEDOT:PSS with chemically modified graphene could be used effectively for hydrogen gas sensing, compared to pure PEDOT:PSS [141].

2.2. Transition Metal Dichalcogenides (TMDs)

Transition metal dichalcogenides are a group of materials with general formula of MX_2, where M is a transition metal element of group IV, V or VI, and X is a chalcogen (S, Se or Te) [68]. Even though bulk crystals of TMDs were known and studied by researchers for decades, their 2D forms have received significant attention from both fundamental and application point of view after the success story of graphene [142]. These materials form two-dimensional layered structures in which the plane of metal atoms is sandwiched between two hexagonal planes of chalcogen atoms. Several 2D TMDs possess bandgap compared to pristine graphene [143,144] and hence these materials are promising candidates for new FETs with high on-off ratios and optoelectronic applications. The success of graphene based sensors greatly inspired scientists to explore the use of 2D TMDs as sensing materials. The electrical, optical and chemical properties of TMDs along with their high surface-to-volume ratio [143] suggest their applications in molecular sensing [68].

Molybdenum disulfide (MoS_2), one of the most popular semiconducting TMDs [144] has shown great promise for a variety of applications in electronics and optoelectronics [145–149]. As expected, gas sensors made from MoS_2 have demonstrated excellent sensing characteristics such as high sensitivity, fast response time and good stability [80–83,85].

Several MoS_2 based gas sensors using micromechanically exfoliated [80,82,83] and liquid phase exfoliated [81] MoS_2 flakes as sensing materials have been reported. Highly sensitive and stable detection of NO gas with detection limit as low as 0.8 ppm had been shown by FET sensors using mechanically exfoliated multilayered (two- three- and four- layered) MoS_2 sheets [80]. They observed strong chemisorption of NO on n-doped MoS_2 flakes, with a slow increase in resistivity of MoS_2 flakes upon exposure to NO gas (over 30 s), due to the induced p-type doping by the NO gas and also slow decrease in resistivity upon removal of NO. Flexible transistor sensor arrays based on 1.5 mm-long MoS_2 channel with rGO electrodes made on a polyethylene terephthalate (PET) substrate displayed much higher NO_2 sensitivity and good reproducibility compared to novel rGO-FET sensors, which can be enhanced up to three times by functionalization of MoS_2 with Pt nanoparticles [81]. But devices made from mechanically and liquid phase exfoliated MoS_2 films suffer from poor scalability and poor electronic quality. MoS_2 based devices made from more scalable approaches have also shown significant changes in electrical conductivity upon NH_3 adsorption [85].

Recently, thin film transistors (TFTs) made from MoS_2 demonstrated selective gas sensing behavior at operating temperature up to 220 °C, with much larger sensitivity [86] compared to similar graphene based gas sensors. TFT structures were fabricated by exfoliation from bulk MoS_2 and were then transferred to Si/SiO_2 substrate with patterned contact electrodes. The working principle of these sensors are based on the generation of positive or negative charges at the MoS_2 surface by the vapor molecules that either enhance or deplete the electron concentration in the channel depending on the type of the gas vapors. As the change in the channel conductance forms the primary basis for gas sensing, different gas species could not be distinguished from each other by looking at this conductance change. Selective gas sensing using graphene was already reported by considering two additional characteristics such as transient time and the peaks in the noise spectrum, when used together with the conductance change [51]. For different vapors such as acetone, acetonitrile, toluene and chloroform, the magnitude, sign of the response, and the response time constant are found to be different. The relative change in the device current and the characteristic transient times of the vapors were used as unique signatures [86]. On comparison, it was observed that for same vapor concentration, the relative resistance changes for graphene were less than 50%. For acetonitrile, the increase in noise under the exposure is nearly an order of magnitude, whereas other gases showed a much smaller change.

Cantalini *et al.* [87] reported linear sensing response of layered MoS_2 films to NO_2 at concentrations below 1 ppm, with a detection limit of around 20 ppb at 200 °C. To 1 ppm NO_2 gas in dry air at temperatures below 250 °C, the chemically exfoliated MoS_2 flakes showed typical p-type behavior with decreasing resistance. The MoS_2 flakes showed reasonable sensitivity of 1.28 (ratio of resistance in air to resistance in NO_2) to 1 ppm NO_2 with fast and reversible response at 150 °C. They did not show any response to CO and H_2 at 150 °C, enabling high selectivity. At temperatures above 250 °C in dry air, the MoS_2 films showed p- to n-type transition of the electrical properties with response to 1 ppm NO_2.

Planar sensor structures consisting of monolayer MoS_2 channel on a SiO_2/Si substrate with Au electrodes (as shown in Figure 8a,b) have shown rapid increase in MoS_2 conductivity on exposure to triethylamine (TEA, a decomposition product of the V-series of nerve gas agent) and acetone. This sensor did not show any response to other analytes such as dichlorobenzene, dichloropentane, nitromethane, nitrotoluene and water vapor [83]. The observed initial response (~5 s) upon TEA exposure was more rapid than the 30 s reported previously for multilayer MoS_2 to NO [80]. The response of the sensor channel to a sequence of 10 TEA pulses, each with a concentration of 0.002% P_0 (~1 ppm), was shown in Figure 9a. This monolayer MoS_2 sensor (green curve in Figure 9a) exhibited no response to water vapor, used as a background constituent (0.025% P_0 ~6 ppm), while a monolayer graphene sensor (purple curve in Figure 9a) showed a pronounced response to water vapor. For each TEA pulse, the change in conductivity followed similar pattern with an initial rapid rise and fall (as shown in Figure 9b). They observed increase in conductivity change with increase in the TEA concentration (as shown in Figure 9c) and the difference in relative sensitivity observed could be attributed to geometric factors and residual contamination during device fabrication. MoS_2 based sensors showed a TEA detection threshold of 10 ppb due to strong response and excellent signal-to-noise ratio. A comparison of the change in conductivity of these sensors with planar sensors fabricated from (a) monolayer graphene grown by chemical vapor deposition on copper and (b) a carbon nanotube (CNT) network consisting of a dense array of CNTs forming an electrically continuous thin film, as the sensor channel for a sequence of TEA pulses (10 s on, 20 s off) of 0.025% P_0 (12 ppm) concentration were plotted in Figure 9d. The response of the sensors based on CNTs and MoS_2 to a single pulse of TEA are comparable, whereas those based on graphene, the response is observed to be small. Upon TEA exposure, the conductivity of both graphene and CNTs decrease, this is marked in contrast with response exhibited by MoS_2. This could be attributed to the facts that TEA is a strong electron donor, CNT networks and graphene show p-type, whereas MoS_2 show n-type character respectively. Hence the transient physisorption of TEA enhances the majority carrier density and the conductivity of MoS_2, whereas these parameters decrease in the case of graphene and CNT network. The MoS_2 sensors exhibited good sensitivity, high selectivity and a complementary response to CNT-network sensor [83].

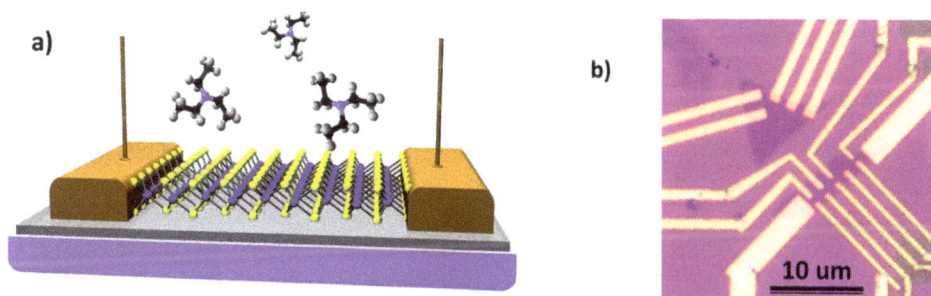

Figure 8. Schematic and image of the MoS_2 monolayer sensor. (**a**) A single monolayer of MoS_2 is supported on a SiO_2/Si substrate and contacted with Au contact pads; (**b**) An optical image of the processed devices showing the monolayer MoS_2 flakes electrically contacted by multiple Au leads. Reprinted with permission from Ref. [83]. Copyright, 2013, American Chemical Society.

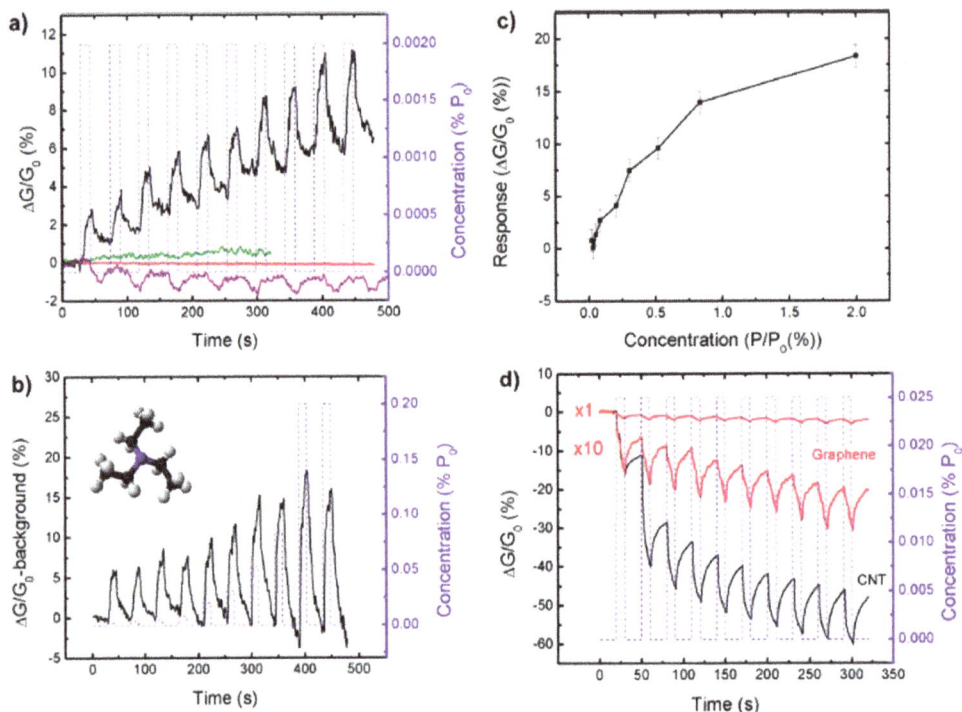

Figure 9. Response of sensors to triethylamine (TEA) exposure. (**a**) Change in conductivity of the monolayer MoS_2 sensor channel upon exposure to a sequence of 0.002% P_0 TEA pulses (black line). The dashed blue lines show the pulse timing (15 s on/ 30 s off) and concentration. The solid red line shows the response to exposure of nitrogen only and serves as a control experiment. The solid green and purple lines show the response of the MoS_2 and graphene sensors to water vapor pulses (0.025% P_0), respectively; (**b**) Same as part a, but for a series of exposure pulses in which the TEA concentration increases from 0.002% P_0 to 0.2% P_0. A positive slope background has been removed. The inset shows a model of the TEA molecule, in which the nitrogen atom is blue, the carbon atoms are black, and the hydrogen atoms are light gray; (**c**) The amplitude of the conductivity change increases with TEA concentration. The vertical axis is the response to each individual pulse (not the time integrated response); (**d**) Change in conductivity of a CVD graphene monolayer (red) and CNT-network sensor (black) upon exposure to a sequence of 0.025% P_0 TEA pulses (10 s on/20 s off). Reprinted with permission from Ref. [83]. Copyright, 2013, American Chemical Society.

Even though atomically thin films of TMDs synthesized by micromechanical cleavage, ultrasonication in organic solvents, aqueous surfactant solutions, or solutions of polymers in solvents, intercalation and exfoliation are well suited for applications in gas sensing [80–83], catalysis [150], composites [151], energy storage and conversion [152,153], the as prepared TMDs are not compatible with standard microlithography techniques in the nanodevice fabrication process. Gatensby and his co-workers [154] proposed a facile route for fabricating devices from MoS_2 and WS_2, grown by vapor phase sulfurization of pre-deposited metal layers which allows the production of highly homogenous TMD films over large areas with fine control of thickness from bulk to monolayer. Using shadow mask lithography, well-defined geometries were produced that could be easily integrated with standard micro-processing methods. The metal was deposited in selective areas by shadow masks, the sample was then sulfurized, interdigitated electrode (IDE) contacts were defined using a second shadow mask and were then sputtered. On exposure of the n-type sensing device to NH_3 (electron donor), the conductivity increased due to the rise in majority carrier concentration. The fabricated sensor responded very quickly to ppb levels of NH_3 with ultrahigh sensitivity down to 400 ppb. Similar to other nanomaterial-based sensors, these sensors also showed slow recovery in pure N_2 flow at room temperature due to the strong binding with NH_3 which thus required UV illumination or high temperature annealing for accelerating the recovery. The ease of device manufacture, cost effectiveness, scalability and compatibility with existing semiconductor fabrication methods make this process favorable for future sensors and other electronic applications [154].

Donarelli *et al.* [84] reported resistive type MoS_2 based gas sensors which showed good response to NO_2, H_2 and relative humidity. The sensor device was fabricated by depositing liquid chemically exfoliated (in N-methyl pyrrolidone, NMP) MoS_2 flakes on Si_3N_4 substrate with pre-patterned IDEs on the front side, and on the back side with heater circuit for thermal annealing and Pt sensor for temperature control. The MoS_2 based sensing device with 150 °C thermal annealing of exfoliated MoS_2 flakes did not respond to 1 ppm NO_2 at room temperature, but these MoS_2 flakes responded at higher operating temperatures. The "150 °C annealed" device, with p-type semiconducting behavior showed a decrease in resistance upon exposure to NO_2 (oxidizing gas) in the 25–200 °C operating temperature range [84]. This "150 °C annealed" device, exhibited good NO_2 response either at 150 °C (ratio of resistance in air to that in NO_2 is 1.29) or 200 °C (ratio of resistance in air to that in NO_2 is 1.15). The "250 °C annealed" device, with n-type semiconducting behavior showed an increase in resistance upon exposure to NO_2, with significant resistance change at room temperature. The "150 °C annealed" device was found to be faster than the "250 °C annealed" device during adsorption. The response intensity of n-type MoS_2 device exceeded by a factor of 5.0 (2.4) than that measured for the p-type device at 200 °C (250 °C). "250 °C annealed" device set at 200 °C showed outstanding performance with detection limit of 20 ppb and response intensity equal to 5.80 at 1 ppm concentration, compared to previously reported sensors. The n-MoS_2 has higher responses than GO based sensors, but with comparable responses to metal oxide and multi-walled CNT based sensors. The Raman and X-ray Photoelectron Spectroscopy (XPS) analyses proved that increase of sulfur vacancies after thermal annealing at 250 °C and the partial surface oxidation of the MoS_2 upper layers in the form of MoO_3 resulted in the n-type behavior and also contributed to the outstanding performance of the "250 °C annealed" MoS_2 sensor [84].

MoS_2 sheets have proved to be exciting candidates for high performance gas sensing in which all the sensing experiments were performed in an inert atmosphere. Under practical environmental conditions that includes oxygen, the properties of MoS_2 gets strongly affected by the adsorption of oxygen which leads to cross sensitivity effects and limits its practical applicability. Recently, room temperature sensing of NO_2 was reported using MoS_2 NS decorated with SnO_2 nanocrystals (MoS_2/SnO_2) due to the improvement in the stability of the MoS_2 sheets in practical environment by functionalization with SnO_2 nanocrystals. High sensitivity, good selectivity and repeatability to NO_2 in practical dry air were also exhibited by this hybrid sensor, as compared to other MoS_2 sensors [155].

Sarkar *et al.* investigated the hydrogen gas sensing performance of a MoS_2-based FET functionalized with Pd NPs [156]. The change in the work function of Pd NPs induced by the adsorption of H_2 led to decrease in p-type doping which was measured by the change in the current of the Pd NP functionalized MoS_2 FET. There was only a negligible change in current upon exposure of the MoS_2 FET without NPs to 3 ppm H_2 (as shown in Figure 10a), but upon the incorporation of Pd NPs, there was large increase in current level of the n-type MoS_2 device (as shown in Figure 10b). For 3 ppm H_2 gas at room temperature, the sensitivity (ratio of change in conductivity/ current to the initial conductivity/ current) of bulk MoS_2 reported earlier was much less than 1 [157], whereas the sensitivity value increased to about 5 for MoS_2 decorated with Pd NPs, which could be attributed to the use of few layer MoS_2, around 8 nm thick and also due to the operation of FET in the subthreshold region [156].

Even though many experimental reports on MoS_2 based gas sensors exist, there have been very few reports on theoretical studies of the adsorption of gas molecules on MoS_2 surface. In this direction, Yue *et al.* [158] analyzed the most stable adsorption position, orientation, associated charge transfer and the modification of electronic properties of the monolayer MoS_2 surface due to the adsorption of H_2, O_2, H_2O, NH_3, NO, NO_2, and CO using first-principles calculation based on DFT. They found that all these molecules acting as either electron donors or acceptors are only physisorbed on MoS_2 surface, with small charge transfer and no significant alteration of band structure upon molecule adsorption. They also observed significant modulation of the charge transfer by the application of a perpendicular electric field [158]. Similarly, first principles simulation have also been employed by Zhao *et al.* [159] to investigate the adsorption of various gas molecules such as CO, CO_2, NH_3, NO, NO_2, CH_4, H_2O, N_2, O_2 and SO_2 on monolayer MoS_2 by including van der Waals interactions between the gas molecules and MoS_2. They found that only NO, NO_2 and SO_2 could bind strongly to MoS_2 surface compared to other gas molecules, which was found to be in good agreement with experimental observations. Hence they suggested that MoS_2 is more sensitive to NO, NO_2 and SO_2 due to the observed charge transfer, variations in the electronic band structure and density of states (DOS) of MoS_2 after gas adsorption [159].

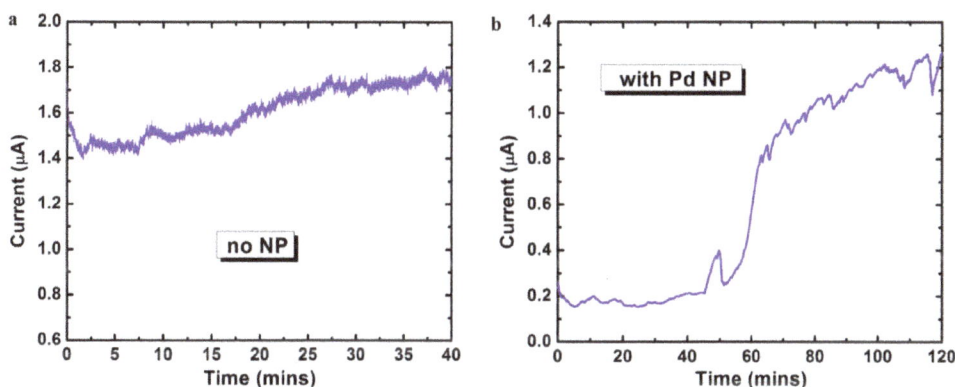

Figure 10. (a) Real-time measurement of current of MoS_2 FET without any NPs. Thickness of MoS_2 used was around 8 nm; (b) Real-time measurement of current of the same MoS_2 FET after incorporation of Pd NPs. Current increases substantially upon exposure to hydrogen (3 ppm from time = 45 min onwards) from 0.2 μA to about 1 μA. Reprinted with permission from Ref. [156]. Copyright, 2015, American Chemical Society.

Cho *et al.* developed a highly sensitive and selective gas sensor using uniform atomic-layered MoS_2 synthesized by thermal CVD for the detection of NO_2 and NH_3 [160]. The charge transfer mechanism was evident from *in situ* photoluminescence characterizations and theoretical calculations. The first-principles DFT based calculations showed that adsorption processes are exothermic and thus probable adsorption of NO_2 and NH_3 onto MoS_2 surface due to the negative adsorption energies for NO_2 and NH_3 gas molecules. The adsorption of NO_2 onto MoS_2 surface led to an increase in the intensity of the positively charged trion (A^+) and suppression of the intensity of neutral exciton (A°) due to the electron depletion of MoS_2 by NO_2 adsorption, whereas in the case of NH_3 adsorption, the A° peak intensity increased and the A^+ peak intensity decreased due to the electron accumulation of MoS_2 by NH_3 adsorption. The resistance of the MoS_2 gas sensor exhibited an increase in the NO_2 gas exposure mode, whereas the resistance exhibited a decrease during the NH_3 gas exposure mode [160].

Recently, Cho *et al.* investigated the gas sensing performance of a 2D heterostructure-based gas sensor via the combination of mechanically exfoliated MoS_2 and CVD grown graphene [161]. MoS_2 flake based gas sensor with Au/Ti metal electrodes exhibited excellent gas sensing stability over many sensing cycle tests with a detection limit of about 1.2 ppm to NO_2. After NO_2 injection, positive sensitivity *i.e.*, increase in resistance was experienced by the Au/Ti/MoS_2 based sensor, whereas in the case of NH_3 injection, negative resistivity was observed. But the sensitivity of MoS_2 based sensor to NH_3 was found to be lower than that to NO_2 due to the small charge transfer from NH_3 and MoS_2. The band structures of MoS_2 after NO_2 and NH_3 adsorption also validated the high sensitivity of MoS_2 to NO_2. In the atomically thin heterostructure-based gas sensor, patterned graphene film was used instead of metal electrodes for the charge collection of MoS_2. They gave a detailed explanation on the gas sensing mechanism of the graphene/MoS_2 device based on an equation having several variable resistance terms. 2D heterostructure-based structure fabricated on a flexible polyimide substrate retained its gas sensing characteristics without any serious performance degradation, even after harsh bending tests of about 5000 bending cycles. The flexible graphene/MoS_2 based sensor exhibited extraordinary long-term stability after 19 months. The graphene/MoS_2 based gas sensor experienced a decrease in resistance upon NO_2 exposure and increase in resistance upon NH_3 exposure, which is just the opposite trend of resistance change observed for Au/Ti/MoS_2 based sensor due to the n-type behavior of Au/Ti/MoS_2 based device and p-type behavior of graphene/MoS_2 heterojunction based gas sensor [161].

Cho *et al.* reported the bifunctional sensing characteristics of CVD synthesized MoS_2 film to detect gas molecules and photons in a sequence [162]. The single MoS_2 based device had demonstrated highly sensitive, selective detection of NO_2 and also good photo sensing performance such as reasonable photoresponsivity, reliable photoresponse and rapid photoswitching. The observed sensing behavior of the MoS_2 based device could be attributed to the charge transfer between NO_2 and MoS_2 sensing film and the detection limit was measured to be 120 ppb. The devices based on atomic-scale MoS_2 films with bifunctional capability would pave the route towards the development of the futuristic multifunctional sensors [162].

Apart from MoS_2, other members of TMDs such as WS_2 and $MoSe_2$ have also shown excellent gas sensing properties [88–91]. Researchers have explored the possibility of employing WS_2 in gas sensing, as WS_2 possess several advantages such as higher thermal stability, wider operation temperature range and favorable band structure as compared to MoS_2. Huo *et al.* for the first time systematically studied the photoelectrical and gas sensing properties of transistors based on multilayer

WS$_2$ nanoflakes exfoliated from WS$_2$ crystals [88]. The photoelectrical properties of multilayer WS$_2$ nanoflake based FETs got strongly influenced by the gas molecules. Upon exposure to reducing gases such as ethanol and NH$_3$, they observed strong and prolonged response with enhanced photo-responsivity and external quantum efficiency, due to the charge transfer between the physically adsorbed gas molecules and multilayer WS$_2$ nanoflake based FETs [88]. High sensitivity detection of WS$_2$ thin films towards NH$_3$ was achieved by WS$_2$ devices synthesized using inductively coupled plasma (ICP) source [89].

The interactions of NH$_3$ and H$_2$O molecules with monolayer WS$_2$ investigated by means of first-principles calculations indicated that both NH$_3$ and H$_2$O molecules are physisorbed on monolayer WS$_2$ [90]. The results from Bader charge analysis and plane-averaged differential charge density showed that NH$_3$ and H$_2$O act as electron donor and acceptor, leading to n- and p-type doping respectively. The charge transfer between the gas molecules and single-layer WS$_2$ was primarily determined from the mixing of the highest occupied and lowest unoccupied molecular orbital with the underlying WS$_2$ orbitals. They also provided detailed explanation about the sensing mechanism of the WS$_2$-FET based gas sensor towards NH$_3$ and H$_2$O. They found enhanced photoresponsivity and external quantum efficiency due to the increase in the total conduction electron density as more electrons are transferred from NH$_3$ to the n-type WS$_2$ channel, upon the adsorption of NH$_3$ gas molecules. In the case of H$_2$O adsorption on monolayer WS$_2$, the source-drain current of WS$_2$-FET based gas sensor got suppressed by the electron trapping of H$_2$O molecule from the WS$_2$ channel [90]. Their theoretical results were found to be in good agreement with the experimental results by Huo et $al.$ [88]. After a single NH$_3$ molecule adsorption on a 4×4 WS$_2$ supercell, the saturation source-drain current (I_{Dsat}) increased by 9.6×10^{-6} A from its value under vacuum, which was found to be comparable to the experimental increment of dark drain current by ~6.9×10^{-7} A in NH$_3$. Similarly I_{Dsat} decreased by 2.8×10^{-6} A, consistent with the experimental decrement of dark drain current to ~1.0×10^{-8} A in air [90].

Late et $al.$ developed a gas sensor based on single layer MoSe$_2$ and the shift in the Raman spectra observed before and after exposing the MoSe$_2$ based gas sensing device to NH$_3$ confirmed the gas detection ability of single layer MoSe$_2$. The detection limit of the MoSe$_2$ based gas sensor was found to be 50 ppm. Their findings proved the potential of MoSe$_2$ and other TMDs as excellent gas sensors [91].

2.3. Phosphorene

Motivated by the utility of 2D materials such as graphene and TMDs in nanodevice applications, scientists started searching for new 2D materials. Phosphorene, a single-atomic layer of black phosphorus (BP), arranged in a puckered honeycomb lattice [163] isolated recently, possess several advantageous properties such as existence of a finite band gap [164] and high charge carrier mobility of around $1000 \text{ cm}^2/\text{Vs}$ [165] over previously isolated 2D materials such as graphene and MoS_2 with zero band gap and low mobility respectively. In addition to these, the anisotropic electrical conductance, high current on/off ratios, high operating frequencies, ambipolar behavior, fast and broadband photo detection of phosphorene [166,167] have already been exploited for FETs, PN junctions, photodetectors, solar cells, *etc.*, [163,168–170] within the last year. Similar to graphene and MoS_2, it was found that gas adsorption strongly influences the electrical properties of phosphorene, which could be effectively employed for gas sensing applications [92,93].

Kou *et al.* presented first-principles study of the adsorption of CO, CO_2, NO, NO_2 and NH_3 on monolayer phosphorene. Their results proved superior gas sensing performance of phosphorene which even surpasses other 2D materials such as graphene and MoS_2. Molecular doping of phosphorene by gas molecules due to the charge transfer between gas molecules and phosphorene resulted in large binding energies, similar to that observed in graphene and MoS_2 [92]. The adsorption of gas molecules was observed to be much stronger than that in graphene and MoS_2, which implies that monolayer phosphorene, can be used to make more sensitive gas sensors. The current-voltage (I-V) characteristics of phosphorene calculated using non-equilibrium Green's function (NEGF) formalism showed sensitive changes to adsorption, depending on the type of the gas molecule and thus leading to high selectivity.

The first experimental verification of gas sensors based on BP was reported by Abbas *et al.* recently [93]. FETs based on multilayer BP (as shown in Figure 11) showed increased conduction upon NO_2 exposure (as shown in Figure 12a) due to the hole doping of the multilayer BP by NO_2 and high sensitivity detection down to 5 ppb, which could be compared with the performance of other 2D materials. The relative conductance change for the multilayer BP based FET on exposure to varying NO_2 concentrations of 5, 10, 20 and 40 ppb followed the Langmuir isotherm for molecules adsorbed on a surface (as shown in Figure 12b). Moreover, the sensor had good stability even after repeated sensing cycles and better recovery after flushing with argon, which presented reversible adsorption and desorption process. Compared to other multilayer 2D materials such as MoS_2 [81,82], the sensitivity of multilayer BP was found to be surprisingly very high [93]. The relative change in conductance of an 18 nm thick MoS_2 flake to 1200 ppb NO_2 was ~1%, whereas the conductance change for 55 nm thick BP to 5 ppb NO_2 was 2.9%. The high sensitivity of BP could be attributed to the high adsorption energies of NO_2 on BP and the less out-of-plane conductance of BP compared to graphene and MoS_2.

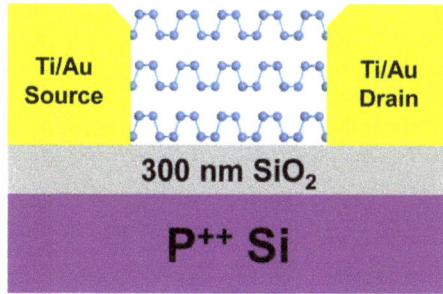

Figure 11. Scheme of a multilayer black phosphorus (BP) FET used for chemical sensing. Reprinted with permission from Ref. [93]. Copyright, 2015, American Chemical Society.

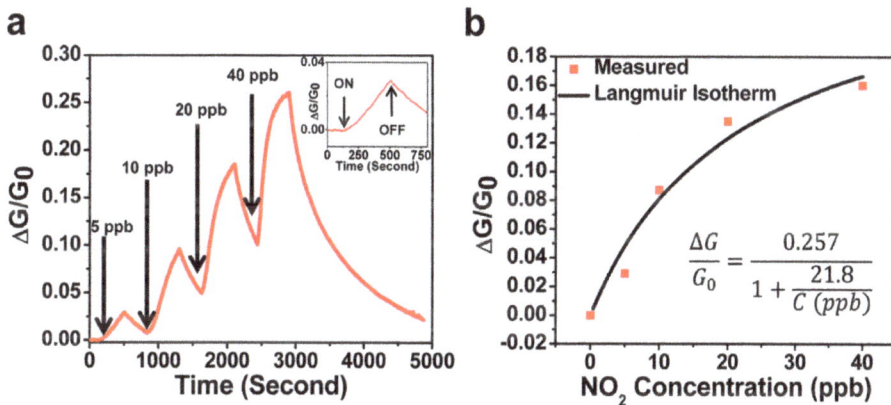

Figure 12. NO_2 gas sensing performance of multilayer BP FET. (**a**) Relative conductance change ($\Delta G/G_0$) *vs.* time in seconds for a multilayer BP sensor. Inset shows a zoomed in image of a 5 ppb NO_2 exposure response with identification of points in time where the NO_2 gas is switched on and off; (**b**) $\Delta G/G_0$ plotted *vs.* NO_2 concentration applied to the BP FET showing an agreement between the measured values (red squares) and the fitted Langmuir isotherm. Reprinted with permission from Ref. [93]. Copyright, 2015, American Chemical Society.

2.4. 2D Semiconducting Metal Oxide Based Nanostructures

Gas sensors based on metal oxide semiconductors such as ZnO, SnO_2, WO_3, Fe_2O_3, In_2O_3, which are the most commonly used solid-state gas detection devices in domestic, industrial and commercial applications [171,172] operate on the principle of induced change in the electrical conductivity of the metal oxide due to the chemical reactions between gas molecules and semiconductor surface. But the sensitivity values reported for these sensors based on bulk materials or dense films were only moderate [172].

Compared to their bulk counterparts, 1D nanostructures in the form of nanowires, nanotubes, nanorods, nanoribbons, nanofibers, *etc.*, provide large surface-to-volume ratios and high density surface reactive sites for gas sensing applications. The sensing platforms based on 1D nanostructured semiconducting metal oxides possess advantages of small size, light weight, good chemical and thermal stabilities under different operating conditions with minimal power consumption [4,173]. Gas sensors using 2D semiconducting metal oxide nanostructures such as nanosheets, nanowalls, nanoplates, nanoscale films have also attracted the major attention of researchers due to the their comparable or even better performance compared to 1D metal oxide nanostructures.

One of the most important and interesting metal oxide, ZnO, a n-type semiconductor, has many unique optical and electrical properties such as wide band gap (3.37 eV), high exciton binding energy (60 meV) in addition to high temperature operation, high sensitivity to toxic and combustible gases, high thermal and chemical stability [174]. During the past, gas sensors based on ZnO were employed particularly for fuel leak detection in automobiles, space crafts, emissions from industrial processes, *etc.*, Recently, ZnO nanostructures in the form of nanowires, nanotubes, nanorods, nanoflowers, *etc.*, have shown superior performance in nanoscale optoelectronics [175,176], nanoscale piezotronics [177,178], catalysis and sensing devices [179–181].

Gas sensors made from 2D ZnO NSs with thickness of about 10–20 nm and width-to-thickness ratios of nearly one thousand, synthesized by simple mixed hydrothermal method in the presence of cetyltrimethyl ammonium bromide (CTAB) and 1,2-propanidiol, exhibited good selectivity and response to acetone and gasoline at high and low temperatures respectively in the presence of ammonia, ethanol and toluene [94]. For acetone, the ZnO NS based gas sensor showed increase in the response with increase in heating temperature, but for gasoline, the sensing behavior was found to be decreasing with rise in heating temperature. Both the response and recovery times got improved with increasing heating temperature. At 300 °C, the response and recovery times were nearly 13 s and 30 s respectively. At 360 °C, the response to acetone (response magnitude of 31) was much higher than that of the other gases such as ammonia, ethanol, toluene and gasoline. At 180 °C, the response to gasoline (response magnitude of 30) was also found to be much higher than that of the other gases. These results proved the potential of employing 2D ZnO NSs with extremely large specific surface area for selective detection of acetone and gasoline through temperature modulation [94].

ZnO nanowalls also have high surface-to-volume ratios and are highly effective in energy storage, field emission, chemical and biological sensing applications.

Chang *et al.* [95] developed a CO gas sensor based on ZnO nanowalls grown on a glass substrate using fast, low-temperature, catalyst-free process in a tube furnace, compared to other ZnO nanowall fabrication methods which involve costly equipment, complex processes, toxic metal-organic precursors and flammable gases. They observed a linear trend of sensitivity ratio with the CO concentration. The current density-potential curve of the ZnO nanowall based gas sensor showed highest relative sensitivity ratio of 1.05 at 300 °C for 3000 ppm CO [95].

Earlier reports have shown enhancement in the gas sensing properties of the nanomaterials as their size approaches the Debye length, which is about 15 nm for ZnO at 325 °C. Recently, ultrathin hexagonal ZnO NSs as thin as 17 nm synthesized by simple hydrothermal method in the presence of CTAB exhibited the highest gas sensing response of 37.8 (ratio of resistance in air to that in target gas) at an operating temperature of 350 °C with response and recovery times of 9 s and 11 s respectively to 50 ppm formaldehyde gas [96].

As 2D porous metal oxide NSs with single-crystalline structure provide relatively large surface area due to the unique sheet-like morphology, they are successful gas sensing materials due to the enhanced sensing response and good stability. In the case of 2D NSs, the synthesis of single-crystalline structure with numerous pores was found to be easier as compared to 1D NWs [172]. Jing *et al.* [97] reported that gas sensor fabricated from porous ZnO nanoplates synthesized by simple microwave method showed strong response to chlorobenzene and ethanol at different operating temperatures. Liu *et al.* verified the gas sensing properties of novel single-crystalline ZnO NSs composed of porous interiors fabricated by annealing $ZnS(en)_{0.5}$ (en = ethylenediamine) complex precursor [98]. The gas sensors fabricated from ZnO nanostructures exhibited high gas sensing response, short response time, fast recovery to formaldehyde and ammonia and also significant long term stability.

Hierarchically porous semiconducting metal oxide materials are proven to be ideal candidates for gas sensing applications due to the high surface area that they provide thanks to its peculiar structure which greatly facilitate significant enhancement in gas diffusion and mass transport, ultimately improving the sensitivity and response time of the gas sensor [182]. Due to ease of scale up, consistency and low power consumption, microstructure sensors based on thin films are appropriate for practical sensing applications. A microstructure sensor based on hierarchically porous ZnO NS thin films demonstrated for the first time by Zeng *et al.* [99] showed the maximum response of 11.2 (ratio of resistance in dry air to that in target gas) at optimal temperature of 300 °C to 100 ppm CO with a response time of 25 s, recovery time of 36 s and high selectivity to CO among other interfering gases such as SO_2, C_7H_8. The comparison of the response of sensors based on ZnO NSs, ZnO NPs and ZnO seed layer as a function of CO concentration showed the fastest increase for ZnO NSs. Due to the hierarchically porous structure of ZnO NSs with well-defined and well-aligned micro-, meso- and nanoporosites, ZnO NS based sensors showed higher and fast response, selectivity with fast recovery compared to other ZnO based nanostructures [99].

Nanomaterials based on NiO, a p-type semiconductor with excellent optical, electrical properties and high chemical stability have also proven as promising gas sensing materials owing to their high electron transport performance. Gas sensors based on NiO nanomaterials have demonstrated high sensitivity, low cost and good compatibility with micromachining. Zero-dimensional (0D) and 1D-NiO based sensors had excellent sensitivity, fast response and recovery, but suffer from poor stability due to poor interconnection between the sensing materials and electrodes on the sensing platform. NiO nanoscale films and foils, when used as sensing elements could overcome the stability issues associated with 0D- and 1D-NiO. Even though various approaches such as CVD, reactive sputtering, metal evaporation, sol-gel and chemical methods have been employed for preparing 2D NiO nanostructures, NiO thin film prepared by chemical reduction has shown high gas sensing performance in terms of repeatability, sensitivity and stability due to high porosity, uniform morphology, continuity and nanocrystallinity of the films. Wang *et al.* [100] fabricated ammonia gas sensors based on NiO porous films on glass substrate through chemical reaction combined with high-temperature oxidation route in air, which showed fast response and excellent sensitivity (as shown in Figure 13). Without the need for pre-concentration step, the change in the sensor conductance reached up to about 18% upon exposure to 30 ppm NH_3 for about 27 s, which could be attributed to the large specific surface area and the porous surface structure of 2D NiO film [100]. The high electron transport performance and good connection between the 2D NiO films and the electrodes improved the recovery and stability of the sensor (as shown in Figure 13). They also observed excellent selectivity towards NH_3 over other organic gases for the 2D NiO grainy film based sensor (as shown in Figure 14).

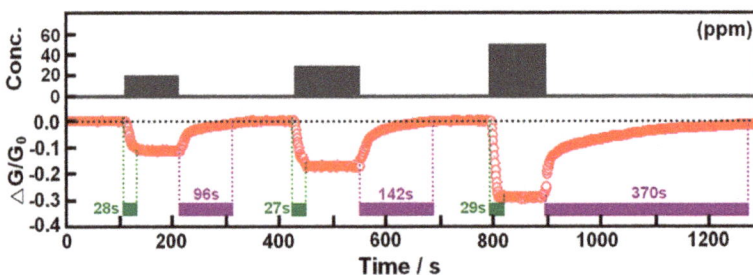

Figure 13. Stability and sensitivity test of gas sensors based on NiO films. The conductance change of NiO films on the glass substrates to 20, 30, and 50 ppm of NH_3 at room temperature [100].

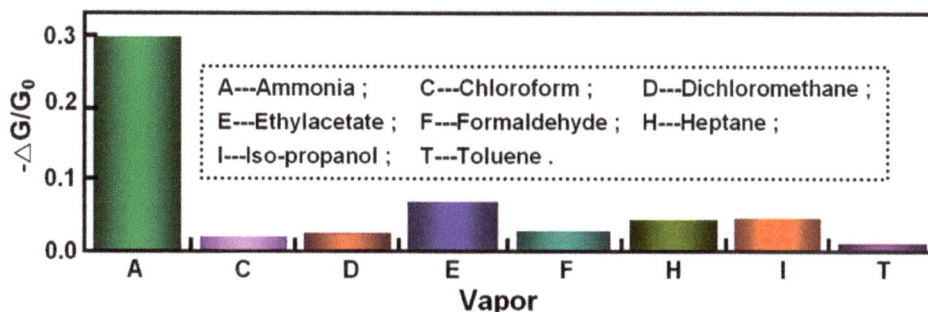

Figure 14. Selectivity tests of gas sensors based on NiO films. The concentration of NH_3 is 50 ppm, and the concentration of other organic gases is higher than 300 ppm [100].

2D nanostructures made from CuO, WO_3, SnO_2 have also shown good gas sensing characteristics. Uniform CuO nanosheets synthesized by mild hydrothermal method in the presence of CTAB showed stable and similar gas sensing response to combustible gases such as ethanol, acetone and gasoline [101]. 2D WO_3 nanoplate based sensors showed superhigh sensitivity to different alcohols such as methanol, ethanol, isopropanol and butanol at operating temperatures of 260–360 °C [102]. The high specific surface area arising from the ultrathin plate like structure and the high crystallinity of WO_3 nanoplates led to fast and effective adsorption of alcohol molecules. The response and recovery times for the 2D WO_3 nanoplate sensor were reported to be less than 15 s for all the tested alcohols.

A highly sensitive and fast responding CO sensor was demonstrated by SnO_2 NSs prepared by a facile chemical route at room temperature with subsequent high temperature annealing [103]. The reported response value of 2.34 and response time of 6 s were found to be significantly larger and less than those of SnO_2 powders (1.57, 88 s) respectively [103]. Sun *et al.* [106] have reported high response, good repeatability, short response and recovery periods, for 10 nm thick SnO_2 NSs prepared by simple and efficient hydrothermal method on exposure to ethanol. The response (ratio of resistance in air to that in tested gas atmosphere) of the SnO_2 NS sensor to 100 ppm ethanol increased with the rise in the temperature from 200 to 250 °C which is followed by decrease in response with increase in operating temperature. The response had the maximum value of 73.3 at 250 °C for ethanol, much greater than the response values to other volatile organic compounds such as acetone, methanol, toluene, butanone and isopropanol. To 200 ppm CO, the SnO_2 NS sensor showed the highest response at 300 °C with good repeatability and stability. Compared to other SnO_2 nanostructures with longer response and recovery times, the response and recovery times of the SnO_2 NS based sensor were about 1 and 3 s respectively, believed to be due to the ease of gas diffusion towards the surface of SnO_2 NSs through the channels and pores and its reaction with the chemisorbed oxygen [106]. The CO gas sensing performance of different SnO_2 NSs prepared through polyvinylpyrrolidone (PVP)-assisted hydrothermal method was investigated by Zeng *et al.* [108]. They found that the as prepared SnO_2 NSs showed a linear increase in sensing response with CO gas concentrations ranging from 100 to 450 ppm at 300 °C. Out of the different SnO_2 NSs that vary in their PVP concentration, SnO_2 NS with the highest PVP concentration of 15 mM exhibited good CO sensing performance such as higher response, fast response and recovery, due to the large contact surface area on the NSs [108].

3. Conclusions and Future Perspectives

The exfoliation of graphene and other 2D materials from their 3D counterparts and the recent utilization of their fascinating properties in various fields have been the main breakthrough in the realm of materials science. 2D materials possess great potential to play an important role in electronics, optoelectronics, energy conversion and storage, chemical and biological sensing due to their unique structural, electrical, physical and chemical properties. The success of new and improved nanodevices based on 2D materials such as MoS_2, graphene and its derivatives in optoelectronics and nanoelectronics have inspired researchers to explore more 2D materials with better performance than those already isolated.

Gas sensors based on 2D materials have demonstrated high sensitivity detection of a wide variety of gas molecules at low concentrations, due to the maximum sensor surface area per unit volume, low noise and superior ability to screen charge fluctuations compared to 0D and 1D systems, in addition to their favorable electrical properties for gas adsorption. Even though pristine graphene, GO and rGO present a very promising gas sensing platform for room temperature gas detection, they get strongly affected by a range of different gas species and also suffer from slow recovery and poor electrical stability under different environmental conditions. Chemical modification of graphene or rGO, functionalized with defects, dopants, metal, metal oxide nanoparticles and polymers have proven to provide promising solution for improving the sensing performance with high specificity, enhanced sensitivity and low detection limit. Over the past two to three years, gas sensors based on other 2D materials such as TMDs and phosphorene have started gaining a lot of research interest due to their extraordinary and superior sensing capability over carbon based nanomaterials.

The desorption of gases on 2D materials such as rGO, MoS_2, phosphorene, *etc.*, is observed to be slow due to the strong adsorption of gas molecules on these materials, which usually requires external assistance such as UV light exposure or high temperature annealing after gas exposure. This drawback of 2D material based gas sensors, which limits their practical use need to be addressed as it degrades the sensing performance in terms of sensitivity, detection limit and repeatability. Most of the gas sensing experiments already reported using 2D materials had to be performed under a controlled environment, as the sensitivity values were found to be affected by the presence of other gases in practical environmental conditions. Hence future work need to concentrate on achieving ultra-high selective gas sensing devices under practical conditions using 2D materials. This issue can be addressed by modification of the material surface with suitable modifiers such as metal or metal oxide nanoparticles. The functionalization scheme could be extended to TMDs and phosphorene, which had already proven to be highly successful for graphene based materials. Lack of large scale manufacturing of 2D materials with large area, high and uniform quality is yet another challenge. Flexible electronics could benefit from the mechanical compatibility of 2D materials with the device fabrication due to their excellent mechanical properties. 2D heterostructures obtained by the combination of different 2D materials provide a suitable sensing platform for developing future wearable electronics.

Modeling of nano sensing devices based on 2D materials is receiving great importance now-a-days, as it could enable the study of underlying sensing mechanisms and the analysis of the sensor performance before going for expensive experimentation. The electronic structure and quantum transport calculations of 2D material based electronic sensors before and after interaction with gas molecules are of great importance to the fabrication and development of novel nano sensors. Even though many reports on theoretical simulations of gas adsorption on graphene and modified graphene exist which give insights into the change in the electrical properties of graphene by gas molecule interactions, similar works on TMDs such as MoS_2, WS_2, $MoSe_2$ and phosphorene are found to be very limited or lacking. Future progress in this field necessitates detailed understanding of the intrinsic properties of new 2D materials at the nano-scale and also the effect of adsorption of gas molecule on the electronic properties for evaluating the gas detection capability of these materials. First-principles calculations on the adsorption of gas molecules using different 2D materials could enable the identification of the suitable modification of these materials for enhanced sensor performance. The gas sensing mechanisms of devices based on newly discovered 2D materials are not completely explained till now. Once the above mentioned shortcomings are addressed, 2D materials could revolutionize the field of gas sensing so that they could emerge as ideal sensing elements with ultrahigh sensitivity, surprisingly high selectivity, fast response, good reversibility, room temperature operation, outstanding stability, *etc.*, for gas sensing systems in the near future.

Conflicts of Interest: The authors declare no conflicts of interest.

References

1. Endres, H.E.; Göttler, W.; Hartinger, R.; Drost, S.; Hellmich, W.; Müller, G.; Braunmühl, C.B.-V.; Krenkow, A.; Perego, C.; Sberveglieri, G. A thin-film SnO_2 sensor system for simultaneous detection of CO and NO_2 with neural signal evaluation. *Sens. Actuators B* **1996**, *36*, 353–357.

2. Siyama, T.; Kato, A. A new detector for gaseous components using semiconductor thin film. *Anal. Chem.* **1962**, *34*, 1502–1503.

3. Tomchenko, A.A.; Harmer, G.P.; Marquis, B.T.; Allen, J.W. Semiconducting metal oxide sensor array for the selective detection of combustion gases. *Sens. Actuators B* **2003**, *93*, 126–134.

4. Arafat, M.; Dinan, B.; Akbar, S.A.; Haseeb, A. Gas sensors based on one dimensional nanostructured metal-oxides: A review. *Sensors* **2012**, *12*, 7207–7258.

5. Barsan, N.; Weimar, U. Conduction model of metal oxide gas sensors. *J. Electroceramics* **2001**, *7*, 143–167.

6. Capone, S.; Forleo, A.; Francioso, L.; Rella, R.; Siciliano, P.; Spadavecchia, J.; Presicce, D.; Taurino, A. Solid state gas sensors: State of the art and future activities. *J. Optoelectron. Adv. Mater.* **2003**, *5*, 1335–1348.

7. Jimenez-Cadena, G.; Riu, J.; Rius, F.X. Gas sensors based on nanostructured materials. *Analyst* **2007**, *132*, 1083–1099.

8. Hanna Varghese, S.; Nair, R.; G. Nair, Baiju; Hanajiri, T.; Maekawa, T.; Yoshida, Y.; Sakthi Kumar, D. Sensors based on carbon nanotubes and their applications: A review. *Curr. Nanosci.* **2010**, *6*, 331–346.

9. Kong, J.; Franklin, N.R.; Zhou, C.; Chapline, M.G.; Peng, S.; Cho, K.; Dai, H. Nanotube molecular wires as chemical sensors. *Science* **2000**, *287*, 622–625.

10. Tabib-Azar, M.; Yan, X. Sensitive NH_3OH and HCl gas sensors using self-aligned and self-welded multiwalled carbon nanotubes. *IEEE Sens. J.* **2007**, *7*, 1435–1439.

11. Li, J.; Lu, Y.; Ye, Q.; Cinke, M.; Han, J.; Meyyappan, M. Carbon nanotube sensors for gas and organic vapor detection. *Nano Lett.* **2003**, *3*, 929–933.

12. Varghese, O.K.; Kichambre, P.D.; Gong, D.; Ong, K.G.; Dickey, E.C.; Grimes, C.A. Gas sensing characteristics of multi-wall carbon nanotubes. *Sens. Actuators B* **2001**, *81*, 32–41.

13. Comini, E.; Sberveglieri, G. Metal oxide nanowires as chemical sensors. *Mater. Today* **2010**, *13*, 36–44.

14. Cui, Y.; Wei, Q.; Park, H.; Lieber, C.M. Nanowire nanosensors for highly sensitive and selective detection of biological and chemical species. *Science* **2001**, *293*, 1289–1292.

15. Chen, X.; Wong, C.K.Y.; Yuan, C.A.; Zhang, G. Nanowire-based gas sensors. *Sens. Actuators B* **2013**, *177*, 178–195.

16. Liu, X.; Cheng, S.; Liu, H.; Hu, S.; Zhang, D.; Ning, H. A survey on gas sensing technology. *Sensors* **2012**, *12*, 9635–9665.

17. Dan, Y.; Evoy, S.; Johnson, A. Chemical gas sensors based on nanowires. In *Nanowire Research Progress*; Nova Science Publisher: Hauppauge, NY, USA, 2008; pp. 95–128.

18. Huang, X.-J.; Choi, Y.-K. Chemical sensors based on nanostructured materials. *Sens. Actuators B* **2007**, *122*, 659–671.

19. Comini, E. Metal oxide nano-crystals for gas sensing. *Anal. Chim. Acta* **2006**, *568*, 28–40.

20. Novoselov, K.S.; Geim, A.K.; Morozov, S.V.; Jiang, D.; Zhang, Y.; Dubonos, S.V.; Grigorieva, I.V.; Firsov, A.A. Electric field effect in atomically thin carbon films. *Science* **2004**, *306*, 666–669.

21. Geim, A.K.; Novoselov, K.S. The rise of graphene. *Nat. Mater.* **2007**, *6*, 183–191.

22. Avouris, P. Graphene: Electronic and photonic properties and devices. *Nano Lett.* **2010**, *10*, 4285–4294.

23. Schwierz, F. Graphene transistors. *Nat. Nano* **2010**, *5*, 487–496.

24. Bonaccorso, F.; Sun, Z.; Hasan, T.; Ferrari, A.C. Graphene photonics and optoelectronics. *Nat. Photonics* **2010**, *4*, 611–622.

25. Pumera, M. Graphene-based nanomaterials for energy storage. *Energy Environ. Sci.* **2011**, *4*, 668–674.

26. Brownson, D.A.C.; Kampouris, D.K.; Banks, C.E. An overview of graphene in energy production and storage applications. *J. Power Sources* **2011**, *196*, 4873–4885.

27. Gwon, H.; Kim, H.-S.; Lee, K.U.; Seo, D.-H.; Park, Y.C.; Lee, Y.-S.; Ahn, B.T.; Kang, K. Flexible energy storage devices based on graphene paper. *Energy Environ. Sci.* **2011**, *4*, 1277–1283.

28. Huang, X.; Qi, X.; Boey, F.; Zhang, H. Graphene-based composites. *Chem. Soc. Rev.* **2012**, *41*, 666–686.

29. Choi, H.-J.; Jung, S.-M.; Seo, J.-M.; Chang, D.W.; Dai, L.; Baek, J.-B. Graphene for energy conversion and storage in fuel cells and supercapacitors. *Nano Energy* **2012**, *1*, 534–551.

30. Wujcik, E.K.; Monty, C.N. Nanotechnology for implantable sensors: Carbon nanotubes and graphene in medicine. *Wiley Interdiscip. Rev.* **2013**, *5*, 233–249.

31. Feng, L.; Liu, Z. Graphene in biomedicine: Opportunities and challenges. *Nanomedicine* **2011**, *6*, 317–324.

32. Shen, H.; Zhang, L.; Liu, M.; Zhang, Z. Biomedical applications of graphene. *Theranostics* **2012**, *2*, 283–294.

33. Chung, C.; Kim, Y.-K.; Shin, D.; Ryoo, S.-R.; Hong, B.H.; Min, D.-H. Biomedical applications of graphene and graphene oxide. *Acc. Chem. Res.* **2013**, *46*, 2211–2224.

34. Feng, L.; Wu, L.; Qu, X. New horizons for diagnostics and therapeutic applications of graphene and graphene oxide. *Adv. Mater.* **2013**, *25*, 168–186.

35. Ratinac, K.R.; Yang, W.; Ringer, S.P.; Braet, F. Toward ubiquitous environmental gas sensors—Capitalizing on the promise of graphene. *Environ. Sci. Technol.* **2010**, *44*, 1167–1176.

36. Yavari, F.; Koratkar, N. Graphene-based chemical sensors. *J. Phys. Chem. Lett.* **2012**, *3*, 1746–1753.

37. Zhou, M.; Zhai, Y.; Dong, S. Electrochemical sensing and biosensing platform based on chemically reduced graphene oxide. *Anal. Chem.* **2009**, *81*, 5603–5613.

38. Liu, Y.; Dong, X.; Chen, P. Biological and chemical sensors based on graphene materials. *Chem. Soci. Rev.* **2012**, *41*, 2283–2307.

39. Castro Neto, A.H.; Guinea, F.; Peres, N.M.R.; Novoselov, K.S.; Geim, A.K. The electronic properties of graphene. *Rev. Mod. Phys.* **2009**, *81*, 109–162.

40. Balandin, A.A. Thermal properties of graphene and nanostructured carbon materials. *Nat. Mater.* **2011**, *10*, 569–581.

41. Sandeep Kumar, V.; Venkatesh, A. Advances in graphene-based sensors and devices. *J. Nanomed. Nanotechol.* **2013**, *4*, e127.

42. Lee, C.; Wei, X.; Kysar, J.W.; Hone, J. Measurement of the elastic properties and intrinsic strength of monolayer graphene. *Science* **2008**, *321*, 385–388.

43. Wehling, T.O.; Novoselov, K.S.; Morozov, S.V.; Vdovin, E.E.; Katsnelson, M.I.; Geim, A.K.; Lichtenstein, A.I. Molecular doping of graphene. *Nano Lett.* **2008**, *8*, 173–177.

44. Schedin, F.; Geim, A.K.; Morozov, S.V.; Hill, E.W.; Blake, P.; Katsnelson, M.I.; Novoselov, K.S. Detection of individual gas molecules adsorbed on graphene. *Nat. Mater.* **2007**, *6*, 652–655.

45. Leenaerts, O.; Partoens, B.; Peeters, F.M. Adsorption of H_2O, NH_3, CO, NO_2, and NO on graphene: A first-principles study. *Phys. Rev. B* **2008**, *77*, 125416.

46. Leenaerts, O.; Partoens, B.; Peeters, F.M. Adsorption of small molecules on graphene. *Microelectron. J.* **2009**, *40*, 860–862.

47. Ko, G.; Kim, H.Y.; Ahn, J.; Park, Y.M.; Lee, K.Y.; Kim, J. Graphene-based nitrogen dioxide gas sensors. *Curr. Appl. Phys.* **2010**, *10*, 1002–1004.

48. Yoon, H.J.; Jun, D.H.; Yang, J.H.; Zhou, Z.; Yang, S.S.; Cheng, M.M.-C. Carbon dioxide gas sensor using a graphene sheet. *Sens. Actuators B* **2011**, *157*, 310–313.

49. Romero, H.E.; Joshi, P.; Gupta, A.K.; Gutierrez, H.R.; Cole, M.W.; Tadigadapa, S.A.; Eklund, P.C. Adsorption of ammonia on graphene. *Nanotechnology* **2009**, *20*, 245501.

50. Chen, C.W.; Hung, S.C.; Yang, M.D.; Yeh, C.W.; Wu, C.H.; Chi, G.C.; Ren, F.; Pearton, S.J. Oxygen sensors made by monolayer graphene under room temperature. *Appl. Phys. Lett.* **2011**, *99*, 243502.

51. Rumyantsev, S.; Liu, G.; Shur, M.S.; Potyrailo, R.A.; Balandin, A.A. Selective gas sensing with a single pristine graphene transistor. *Nano Lett.* **2012**, *12*, 2294–2298.

52. Chen, G.; Paronyan, T.M.; Harutyunyan, A.R. Sub-ppt gas detection with pristine graphene. *Appl. Phys. Lett.* **2012**, *101*, 053119.

53. Fattah, A.; Khatami, S. Selective H_2S gas sensing with a graphene/n-si schottky diode. *IEEE Sens. J.* **2014**, *14*, 4104–4108.

54. Nemade, K.R.; Waghuley, S.A. Chemiresistive gas sensing by few-layered graphene. *J. Electron. Mater.* **2013**, *42*, 2857–2866.

55. Kumar, S.; Kaushik, S.; Pratap, R.; Raghavan, S. Graphene on paper: A simple, low-cost chemical sensing platform. *ACS Appl. Mater. Interfaces* **2015**, *7*, 2189–2194.

56. Ricciardella, F.; Alfano, B.; Loffredo, F.; Villani, F.; Polichetti, T.; Miglietta, M.L.; Massera, E.; di Francia, G. Inkjet Printed Graphene-Based Chemi-Resistors for Gas Detection in Environmental Conditions. In Proceedings of the AISEM Annual Conference, 2015 XVIII, Trento, Italy, 3–5 February 2015; pp. 1–4.

57. Ganji, M.D.; Hosseini-Khah, S.; Amini-Tabar, Z. Theoretical insight into hydrogen adsorption onto graphene: A first-principles B3LYP-D3 study. *Phys. Chem. Chem. Phys.* **2015**, *17*, 2504–2511.

58. Prezioso, S.; Perrozzi, F.; Giancaterini, L.; Cantalini, C.; Treossi, E.; Palermo, V.; Nardone, M.; Santucci, S.; Ottaviano, L. Graphene oxide as a practical solution to high sensitivity gas sensing. *J. Phys. Chem. C* **2013**, *117*, 10683–10690.

59. Bi, H.; Yin, K.; Xie, X.; Ji, J.; Wan, S.; Sun, L.; Terrones, M.; Dresselhaus, M.S. Ultrahigh humidity sensitivity of graphene oxide. *Sci. Rep.* **2013**.

60. Peng, Y.; Li, J. Ammonia adsorption on graphene and graphene oxide: A first-principles study. *Front. Environ. Sci. Eng.* **2013**, *7*, 403–411.

61. Tang, S.; Cao, Z. Adsorption of nitrogen oxides on graphene and graphene oxides: Insights from density functional calculations. *J. Chem. Phys.* **2011**, *134*, 044710.

62. Robinson, J.T.; Perkins, F.K.; Snow, E.S.; Wei, Z.; Sheehan, P.E. Reduced graphene oxide molecular sensors. *Nano Lett.* **2008**, *8*, 3137–3140.

63. Lu, G.; Park, S.; Yu, K.; Ruoff, R.S.; Ocola, L.E.; Rosenmann, D.; Chen, J. Toward practical gas sensing with highly reduced graphene oxide: A new signal processing method to circumvent run-to-run and device-to-device variations. *ACS Nano* **2011**, *5*, 1154–1164.

64. Lipatov, A.; Varezhnikov, A.; Wilson, P.; Sysoev, V.; Kolmakov, A.; Sinitskii, A. Highly selective gas sensor arrays based on thermally reduced graphene oxide. *Nanoscale* **2013**, *5*, 5426–5434.

65. Hassinen, J.; Kauppila, J.; Leiro, J.; Määttänen, A.; Ihalainen, P.; Peltonen, J.; Lukkari, J. Low-cost reduced graphene oxide-based conductometric nitrogen dioxide-sensitive sensor on paper. *Anal. Bioanal. Chem.* **2013**, *405*, 3611–3617.

66. Wang, D.; Hu, Y.; Zhao, J.; Zeng, L.; Tao, X.; Chen, W. Holey reduced graphene oxide nanosheets for high performance room temperature gas sensing. *J. Mater. Chem. A* **2014**, *2*, 17415–17420.

67. Miró, P.; Audiffred, M.; Heine, T. An atlas of two-dimensional materials. *Chem. Soc. Rev.* **2014**, *43*, 6537–6554.

68. Wang, Q.H.; Kalantar-Zadeh, K.; Kis, A.; Coleman, J.N.; Strano, M.S. Electronics and optoelectronics of two-dimensional transition metal dichalcogenides. *Nat. Nano* **2012**, *7*, 699–712.

69. Butler, S.Z.; Hollen, S.M.; Cao, L.; Cui, Y.; Gupta, J.A.; Gutiérrez, H.R.; Heinz, T.F.; Hong, S.S.; Huang, J.; Ismach, A.F.; *et al.* Progress, challenges, and opportunities in two-dimensional materials beyond graphene. *ACS Nano* **2013**, *7*, 2898–2926.

70. Schwierz, F.; Pezoldt, J.; Granzner, R. Two-dimensional materials and their prospects in transistor electronics. *Nanoscale* **2015**, *7*, 8261–8283.

71. Fiori, G.; Bonaccorso, F.; Iannaccone, G.; Palacios, T.; Neumaier, D.; Seabaugh, A.; Banerjee, S.K.; Colombo, L. Electronics based on two-dimensional materials. *Nat. Nano* **2014**, *9*, 768–779.

72. Das, S.; Robinson, J.A.; Dubey, M.; Terrones, H.; Terrones, M. Beyond graphene: Progress in novel two-dimensional materials and van der waals solids. *Annu. Rev. Mater. Res.* **2015**, *45*, 1–27.

73. Yazyev, O.V.; Chen, Y.P. Polycrystalline graphene and other two-dimensional materials. *Nat. Nano* **2014**, *9*, 755–767.

74. Das, S.; Kim, M.; Lee, J.-W.; Choi, W. Synthesis, properties, and applications of 2-D materials: A comprehensive review. *Crit. Rev. Solid State Mater. Sci.* **2014**, *39*, 231–252.

75. Rao, C.N.R.; Gopalakrishnan, K.; Maitra, U. Comparative study of potential applications of graphene, MoS$_2$, and other two-dimensional materials in energy devices, sensors, and related areas. *ACS Appl. Mater. Interfaces* **2015**, *7*, 7809–7832.

76. Varghese, S.S.; Lonkar, S.; Singh, K.K.; Swaminathan, S.; Abdala, A. Recent advances in graphene based gas sensors. *Sens. Actuators B* **2015**, *218*, 160–183.

77. Yuan, W.; Shi, G. Graphene-based gas sensors. *J. Mater. Chem. A* **2013**, *1*, 10078–10091.

78. Basu, S.; Bhattacharyya, P. Recent developments on graphene and graphene oxide based solid state gas sensors. *Sens. Actuators B* **2012**, *173*, 1–21.

79. Novoselov, K.S.; Jiang, D.; Schedin, F.; Booth, T.J.; Khotkevich, V.V.; Morozov, S.V.; Geim, A.K. Two-dimensional atomic crystals. *Proc. Natl. Acad. Sci. USA* **2005**, *102*, 10451–10453.

80. Li, H.; Yin, Z.; He, Q.; Li, H.; Huang, X.; Lu, G.; Fam, D.W.H.; Tok, A.I.Y.; Zhang, Q.; Zhang, H. Fabrication of single- and multilayer MoS$_2$ film-based field-effect transistors for sensing no at room temperature. *Small* **2012**, *8*, 63–67.

81. He, Q.; Zeng, Z.; Yin, Z.; Li, H.; Wu, S.; Huang, X.; Zhang, H. Fabrication of flexible MoS$_2$ thin-film transistor arrays for practical gas-sensing applications. *Small* **2012**, *8*, 2994–2999.

82. Late, D.J.; Huang, Y.-K.; Liu, B.; Acharya, J.; Shirodkar, S.N.; Luo, J.; Yan, A.; Charles, D.; Waghmare, U.V.; Dravid, V.P.; *et al.* Sensing behavior of atomically thin-layered MoS$_2$ transistors. *ACS Nano* **2013**, *7*, 4879–4891.

83. Perkins, F.K.; Friedman, A.L.; Cobas, E.; Campbell, P.M.; Jernigan, G.G.; Jonker, B.T. Chemical vapor sensing with monolayer MoS$_2$. *Nano Lett.* **2013**, *13*, 668–673.

84. Donarelli, M.; Prezioso, S.; Perrozzi, F.; Bisti, F.; Nardone, M.; Giancaterini, L.; Cantalini, C.; Ottaviano, L. Response to NO$_2$ and other gases of resistive chemically exfoliated MoS$_2$-based gas sensors. *Sens. Actuators B* **2015**, *207 Part A*, 602–613.

85. Lee, K.; Gatensby, R.; McEvoy, N.; Hallam, T.; Duesberg, G.S. High-performance sensors based on molybdenum disulfide thin films. *Adv. Mater.* **2013**, *25*, 6699–6702.

86. Shur, M.; Rumyantsev, S.; Jiang, C.; Samnakay, R.; Renteria, J.; Balandin, A.A. Selective Gas Sensing with MoS$_2$ Thin Film Transistors. In Proceedings of the SENSORS, 2014 IEEE, Valencia, Spain, 2–5 November 2014; pp. 55–57.

87. Cantalini, C.; Giancaterini, L.; Donarelli, M.; Santucci, S.; Ottaviano, L. NO$_2$ Response to Few-Layers MoS$_2$. In Proceedings of the IMCS 2012–The 14th International Meeting on Chemical Sensors, Nuremberg, Germany, 20–23 May 2012; pp. 1656–1659.

88. Huo, N.; Yang, S.; Wei, Z.; Li, S.-S.; Xia, J.-B.; Li, J. Photoresponsive and gas sensing field-effect transistors based on multilayer WS$_2$ nanoflakes. *Sci. Rep.* **2014**, *4*, 5209.

89. O'Brien, M.; Lee, K.; Morrish, R.; Berner, N.C.; McEvoy, N.; Wolden, C.A.; Duesberg, G.S. Plasma assisted synthesis of WS$_2$ for gas sensing applications. *Chem. Phys. Lett.* **2014**, *615*, 6–10.

90. Zhou, C.J.; Yang, W.H.; Wu, Y.P.; Lin, W.; Zhu, H.L. Theoretical study of the interaction of electron donor and acceptor molecules with monolayer WS$_2$. *J. Phys. D* **2015**, *48*, 285303.

91. Late, D.J.; Doneux, T.; Bougouma, M. Single-layer MoSe$_2$ based NH$_3$ gas sensor. *Appl. Phys. Lett.* **2014**, *105*, 233103.

92. Kou, L.; Frauenheim, T.; Chen, C. Phosphorene as a superior gas sensor: Selective adsorption and distinct I–V response. *J. Phys. Chem. Lett.* **2014**, *5*, 2675–2681.

93. Abbas, A.N.; Liu, B.; Chen, L.; Ma, Y.; Cong, S.; Aroonyadet, N.; Köpf, M.; Nilges, T.; Zhou, C. Black phosphorus gas sensors. *ACS Nano* **2015**, *9*, 5618–5624.

94. Fan, H.; Jia, X. Selective detection of acetone and gasoline by temperature modulation in zinc oxide nanosheets sensors. *Solid State Ionics* **2011**, *192*, 688–692.

95. Chang, S.-P.; Wen, C.-H.; Chang, S.-J. Two-dimensional ZnO nanowalls for gas sensor and photoelectrochemical applications. *Electron. Mater. Lett.* **2014**, *10*, 693–697.

96. Guo, W.; Fu, M.; Zhai, C.; Wang, Z. Hydrothermal synthesis and gas-sensing properties of ultrathin hexagonal ZnO nanosheets. *Ceram. Int.* **2014**, *40*, 2295–2298.

97. Jing, Z.; Zhan, J. Fabrication and gas-sensing properties of porous ZnO nanoplates. *Adv. Mater.* **2008**, *20*, 4547–4551.

98. Liu, J.; Guo, Z.; Meng, F.; Luo, T.; Li, M.; Liu, J. Novel porous single-crystalline zno nanosheets fabricated by annealing ZnS(en)$_{0.5}$ (en = ethylenediamine) precursor. Application in a gas sensor for indoor air contaminant detection. *Nanotechnology* **2009**, *20*, 125501.

99. Zeng, Y.; Qiao, L.; Bing, Y.; Wen, M.; Zou, B.; Zheng, W.; Zhang, T.; Zou, G. Development of microstructure CO sensor based on hierarchically porous ZnO nanosheet thin films. *Sens. Actuators B* **2012**, *173*, 897–902.

100. Wang, J.; Yang, P.; Wei, X.; Zhou, Z. Preparation of NiO two-dimensional grainy films and their high-performance gas sensors for ammonia detection. *Nanoscale Res. Lett.* **2015**, *10*, 1–6.

101. Jia, X.; Fan, H.; Yang, W. Hydrothermal synthesis and primary gas sensing properties of CuO nanosheets. *J. Dispers. Sci. Technol.* **2010**, *31*, 866–869.

102. Chen, D.; Hou, X.; Wen, H.; Wang, Y.; Wang, H.; Li, X.; Zhang, R.; Lu, H.; Xu, H.; Guan, S. The enhanced alcohol-sensing response of ultrathin WO$_3$ nanoplates. *Nanotechnology* **2010**, *21*, 035501.

103. Moon, C.S.; Kim, H.-R.; Auchterlonie, G.; Drennan, J.; Lee, J.-H. Highly sensitive and fast responding CO sensor using SnO$_2$ nanosheets. *Sens. Actuators B* **2008**, *131*, 556–564.

104. Li, K.-M.; Li, Y.-J.; Lu, M.-Y.; Kuo, C.-I.; Chen, L.-J. Direct conversion of single-layer sno nanoplates to multi-layer SnO$_2$ nanoplates with enhanced ethanol sensing properties. *Adv. Funct. Mater.* **2009**, *19*, 2453–2456.

105. Xu, M.-H.; Cai, F.-S.; Yin, J.; Yuan, Z.-H.; Bie, L.-J. Facile synthesis of highly ethanol-sensitive SnO$_2$ nanosheets using homogeneous precipitation method. *Sens. Actuators B* **2010**, *145*, 875–878.

106. Sun, P.; Cao, Y.; Liu, J.; Sun, Y.; Ma, J.; Lu, G. Dispersive SnO$_2$ nanosheets: Hydrothermal synthesis and gas-sensing properties. *Sens. Actuators B* **2011**, *156*, 779–783.

107. Lou, Z.; Wang, L.; Wang, R.; Fei, T.; Zhang, T. Synthesis and ethanol sensing properties of SnO$_2$ nanosheets via a simple hydrothermal route. *Solid-State Electron.* **2012**, *76*, 91–94.

108. Zeng, W.; Wu, M.; Li, Y.; Wu, S. Hydrothermal synthesis of different SnO_2 nanosheets with co gas sensing properties. *J. Mater. Sci.: Mater. Electron.* **2013**, *24*, 3701–3706.

109. Du, X.; Skachko, I.; Barker, A.; Andrei, E.Y. Approaching ballistic transport in suspended graphene. *Nat Nano* **2008**, *3*, 491–495.

110. de Heer, W.A.; Berger, C.; Wu, X.; First, P.N.; Conrad, E.H.; Li, X.; Li, T.; Sprinkle, M.; Hass, J.; Sadowski, M.L.; *et al.* Epitaxial graphene. *Solid State Commun.* **2007**, *143*, 92–100.

111. Park, S.; Ruoff, R.S. Chemical methods for the production of graphenes. *Nat Nano* **2009**, *4*, 217–224.

112. Balandin, A.A.; Ghosh, S.; Bao, W.; Calizo, I.; Teweldebrhan, D.; Miao, F.; Lau, C.N. Superior thermal conductivity of single-layer graphene. *Nano Lett.* **2008**, *8*, 902–907.

113. Blake, P.; Hill, E.W.; Castro Neto, A.H.; Novoselov, K.S.; Jiang, D.; Yang, R.; Booth, T.J.; Geim, A.K. Making graphene visible. *Appl. Phys. Lett.* **2007**, *91*, 063124.

114. Dutta, P.; Horn, P.M. Low-frequency fluctuations in solids: 1/f Noise. *Rev. Mod. Phys.* **1981**, *53*, 497–516.

115. McAllister, M.J.; Li, J.-L.; Adamson, D.H.; Schniepp, H.C.; Abdala, A.A.; Liu, J.; Herrera-Alonso, M.; Milius, D.L.; Car, R.; Prud'homme, R.K.; *et al.* Single sheet functionalized graphene by oxidation and thermal expansion of graphite. *Chem. Mater.* **2007**, *19*, 4396–4404.

116. Zhang, C.; Lv, W.; Xie, X.; Tang, D.; Liu, C.; Yang, Q.-H. Towards low temperature thermal exfoliation of graphite oxide for graphene production. *Carbon* **2013**, *62*, 11–24.

117. Stankovich, S.; Dikin, D.A.; Piner, R.D.; Kohlhaas, K.A.; Kleinhammes, A.; Jia, Y.; Wu, Y.; Nguyen, S.T.; Ruoff, R.S. Synthesis of graphene-based nanosheets via chemical reduction of exfoliated graphite oxide. *Carbon* **2007**, *45*, 1558–1565.

118. Si, Y.; Samulski, E.T. Synthesis of water soluble graphene. *Nano Letters* **2008**, *8*, 1679–1682.

119. Zhang, Y.-H.; Chen, Y.-B.; Zhou, K.-G.; Liu, C.-H.; Zeng, J.; Zhang, H.-L.; Peng, Y. Improving gas sensing properties of graphene by introducing dopants and defects: A first-principles study. *Nanotechnology* **2009**, *20*, 185504.

120. Shao, L.; Chen, G.; Ye, H.; Wu, Y.; Qiao, Z.; Zhu, Y.; Niu, H. Sulfur dioxide adsorbed on graphene and heteroatom-doped graphene: A first-principles study. *Eur. Phys. J. B* **2013**, *86*, 1–5.

121. Zhang, H.; Luo, X.; Song, H.; Lin, X.; Lu, X.; Tang, Y. DFT study of adsorption and dissociation behavior of H_2S on Fe-doped graphene. *Appl. Surface Sci.* **2014**, *317*, 511–516.

122. Ma, C.; Shao, X.; Cao, D. Nitrogen-doped graphene as an excellent candidate for selective gas sensing. *Sci. China Chem.* **2014**, *57*, 911–917.

123. Liu, X.-Y.; Zhang, J.-M.; Xu, K.-W.; Ji, V. Improving SO_2 gas sensing properties of graphene by introducing dopant and defect: A first-principles study. *Appl. Surface Sci.* **2014**, *313*, 405–410.

124. Zhou, Q.; Yuan, L.; Yang, X.; Fu, Z.; Tang, Y.; Wang, C.; Zhang, H. DFT study of formaldehyde adsorption on vacancy defected graphene doped with B, N, and S. *Chem. Phys.* **2014**, *440*, 80–86.

125. Wang, X.; Sun, G.; Routh, P.; Kim, D.-H.; Huang, W.; Chen, P. Heteroatom-doped graphene materials: Syntheses, properties and applications. *Chem. Soc. Rev.* **2014**, *43*, 7067–7098.

126. Lv, R.; Li, Q.; Botello-Méndez, A.R.; Hayashi, T.; Wang, B.; Berkdemir, A.; Hao, Q.; Elías, A.L.; Cruz-Silva, R.; Gutiérrez, H.R.; *et al.* Nitrogen-doped graphene: Beyond single substitution and enhanced molecular sensing. *Sci. Rep.* **2012**, *2*, 586.

127. Niu, F.; Liu, J.-M.; Tao, L.-M.; Wang, W.; Song, W.-G. Nitrogen and silica co-doped graphene nanosheets for NO_2 gas sensing. *J. Mater. Chem. A* **2013**, *1*, 6130–6133.

128. Chung, M.G.; Kim, D.-H.; Seo, D.K.; Kim, T.; Im, H.U.; Lee, H.M.; Yoo, J.-B.; Hong, S.-H.; Kang, T.J.; Kim, Y.H. Flexible hydrogen sensors using graphene with palladium nanoparticle decoration. *Sens. Actuators B* **2012**, *169*, 387–392.

129. Cho, B.; Yoon, J.; Hahm, M.G.; Kim, D.-H.; Kim, A.R.; Kahng, Y.H.; Park, S.-W.; Lee, Y.-J.; Park, S.-G.; Kwon, J.-D. Graphene-based gas sensor: Metal decoration effect and application to a flexible device. *J. Mater. Chem. C* **2014**, *2*, 5280–5285.

130. Wang, J.; Rathi, S.; Singh, B.; Lee, I.; Maeng, S.; Joh, H.-I.; Kim, G.-H. Dielectrophoretic assembly of Pt nanoparticle-reduced graphene oxide nanohybrid for highly-sensitive multiple gas sensor. *Sens. Actuators B* **2015**, *220*, 755–761.

131. Zhang, Z.; Zou, X.; Xu, L.; Liao, L.; Liu, W.; Ho, J.; Xiao, X.; Jiang, C.; Li, J. Hydrogen gas sensor based on metal oxide nanoparticles decorated graphene transistor. *Nanoscale* **2015**, *7*, 10078–10084.

132. Liu, S.; Yu, B.; Zhang, H.; Fei, T.; Zhang, T. Enhancing NO_2 gas sensing performances at room temperature based on reduced graphene oxide-ZnO nanoparticles hybrids. *Sens. Actuators B* **2014**, *202*, 272–278.

133. Zhou, L.; Shen, F.; Tian, X.; Wang, D.; Zhang, T.; Chen, W. Stable Cu_2O nanocrystals grown on functionalized graphene sheets and room temperature H_2S gas sensing with ultrahigh sensitivity. *Nanoscale* **2013**, *5*, 1564–1569.

134. Su, P.-G.; Peng, S.-L. Fabrication and NO_2 gas-sensing properties of reduced graphene oxide/WO_3 nanocomposite films. *Talanta* **2015**, *132*, 398–405.

135. Jiang, Z.; Li, J.; Aslan, H.; Li, Q.; Li, Y.; Chen, M.; Huang, Y.; Froning, J.P.; Otyepka, M.; Zbořil, R. A high efficiency H_2S gas sensor material: Paper like Fe_2O_3/graphene nanosheets and structural alignment dependency of device efficiency. *J. Mater. Chem. A* **2014**, *2*, 6714–6717.

136. Lei, W.; Si, W.; Xu, Y.; Gu, Z.; Hao, Q. Conducting polymer composites with graphene for use in chemical sensors and biosensors. *Microchim. Acta* **2014**, *181*, 707–722.

137. Al-Mashat, L.; Shin, K.; Kalantar-zadeh, K.; Plessis, J.D.; Han, S.H.; Kojima, R.W.; Kaner, R.B.; Li, D.; Gou, X.; Ippolito, S.J.; *et al.* Graphene/polyaniline nanocomposite for hydrogen sensing. *J. Phys. Chem. C* **2010**, *114*, 16168–16173.

138. Bai, H.; Sheng, K.; Zhang, P.; Li, C.; Shi, G. Graphene oxide/conducting polymer composite hydrogels. *J. Mater. Chem.* **2011**, *21*, 18653–18658.

139. Jang, W.-K.; Yun, J.; Kim, H.-I.; Lee, Y.-S. Improvement of ammonia sensing properties of polypyrrole by nanocomposite with graphitic materials. *Colloid Polym. Sci.* **2013**, *291*, 1095–1103.

140. Ranola, R.A.G.; Concina, I.; Sevilla, F.B.; Ferroni, M.; Sangaletti, L.; Sberveglieri, G.; Comini, E. Room Temperature Trimethylamine Gas Sensor Based on Aqueous Dispersed Graphene, Proceedings of the 2015 XVIII AISEM Annual Conference, Trento, Italy, 3–5 February 2015; pp. 1–4.

141. Zheng, Y.; Lee, D.; Koo, H.Y.; Maeng, S. Chemically modified graphene/PEDOT:PSS nanocomposite films for hydrogen gas sensing. *Carbon* **2015**, *81*, 54–62.

142. Ayari, A.; Cobas, E.; Ogundadegbe, O.; Fuhrer, M.S. Realization and electrical characterization of ultrathin crystals of layered transition-metal dichalcogenides. *J. Appl. Phys.* **2007**, *101*, 14507–14507.

143. Wilson, J.A.; Yoffe, A.D. The transition metal dichalcogenides discussion and interpretation of the observed optical, electrical and structural properties. *Adv. Phys.* **1969**, *18*, 193–335.

144. Mak, K.F.; Lee, C.; Hone, J.; Shan, J.; Heinz, T.F. Atomically thin MoS_2: A new direct-gap semiconductor. *Phys. Rev. Lett.* **2010**, *105*, 136805.

145. Izyumskaya, N.; Demchenko, D.O.; Avrutin, V.; Özgur, U.; Morkoc, H. Two-dimensional MoS_2 as a new material for electronic devices. *Turkish J. Phys.* **2014**, *38*, 478–496.

146. Li, X.; Zhu, H. Two-dimensional MoS_2: Properties, preparation, and applications. *J. Materiomics* **2015**, *1*, 33–44.

147. Radisavljevic, B; Radenovic, A; Brivio, J; Giacometti, V; Kis, A. Single-layer MoS_2 transistors. *Nat. Nano* **2011**, *6*, 147–150.

148. Pu, J.; Yomogida, Y.; Liu, K.-K.; Li, L.-J.; Iwasa, Y.; Takenobu, T. Highly flexible MoS_2 thin-film transistors with ion gel dielectrics. *Nano Lett.* **2012**, *12*, 4013–4017.

149. Lopez-Sanchez, O.; Lembke, D.; Kayci, M.; Radenovic, A.; Kis, A. Ultrasensitive photodetectors based on monolayer MoS_2. *Nat. Nano* **2013**, *8*, 497–501.

150. Lukowski, M.A.; Daniel, A.S.; Meng, F.; Forticaux, A.; Li, L.; Jin, S. Enhanced hydrogen evolution catalysis from chemically exfoliated metallic MoS_2 nanosheets. *J. Am. Chem. Soc.* **2013**, *135*, 10274–10277.

151. Tan, C.; Zhang, H. Two-dimensional transition metal dichalcogenide nanosheet-based composites. *Chem. Soc. Rev.* **2015**, *44*, 2713–2731.

152. Cao, X.; Shi, Y.; Shi, W.; Rui, X.; Yan, Q.; Kong, J.; Zhang, H. Preparation of MoS_2-coated three-dimensional graphene networks for high-performance anode material in lithium-ion batteries. *Small* **2013**, *9*, 3433–3438.

153. Cao, L.; Yang, S.; Gao, W.; Liu, Z.; Gong, Y.; Ma, L.; Shi, G.; Lei, S.; Zhang, Y.; Zhang, S.; *et al.* Direct laser-patterned micro-supercapacitors from paintable MoS_2 films. *Small* **2013**, *9*, 2905–2910.

154. Gatensby, R.; McEvoy, N.; Lee, K.; Hallam, T.; Berner, N.C.; Rezvani, E.; Winters, S.; O'Brien, M.; Duesberg, G.S. Controlled synthesis of transition metal dichalcogenide thin films for electronic applications. *Appl. Surface Sci.* **2014**, *297*, 139–146.

155. Cui, S.; Wen, Z.; Huang, X.; Chang, J.; Chen, J. Stabilizing MoS_2 nanosheets through SnO_2 nanocrystal decoration for high-performance gas sensing in air. *Small* **2015**, *11*, 2305–2313.

156. Sarkar, D.; Xie, X.; Kang, J.; Zhang, H.; Liu, W.; Navarrete, J.; Moskovits, M.; Banerjee, K. Functionalization of transition metal dichalcogenides with metallic nanoparticles: Implications for doping and gas-sensing. *Nano Lett.* **2015**, *15*, 2852–2862.

157. Miremadi, B.; Singh, R.; Morrison, S.R.; Colbow, K. A highly sensitive and selective hydrogen gas sensor from thick oriented films of MoS_2. *Appl. Phys. A* **1996**, *63*, 271–275.

158. Yue, Q.; Shao, Z.; Chang, S.; Li, J. Adsorption of gas molecules on monolayer MoS_2 and effect of applied electric field. *Nanoscale Res. Lett.* **2013**, *8*, 1–7.

159. Zhao, S.; Xue, J.; Kang, W. Gas adsorption on MoS_2 monolayer from first-principles calculations. *Chem. Phys. Lett.* **2014**, *595–596*, 35–42.

160. Cho, B.; Hahm, M.G.; Choi, M.; Yoon, J.; Kim, A.R.; Lee, Y.-J.; Park, S.-G.; Kwon, J.-D.; Kim, C.S.; Song, M.; et al. Charge-transfer-based gas sensing using atomic-layer MoS_2. *Sci. Rep.* **2015**, *5*, 8052.

161. Cho, B.; Yoon, J.; Lim, S.K.; Kim, A.R.; Kim, D.-H.; Park, S.-G.; Kwon, J.-D.; Lee, Y.-J.; Lee, K.-H.; Lee, B.H.; et al. Chemical sensing of 2d graphene/MoS_2 heterostructure device. *ACS Appl. Mater. Interfaces* **2015**, *7*, 16775–16780.

162. Cho, B.; Kim, A.R.; Park, Y.; Yoon, J.; Lee, Y.-J.; Lee, S.; Yoo, T.J.; Kang, C.G.; Lee, B.H.; Ko, H.C.; et al. Bifunctional sensing characteristics of chemical vapor deposition synthesized atomic-layered MoS_2. *ACS Appl. Mater. Interfaces* **2015**, *7*, 2952–2959.

163. Li, L.; Yu, Y.; Ye, G.J.; Ge, Q.; Ou, X.; Wu, H.; Feng, D.; Chen, X.H.; Zhang, Y. Black phosphorus field-effect transistors. *Nat. Nano* **2014**, *9*, 372–377.

164. Zhu, Z.; Tománek, D. Semiconducting layered blue phosphorus: A computational study. *Phys. Rev. Lett.* **2014**, *112*, 176802.

165. Liu, H.; Neal, A.T.; Zhu, Z.; Luo, Z.; Xu, X.; Tománek, D.; Ye, P.D. Phosphorene: An unexplored 2d semiconductor with a high hole mobility. *ACS Nano* **2014**, *8*, 4033–4041.

166. Koenig, S.P.; Doganov, R.A.; Schmidt, H.; Castro Neto, A.H.; Özyilmaz, B. Electric field effect in ultrathin black phosphorus. *Appl. Phys. Lett.* **2014**, *104*, 103106.

167. Buscema, M.; Groenendijk, D.J.; Blanter, S.I.; Steele, G.A.; van der Zant, H.S.J.; Castellanos-Gomez, A. Fast and broadband photoresponse of few-layer black phosphorus field-effect transistors. *Nano Lett.* **2014**, *14*, 3347–3352.

168. Dai, J.; Zeng, X.C. Bilayer phosphorene: Effect of stacking order on bandgap and its potential applications in thin-film solar cells. *J. Phys. Chem. Lett.* **2014**, *5*, 1289–1293.

169. Xia, F.; Wang, H.; Jia, Y. Rediscovering black phosphorus as an anisotropic layered material for optoelectronics and electronics. *Nat. Commun.* **2014**.

170. Youngblood, N.; Chen, C.; Koester, S.J.; Li, M. Waveguide-integrated black phosphorus photodetector with high responsivity and low dark current. *Nat. Photonics* **2015**, *9*, 247–252.

171. Morrison, S.R. Semiconductor gas sensors. *Sens. Actuators* **1981**, *2*, 329–341.

172. Sun, Y.-F.; Liu, S.-B.; Meng, F.-L.; Liu, J.-Y.; Jin, Z.; Kong, L.-T.; Liu, J.-H. Metal oxide nanostructures and their gas sensing properties: A review. *Sensors* **2012**, *12*, 2610–2631.

173. Kolmakov, A.; Moskovits, M. Chemical sensing and catalysis by one-dimensional metal-oxide nanostructures. *Annu. Rev. Mater. Res.* **2004**, *34*, 151–180.

174. Brown, H.E. *Zinc Oxide—Properties and Applications*; International Lead Zinc Research Organization, Inc.: New York, NY, USA, 1978; p. 218.

175. Djurišić, A.B.; Ng, A.M.C.; Chen, X.Y. ZnO nanostructures for optoelectronics: Material properties and device applications. *Prog. Quantum Electron.* **2010**, *34*, 191–259.

176. Jiang, C.Y.; Sun, X.W.; Lo, G.Q.; Kwong, D.L.; Wang, J.X. Improved dye-sensitized solar cells with a ZnO-nanoflower photoanode. *Appl. Phys. Lett.* **2007**, *90*, 263501.

177. Wang, Z.L. From nanogenerators to piezotronics—A decade-long study of ZnO nanostructures. *MRS Bull.* **2012**, *37*, 814–827.

178. Lu, Y.; Emanetoglu, N.W.; Chen, Y. Chapter 13—ZnO piezoelectric devices. In *Zinc Oxide Bulk, Thin Films and Nanostructures*; Jagadish, C., Pearton, S., Eds.; Elsevier Science Ltd.: Oxford, UK, 2006; pp. 443–489.

179. Chen, Y.-J.; Zhu, C.-L.; Xiao, G. Ethanol sensing characteristics of ambient temperature sonochemically synthesized ZnO nanotubes. *Sens. Actuators B: Chem.* **2008**, *129*, 639–642.

180. Liao, L.; Lu, H.; Shuai, M.; Li, J.; Liu, Y.; Liu, C.; Shen, Z.; Yu, T. A novel gas sensor based on field ionization from ZnO nanowires: Moderate working voltage and high stability. *Nanotechnology* **2008**, *19*, 175501.

181. Heo, Y.-W.; Ren, F.; Norton, D.P. Chapter 14—Gas, chemical and biological sensing with ZnO. In *Zinc Oxide Bulk, Thin Films and Nanostructures*; Jagadish, C., Pearton, S., Eds.; Elsevier Science Ltd.: Oxford, UK, 2006; pp. 491–523.

182. Li, J.; Fan, H.; Jia, X. Multilayered ZnO nanosheets with 3d porous architectures: Synthesis and gas sensing application. *J. Phys. Chem. C* **2010**, *114*, 14684–14691.

On the Stability and Electronic Structure of Transition-Metal Dichalcogenide Monolayer Alloys $Mo_{1-x}X_xS_{2-y}Se_y$ with X = W, Nb

Agnieszka Kuc and Thomas Heine

Abstract: Layered transition-metal dichalcogenides have extraordinary electronic properties, which can be easily modified by various means. Here, we have investigated how the stability and electronic structure of MoS_2 monolayers is influenced by alloying, *i.e.*, by substitution of the transition metal Mo by W and Nb and of the chalcogen S by Se. While W and Se incorporate into the MoS_2 matrix homogeneously, forming solid solutions, the incorporation of Nb is energetically unstable and results in phase separation. However, all three alloying atoms change the electronic band structure significantly. For example, a very small concentration of Nb atoms introduces localized metallic states, while $Mo_{1-x}W_xS_2$ and $MoS_{2-y}Se_y$ alloys exhibit spin-splitting of the valence band of strength that is in between that of the pure materials. Moreover, small, but evident spin-splitting is introduced in the conduction band due to the symmetry breaking. Therefore, transition-metal dichalcogenide alloys are interesting candidates for optoelectronic and spintronic applications.

Reprinted from *Electronics*. Cite as: Kuc, A.; Heine, T. On the Stability and Electronic Structure of Transition-Metal Dichalcogenide Monolayer Alloys $Mo_{1-x}X_xS_{2-y}Se_y$ with X = W, Nb. *Electronics* **2016**, *5*, 1.

1. Introduction

The electronic structure of two-dimensional (2D) layered materials, in particular transition-metal chalcogenides (TMCs), have gained enormous interest in the past five years. Special attention is paid to semiconducting TMCs of the 2H TX_2 type (T, transition metal; X, chalcogen atom), because their electronic properties can be easily tuned by various factors, such as quantum confinement, external electric field, strain modulations or doping [1–11]. Easy tuning of the electronic structure is very important for several applications, e.g., for spin- and opto-electronics.

The interest in spintronic applications of TMCs arises from the hexagonal symmetry and the finite band gap, which, depending on the stoichiometry and number of layers, is in the range of 1–2 eV. Layered TMCs with an even number of layers (e.g., bilayers) have inversion symmetry, which is explicitly broken in systems with an odd number of layers (including monolayers). Hexagonal symmetry of TMCs imposes the existence of inequivalent energy valleys in the Brillouin zone

(K and K'). Due to lack of inversion symmetry in monolayers, a giant spin-orbit (SO) coupling (SOC) exists in the top of the valence band, ranging from about 150 meV (MoS_2) to nearly 500 meV (WTe_2) [12,13].

Quantum confinement of bulk TMCs down to the monolayer limit results in an increase of the band gap and an indirect-direct band gap transition. Although an external electric field does not change the electronic properties of monolayers, it causes band gap closure and spin-orbit splitting in bi- and multi-layer TMCs. The monolayer band structure can, however, be modulated by tensile strain or hydrostatic pressure, leading to semiconductor-metal transition [8–11,14,15].

Results on TMC nanotubes suggest that doping small quantities of Group 13 or Group 15 atoms results in a metallic character [16]; however, Nb doping of MoS_2 tubes was found to be unfavorable energetically [17–19]. The incorporation of a dilute impurity concentration into a host material results in significant changes to the band structure, which was shown in the cases of III-V semiconductors, e.g., the As impurity in GaN leads to a significant band gap reduction, and such a GaNAs alloy can be used for green emission [20,21]. This discovery opens up a new direction in the field of light-emitting diodes. Similarly, the addition of dilute-N impurity into the GaAs material results in low threshold laser devices [22,23].

Several experimental and theoretical studies have also been published up to date on the doping and alloy formation in TMC materials [24–30]. Many of these investigations are focused on the changes of Raman modes, optical or electronic properties in the mixed systems. For example, Raman modes, especially A_{1g} and E_{2g}, are increasing and decreasing, respectively, with W concentration in MoS_2 monolayers [26–28]. There is, however, a limited literature available on the spin-orbit splitting in the electronic structure of TMC alloys.

Among various TMCs, molybdenum disulfide (MoS_2) is the most widely-studied material. TMCs of the $2H$ TX_2 type are composed of two-dimensional $X–T–X$ sheets stacked on top of one another and held together via weak interlayer interactions, allowing easy exfoliation. Each sheet is trilayered with T atoms sandwiched between two chalcogen layers, as shown in Figure 1. Therefore, doping/alloying can occur both in the T and X sites.

In this paper, we have studied the complete transition of substituted MoS_2 as (i) $Mo_{1-x}W_xS_2$, (ii) $Mo_{1-x}Nb_xS_2$ and (iii) $MoS_{2-x}Se_x$ and discuss the stability and electronic properties of these phases. Such ternary alloys are a straightforward way of tuning the electronic properties of TMC materials and might be of great interest in the fields of electronic transport, optoelectronics, as well as spintronics. We have discussed in detail the electronic band structures and projected densities of the states of MoS_2-based alloys, showing that band gaps and the spin-splitting in the valence and conduction bands can be effectively tuned, with values in between those of the pure parental materials. Moreover, a small amount of NbS_2 mixed into the MoS_2

matrix makes it metallic. We show that semiconducting WS_2 and $MoSe_2$ easily form solid solutions with MoS_2, while metallic NbS_2 results in a phase separation.

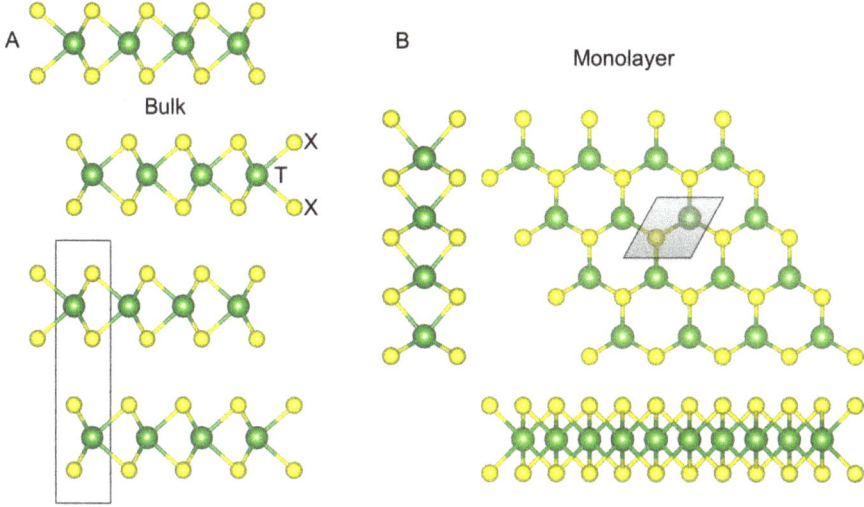

Figure 1. The atomic structure of transition-metal dichalcogenides of the $2H$ TX_2 type (T, transition metal; X, chalcogenide). (**A**) Bulk and (**B**) monolayer systems are shown together with their unit cells. Different sheets of TX_2 are composed of three atomic layers X−T−X, where T and X are covalently bonded. Sheets are held together by weak van der Waals forces.

2. Methods Section

In this work, we have studied atomic substitutions (alloys) in the MoS_2 monolayer. We have considered W and Nb atoms as substitutes of Mo and Se to substitute S atoms. The initial structures have hexagonal symmetry and belong to the $P6_3/mmc$ space group. The monolayers were cut out from the fully-optimized bulk structures as (0 0 1) surfaces. Alloy monolayers were further fully optimized; thus, the final symmetry is lowered. Optimization was performed using analytical energy gradients with respect to atomic coordinates and unit cell parameters within the quasi-Newton scheme combined with the BFGS (Broyden–Fletcher–Goldfarb–Shanno) scheme for Hessian updating. First-principle optimization calculations were performed on the basis of density functional theory (DFT) as implemented in the CRYSTAL09 code [31]. The exchange and correlation terms were described using general gradient approximation (GGA) in the scheme of the PBE (Perdew–Burke–Ernzerhof) [32] functional. For the sulfur and selenium atoms, the all-electron 86-311G* and 976-311d51G bases were chosen, respectively, while for heavier elements, the effective core potential (ECP) approach with large cores was employed, accounting for scalar relativistic effects [33–35]. The electronic structure calculations with the spin-orbit coupling (SOC) were performed using the PBE exchange-correlation functional and with Becke and Johnson damping (BJ-damping), as implemented in the ADF/BAND package [36,37]. Local basis functions (numerical and Slater-type basis functions of valence triple zeta quality with one polarization function (TZP)) were adopted for all atom types, and the frozen core approach (small core) was chosen.

The alloy monolayers were studied using 4×4 supercells, which consist of 16 metal atoms and 32 chalcogen atoms. The shrinking factor for bulk and layered structures was set to 8, which results in the corresponding number of 50 and 30 k-points in the irreducible Brillouin zone, respectively. The mesh of k-points for optimization calculations was obtained according to the scheme proposed by Monkhorst and Pack [38]. The SOC calculations are very time consuming and expensive; therefore, we have carried them out on smaller supercells (2×2) and alloys with 50 at% mixing. Band structures were calculated along the high symmetry points using the following path: $\Gamma - M - K - \Gamma$.

The alloying energies, E_{alloy}, were calculated as follows:

$$E_{alloy} = \frac{E_{mixed} - \left[x_{host} E_{host} + y_{guest} E_{guest} \right]}{y} \tag{1}$$

where E_{mixed} is the energy of the alloy monolayer, E_{host} is the energy of perfect MoS_2 monolayer and E_{guest} is the energy of the dopant perfect monolayer form, namely WS_2, NbS_2 and $MoSe_2$. x and y denote stoichiometric factors and refer to the number of formula units of each species. For example, in the case of $Mo_{12}W_4S_{32}$ with 6.25 at% W substitution, $x = 12$ and $y = 4$, giving:

$$E_{alloy} = \frac{E_{Mo_{12}W_4S_{32}} - \left[12E_{MoS_2} + 4E_{WS_2} \right]}{4} \tag{2}$$

E_{alloy} are given per TX_2 formula unit of the guest species (dopants).

3. Results and Discussion

We have studied three different ternary alloys on the MoS_2 monolayer with W, Nb and Se atoms. The alloying energies were calculated for all materials according to Equation (2) and given per dopant formula unit as TX_2. Band structures and the atom-projected densities of states were calculated with and without spin-orbit coupling correction. We have studied the W-MoS_2 alloy in great detail; however, general conclusions can also be drawn from the less-detailed analysis of Se- and Nb-MoS_2 alloys. The W and Se alloys with the MoS_2 monolayer are two examples of semiconductor-semiconductor alloys, while the one with Nb represents a metal-semiconductor alloy. It is important to note that dopant/alloy concentrations discussed hereafter are given as atomic percent of the metal atoms in the host material, e.g., 1 W atom in the 4×4 MoS_2 supercell (total of 16 Mo atoms) gives 6.25 at% of W.

3.1. MoS_2 Alloys with W

The MoS_2 monolayer, as well as all W-alloys were fully optimized in terms of lattice vectors and atomic positions. The optimized lattice parameters of W(Mo)-based Mo(W)S_2 alloys are given in Table 1 together with the shortest distances between the substituting W (Mo) atom. Notice that below (above) a 50 at% concentration of substituting atoms, we are talking about the $Mo_{1-x}W_xS_2$ ($W_{1-x}Mo_xS_2$) alloys. In both cases, the substituting atoms prefer to distribute homogeneously in the host material, forming solid solutions. This is supported by relatively large distances between W (Mo) of about 6.3 Å for concentrations up to 25 at%. At this concentration, we have also obtained the most favorable alloying energies for both $Mo_{1-x}W_xS_2$ and $W_{1-x}Mo_xS_2$ alloys (see Figure 2).

Table 1. Calculated lattice parameters a and b (Å) of the 4×4 supercell and dopant closest distances (Å) of the most stable $Mo_{1-x}W_xS_2$ (or $Mo_xW_{1-x}S_2$) systems. Experimental data are given in parenthesis [39–41].

Formula	W at%	a, b	W–W	Formula	Mo at%	a, b	Mo–Mo
$Mo_{16}S_{32}$	0.00	12.683 (12.656)	–	$W_{16}S_{32}$	0.00	12.642 (12.616)	–
$Mo_{15}W_1S_{32}$	6.25	12.678	–	$Mo_1W_{15}S_{32}$	6.25	12.643	–
$Mo_{14}W_2S_{32}$	12.50	12.675	6.333	$Mo_2W_{14}S_{32}$	12.50	12.646	6.322
$Mo_{13}W_3S_{32}$	18.75	12.672	6.334	$Mo_3W_{13}S_{32}$	18.75	12.648	6.322
$Mo_{12}W_4S_{32}$	25.00	12.669	6.334	$Mo_4W_{12}S_{32}$	25.00	12.651	6.325
$Mo_{11}W_5S_{32}$	31.25	12.667	3.167	$Mo_5W_{11}S_{32}$	31.25	12.652	3.163
$Mo_{10}W_6S_{32}$	37.50	12.664	3.166	$Mo_6W_{10}S_{32}$	37.50	12.655	3.164
$Mo_9W_7S_{32}$	43.75	12.662	3.166	$Mo_7W_9S_{32}$	43.75	12.657	3.164
$Mo_8W_8S_{32}$	50.00	12.659	3.165	$Mo_8W_8S_{32}$	50.00	12.659	3.165

It has to be stressed at this point that for larger supercell models, it might be possible to obtain more favorable alloys with larger concentrations of W or Mo. For example, Gan et al. [24] and Kutana et al. [25] have reported the maximum formation/mixing energies for the W concentrations in the range of 33%–60%. Nevertheless, our trends agree well with other DFT-based works, showing solid solution formation in Mo/W disulfide alloys [24–26].

The lattice parameters of alloy materials are always in between the pure materials and change linearly with the concentration of W (Mo) doping atoms. This is also shown in Figure 3A.

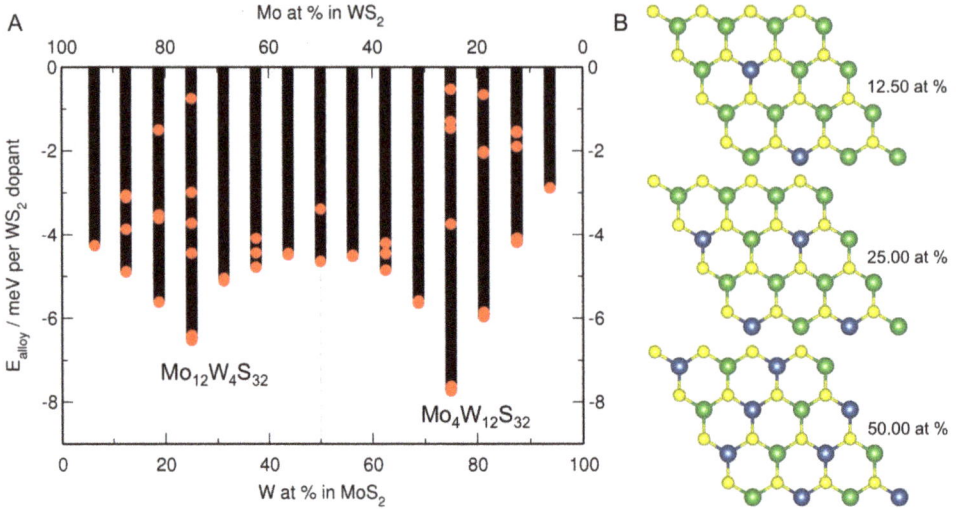

Figure 2. (**A**) Calculated alloying energies of $Mo_{1-x}W_xS_2$ systems ($x = 0 - 1$). Numbers calculated according to the Equation (2); (**B**) Selected alloys with the most stable W arrangement for a given concentration. Green, Mo; blue, W; yellow, S.

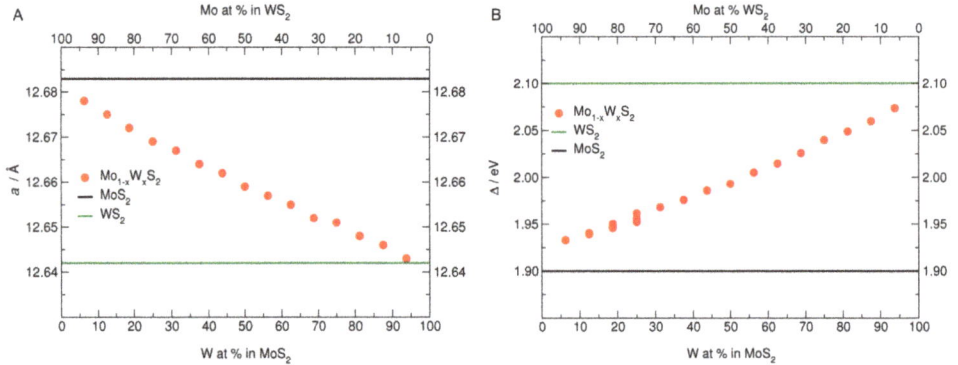

Figure 3. (**A**) The change of lattice parameter a and (**B**) the band gap Δ in $Mo_{1-x}W_xS_2$ systems ($x = 0 - 1$). The values of MoS_2 and WS_2 monolayers are shown with black and green horizontal lines, respectively.

Moreover, the electronic properties are also changed for various alloys (see Figure 3B), namely the band gaps change in a linear manner. Pure MoS_2 and WS_2 monolayers are direct-gap semiconductors with band gaps of 1.84 and 2.01 eV, respectively. These values agree very well with other theoretical and experimental data [1–3,5]. $Mo_{1-x}W_xS_2$ alloys offer mono-layered materials with band gaps in the range 1.84–2.01 eV. Although the range of available band gaps is not large, the idea of ternary alloys is very interesting in the fields of nanoelectronics, and other TMC alloys, with a more distinct difference in the band gap values, might offer an even larger range of properties.

We have analyzed the electronic structures of W/Mo disulfide alloys in more detail. Figures 4 and 5 show the band structures and density of states of pure materials and the 50%–50% mixture. We have calculated the band structures at two different levels of theory, namely with and without relativistic effects of spin-orbit coupling. TMC materials are well know to exhibit giant spin-orbit splitting, which might of course be altered by alloy formation.

Figure 4 shows that the bands of alloys are exactly in between the bands of pure materials. Furthermore, the density of states shows perfect mixing and hybridization between Mo and W atoms in the alloy. Both metal atoms contribute equally to the states close to the Fermi level. The band edges, valence band maximum (VBM) and conduction band minimum (CBM), in all three materials are situated at the high-symmetry K point, showing that alloy formation does not affect the direct-gap characteristics of TMCs. Therefore, alloys might be interesting for optoelectronic applications.

Figure 5 shows a more detailed electronic band structure of the pure materials and the 50%–50% mixture as extreme cases. MoS_2 and WS_2 exhibit spin-orbit splitting (Δ_{SOC}) in the VBMs of 150 and 425 meV, respectively. These results agree perfectly with other reports available in the literature [12,13]. The alloy materials also exhibit strong Δ_{SOC}, which again is between the values of the pure phases. In the discussed case, we have obtained Δ_{SOC} of 279 meV in the VBM. These results are in agreement with DFT calculations by Wang $et\ al.$ [26], who obtained Δ_{SOC} of 182, 452 and 305 meV for MoS_2, WS_2 and $Mo_{0.5}W_{0.5}S_2$, respectively, at the LDA level of theory. However, the same authors performed experimental investigations and reported bowing of Δ_{SOC} for different alloys, rather than a linear trend, which has not yet been anticipated by theory.

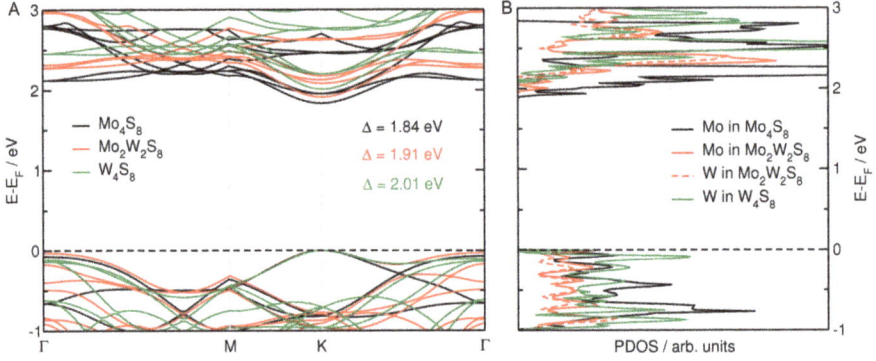

Figure 4. Electronic band structure (**A**) and density of states (**B**) of MoS_2, WS_2 and $Mo_{0.5}W_{0.5}S_2$ systems. The fundamental band gaps are indicated. The projection of states is shown on metal atoms. The Fermi level is shifted to the valence band maximum. No spin-orbit coupling is taken into account.

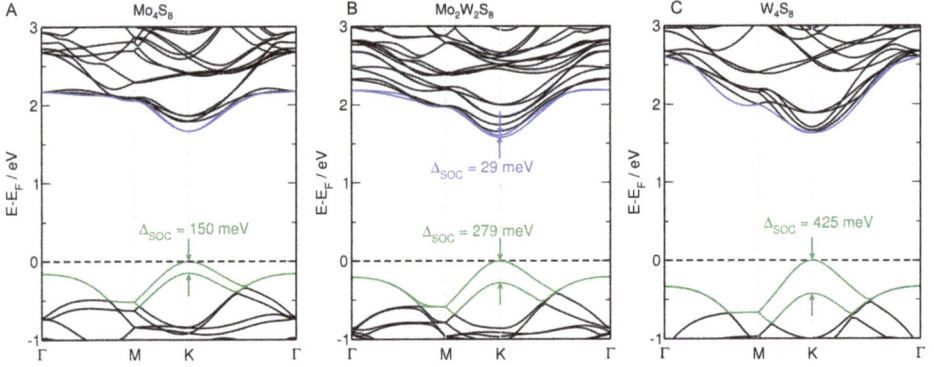

Figure 5. Calculated band structures of MoS_2, WS_2 and $Mo_{0.5}W_{0.5}S_2$ systems with spin-orbit coupling (Δ_{SOC}).

Moreover, alloys offer extra Δ_{SOC} in the CBM, which reaches about 30 meV for the 50%–50% mixture and which is almost suppressed in the parental monolayers. The spin-splitting in CBM reveals also a small Rashba effect (not discussed here). These features make ternary alloys much more interesting for opto- and spintronic applications compared to pure TMC materials.

We have further analyzed the alloy formation effect on the heat capacity and the Raman normal modes. Selected results are shown in Figure 6. It should be noted that the heat capacity values are lower than the 3R (R—gas constant) limit, because they do not include the phonon contribution. Nevertheless, the trends should not be significantly affected. The results show the lowering of the heat capacity of ternary alloys compared to the pure materials, especially at room temperature. This suggests that alloys or small dopings in TMC monolayers might be useful in search for thermoelectric materials. The electronic components of the figure of merit (ZT) for conventional 3D crystals are interrelated in such a way that it is difficult to control them independently to increase ZT. This problem is solved in low-dimensional materials, such as monolayers. Moreover, lowering the heat capacity decreases the thermal conductivity and, as a result, increases ZT.

We have analyzed the normal modes of MoS_2 and alloys of up to a 50 at% W concentration. Phonon calculations in alloys become very complex and are computationally demanding. Our results show that the MoS_2-like A_{1g} modes (which correspond to the out of plane symmetric vibrations of the S atoms) increase slightly with W concentration. On the other hand, the MoS_2-like E_{2g} modes (in-plane symmetric vibrations) decrease with W concentration. At a mixture of 30% (W)–70% (Mo) and larger, the WS_2-like modes start to appear, resulting in a two-mode behavior for each of the two normal modes. This is in a good agreement with experimental data reported by Liu *et al.* [27].

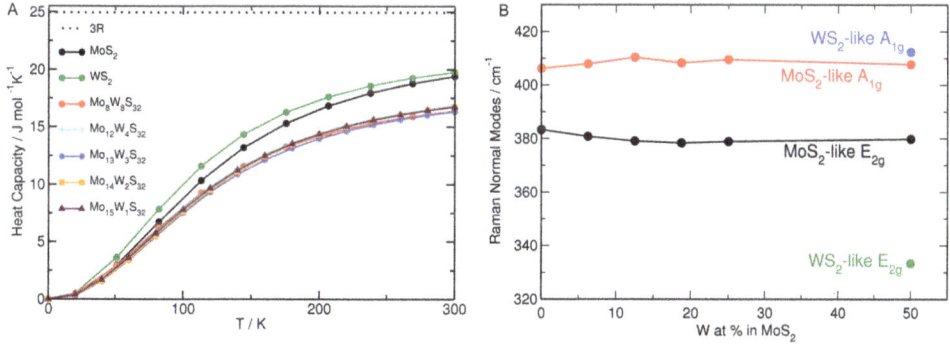

Figure 6. (**A**) Heat capacity and (**B**) Raman normal modes change with the W concentration in the MoS_2 monolayer. The normal modes are indicated.

3.2. MoS_2 Alloys with Nb

NbS_2 is a metallic transition-metal dichalcogenide. When alloyed with MoS_2, even in a very small concentration, it introduces metallic states in the band gap.

The lattice parameters and the smallest Nb–Nb distances of $Mo_{1-x}Nb_xS_2$ systems (with $x = 0$–0.25) are shown in Table 2. The distances between the doping/alloying metal atoms in the MoS_2 monolayer of only about 3.3 Å suggest phase separation, as this distance corresponds to the first metal neighbors. This is also shown in Figure 7, where the alloying energy is plotted together with the most stable structure of the $Mo_{0.75}Nb_{0.25}S_2$ alloy.

Table 2. Calculated lattice parameters a and b (Å) of the 4×4 supercell and dopant closest distances (Å) of the most stable $Mo_{1-x}Nb_xS_2$ systems. Experimental data are given in parenthesis [39–41].

Formula	Nb at%	a, b	Nb–Nb
$Mo_{16}S_{32}$	0.00	12.683 (12.656)	–
$Mo_{15}Nb_1S_{32}$	6.25	12.678	–
$Mo_{14}Nb_2S_{32}$	12.50	12.675	3.325
$Mo_{13}Nb_3S_{32}$	18.75	12.672	3.380
$Mo_{12}Nb_4S_{32}$	25.00	12.669	3.394
$Nb_{16}S_{32}$	100.00	13.348 (13.240)	3.338

The alloying energies are positive, meaning that it is rather unlikely that Nb will easily mix into the matrix of MoS_2 monolayers. These energies, however, become nearly zero for Nb concentration above 15 at%, as at this point, the dopants start to cluster inside the host material. Ivanovskaya *et al.* [17,19] have also reported clustering of Nb atoms in nanotubes and monolayers; however, they obtained negative formation energies from the density functional-based tight-binding (DFTB) calculations.

We have calculated the electronic structure of the $Mo_{0.9375}Nb_{0.0625}S_2$ alloy and compared it to those of pure MoS_2 and NbS_2 materials. To manage the computational complexity, these calculations were performed without spin-orbit correction; however, it should be pointed out that spin-splitting occurs also in the case of metallic NbS_2. Since we have considered only one Nb atom in the supercell, the calculations are spin-polarized, and both alpha and beta electrons are plotted.

The band structures of atom-projected densities of states are shown in Figure 8. These calculations were performed for the largest supercells considered in this work, in order to show the band structure change for the smallest Nb concentration in a given model (6.25 at%). The Nb atom has one electron less than the Mo atom in the valence shell; therefore, an introduction of a single Nb results in an electron hole, which acts as an acceptor impurity. Similar results were obtained from DFTB calculations by Ivanovskaya *et al.* [19].

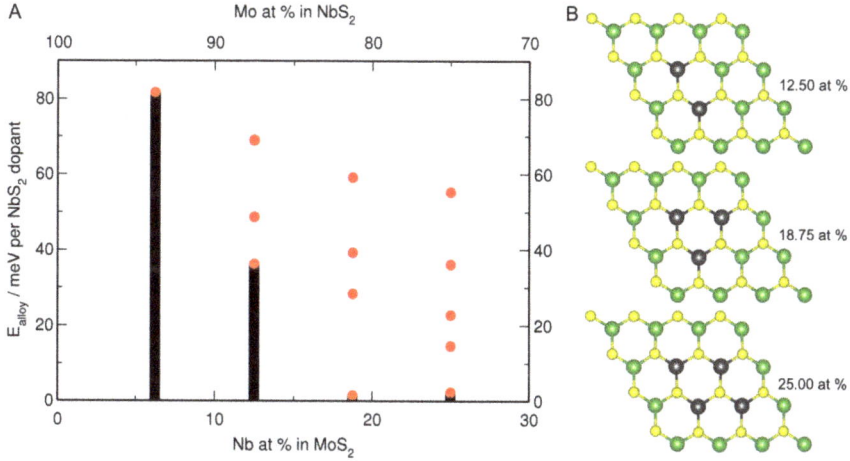

Figure 7. (**A**) The alloying energies (E_{alloy}, in meV per NbS_2 unit) of the $Mo_{1-x}Nb_xS_2$ systems and (**B**) selected alloys with the most stable Nb arrangement for a given concentration. Green, Mo; grey, Nb; yellow, S.

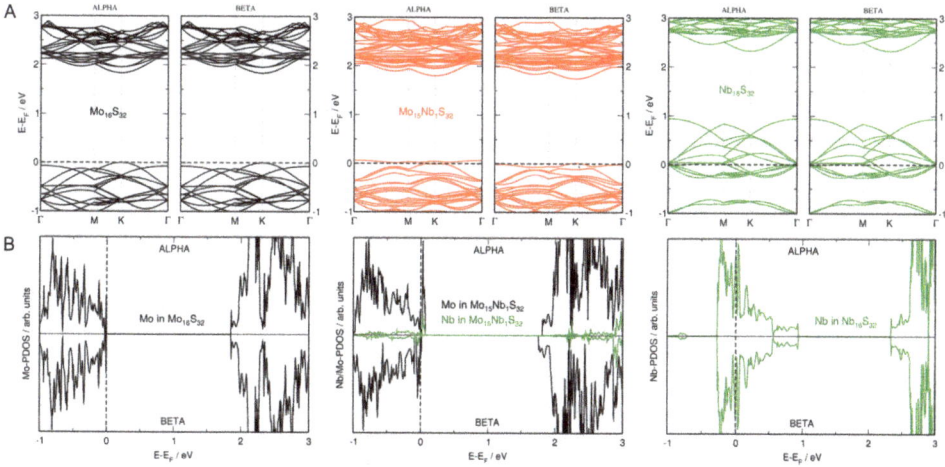

Figure 8. Spin-polarized electronic band structure (**A**) and density of states (**B**) of MoS_2, NbS_2 and $Mo_{0.9275}Nb_{0.0625}S_2$ systems. The projection of states is shown on metal atoms. The Fermi level is indicated by the dashed horizontal line. No spin-orbit coupling is taken into account.

3.3. MoS_2 Alloys with Se

Here, we discuss the substitution of S by Se atoms. In this case, there are several possibilities, as the chalcogen atom may be substituted in only one layer or in both layers. In the latter case, the substitution can be symmetric or asymmetric. Since the substitution is more favorable in two layers, we will show only this case. Moreover, the alloying energies of symmetric and asymmetric alloys are very similar, meaning that polymorphism can be expected. Table 3 shows the lattice parameters and the shortest Se–Se distances in the symmetric substitution cases, unless otherwise stated. Again, the lattice parameters change in a linear manner with the Se concentration, similar to the W-based alloys. Furthermore, Se atoms distribute homogeneously in the MoS_2 monolayer with the shortest distance of about 5.5 Å up to a 20 at% concentration. This means that Se, similar to W, forms a solid solution with the MoS_2 matrix.

Table 3. Calculated lattice parameters a and b (Å) of the 4×4 supercell, dopant closest distances (Å) and the band gaps, Δ (eV), of the most stable $MoSe_{1-x}S_x$ systems with Se atoms included in both chalcogen layers. Experimental data are given in parenthesis [1,39–42].

Formula	$MoSe_2$ at%	a, b	Se–Se	Δ
$Mo_{16}S_{32}$	0.00	12.683 (12.656)	–	1.90 (1.80)
$Mo_{16}S_{30}Se_2$	6.25	12.710	–	1.81
$Mo_{16}S_{28}Se_4$	12.50	12.738	5.520	1.79
$Mo_{16}S_{26}Se_6$	18.75	12.767	5.532	1.77
$Mo_{16}S_{24}Se_8$	25.00	12.795	5.540	1.75
$Mo_{16}S_{16}Se_{16}$	50.00 symmetric	12.915	3.246	1.63
$Mo_{16}S_{16}Se_{16}$	50.00 asymmetric	12.903	3.243	1.68
$Mo_{16}Se_{32}$	100.00	13.261 (13.152)	3.288	1.53 (1.55)

The mixing of Se into MoS_2 is energetically very favorable (see Figure 9) for any concentration of at least up to 25 at%. Various distributions of Se are in principle possible, as alloying energies of different systems are very close in value. Moreover, the energies are one order of magnitude larger than in the case of W-alloys, which means that mixing of MoS_2 with $MoSe_2$ is even more probable. However, we have obtained somewhat lower formation energies than reported by Kang et al. [43]. This might be due to the relation of formation energies to the number of atoms: we are referring to the number of $MoSe_2$ units inside the MoS_2, while Kang et al. refer to the number of anions.

For the Se-based alloys, we have again calculated the electronic band structures with and without spin-orbit corrections. Figure 10 shows the electronic band structure and atom-projected densities of states calculated without spin-orbit coupling. The 50%–50% symmetric alloy is a direct-band gap semiconductor, similar to the pure materials, with a band gap of 1.63 eV just in between MoS_2 (1.84 eV) and $MoSe_2$ (1.53 eV). In the case of Se-based MoS_2 alloys, we can tune the band gap in the range of about 300 meV, which is already larger than for the case of W-based systems (about 150 meV).

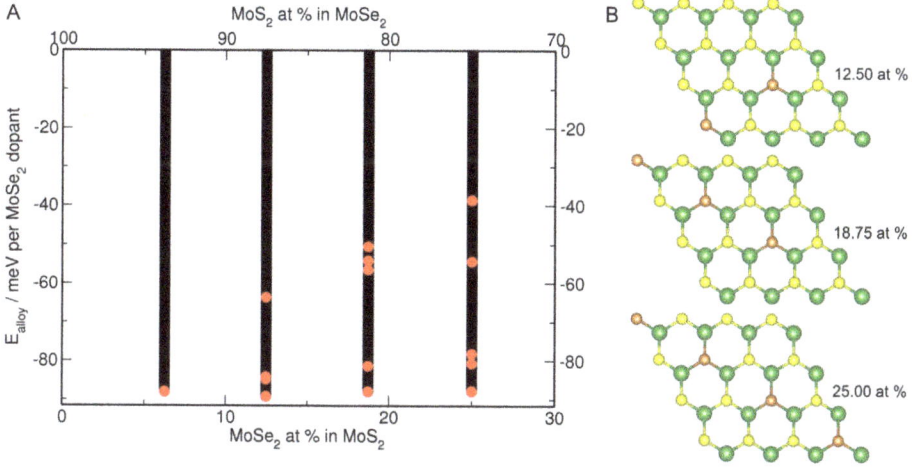

Figure 9. (**A**) The alloying energies (E_{alloy}, in meV per $MoSe_2$ unit) of the $MoSe_{1-x}S_x$ systems with Se atoms included in both chalcogen layers and (**B**) selected alloys with the most stable symmetric Se arrangement for a given concentration. Green, Mo; orange, Se; yellow, S.

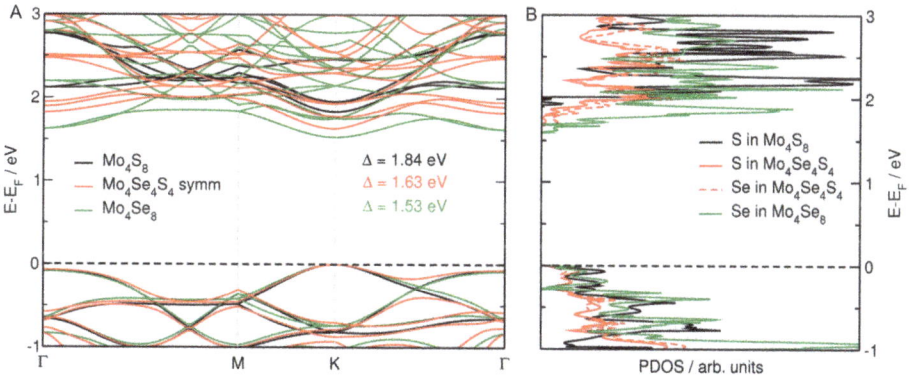

Figure 10. Electronic band structure (**A**) and density of states (**B**) of MoS_2, $MoSe_2$ and MoSeS systems. The fundamental band gaps are indicated. The projection of states is shown on chalcogen atoms. The Fermi level is shifted to the valence band maximum. No spin-orbit coupling is taken into account.

Th projected density of states shows that once again, the doping atoms are well incorporated into the host materials, with both S and Se atoms contributing equally to the states close to the Fermi level.

The spin-orbit corrected calculations show that also Se-based alloys exhibit large spin-orbit splitting of about 170–180 meV in the case of a 50%–50% mixture (see Figure 11). These values are just in between those of the pure MoS_2 (150 meV) and $MoSe_2$ (192 meV) materials.

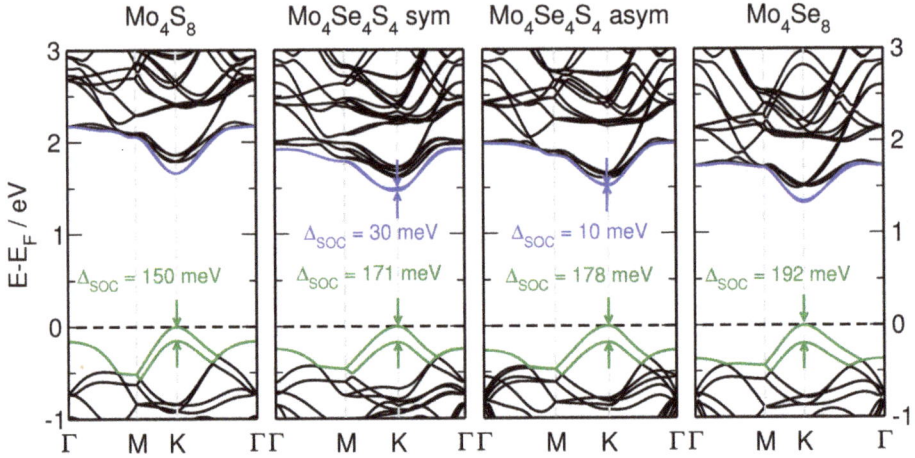

Figure 11. Calculated band structures of MoS_2, $MoSe_2$, and $MoSeS$ systems with spin-orbit coupling (Δ_{SOC}).

The above results show that the MoS_2-based alloys with W and Se are very good candidates for various applications, including optoelectronics, spintronics or even thermoelectrics. The electronic properties can be tuned in a controlled manner with any features that are in between those of the pure systems, which form the alloys. This means that carefully choosing the systems, with given band gaps or given spin-orbit splitting, will result in very promising alloys. These conclusions and predictions might not only be limited to monolayers, but could as well open a field for bi- and multi-layered alloys.

4. Conclusions

We have studied ternary alloys of the MoS_2 monolayer with W, Nb and Se atoms from first-principles. Our results show that W and Se incorporate easily into the host material and distribute homogeneously, resulting in a solid solution. On the other hand, Nb atoms cluster together for larger metal concentrations and form phase separation. Energetically, the Nb alloys are not very favorable for smaller concentrations, while W and Se easily form mixture materials with MoS_2.

All studied materials show, however, significant changes in the electronic structures compared to the pure parental MoS_2 monolayer. Especially in the case of Nb atoms, the semiconducting MoS_2 shows first localized metallic states for a small metal concentration and, later, full metallic bands for a larger amount of Nb. In the cases of W and Se alloys, the band structures change in such a way that the electronic properties are somewhere in between the intrinsic properties of the pure materials. For example, the spin-splitting of the valence band in MoS_2 and WS_2 is calculated to be 150 meV and 425 meV, respectively, while the 50% mixture exhibits 279 meV. Moreover, spin-splitting occurs in the conduction band with the value of about 30 meV for the $Mo_{0.5}W_{0.5}S_2$. Furthermore, the optical/fundamental band gaps of the semiconducting alloys are direct gaps with values in between the pure phases, e.g., the band gaps of MoS_2, $MoSe_2$ and $MoSeS$ are 1.84 eV, 1.63 eV and 1.53 eV, respectively.

Alloys allow tuning of the optical and spintronic properties of transition-metal dichalcogenide materials in a wide range. This feature could be further exploited in the bi- and multi-layered materials, which would probably result in even more interesting properties.

Acknowledgments: The European Commission (FP7-PEOPLE-2012-ITN MoWSeS, GA 317451).

Author Contributions: Agnieszka Kuc and Thomas Heine generated, analyzed and discussed the results. Both authors contributed to writing this paper.

Conflicts of Interest: The authors declare no conflict of interest.

References

1. Mak, K.F.; Lee, C.; Hone, J.; Shan, J.; Heinz, T.F. Atomically Thin MoS_2: A New Direct-Gap Semiconductor. *Phys. Rev. Lett.* **2010**, *105*, 136805.
2. Splendiani, A.; Sun, L.; Zhang, Y.B.; Li, T.S.; Kim, J.; Chim, C.Y.; Galli, G.; Wang, F. Emerging Photoluminescence in Monolayer MoS_2. *Nano Lett.* **2010**, *10*, 1271–1275.
3. Kuc, A.; Zibouche, N.; Heine, T. Influence of quantum confinement on the electronic structure of the transition metal sulfide TS_2. *Phys. Rev. B* **2011**, *83*, 245213.
4. Komsa, H.-P.; Krasheninnikov, A.V. Two-Dimensional Transition Metal Dichalcogenide Alloys: Stability and Electronic Properties. *J. Phys. Chem. Lett.* **2012**, *3*, 3652–3656.
5. Ataca, C.; Sahin, H.; Ciraci, S. Stable, Single-Layer MX_2 Transition-Metal Oxides and Dichalcogenides in a Honeycomb-Like Structure. *J. Phys. Chem. C* **2012**, *116*, 8983–8999.
6. Zibouche, N.; Philipsen, P.; Kuc, A.; Heine, T. Transition-metal dichalcogenide bilayers: Switching materials for spintronic and valleytronic applications. *Phys. Rev. B* **2014**, *90*, 125440.
7. Ramasubramaniam, A.; Naveh, D.; Towe, E. Tunable band gaps in bilayer transition-metal dichalcogenides. *Phys. Rev. B* **2011**, *84*, 205325.
8. Ghorbani-Asl, M.; Zibouche, N.; Wahiduzzaman, M.; Oliveira, A.F.; Kuc, A.; Heine, T. Electromechanics in MoS_2 and WS_2: Nanotubes vs. monolayers. *Sci. Rep.* **2013**, *3*, 2961.
9. Johari, P.; Shenoy, V.B. Tuning the Electronic Properties of Semiconducting Transition Metal Dichalcogenides by Applying Mechanical Strains. *ACS Nano* **2012**, *6*, 5449–5456.

10. Yue, Q.; Kang, J.; Shao, Z.; Zhang, X.; Chang, S.; Wang, G.; Qin, S.; Li, J. Mechanical and electronic properties of monolayer MoS_2 under elastic strain. *Phys. Lett. A* **2012**, *376*, 1166–1170.

11. Yun, W.S.; Han, S.W.; Hong, S.C.; Kim, I.G.; Lee, J.D. Thickness and strain effects on electronic structures of transition metal dichalcogenides: $2H$-MX_2 semiconductors (M = Mo, W; X = S, Se, Te). *Phys. Rev. B* **2012**, *85*, 033305.

12. Zhu, Z.Y.; Cheng, Y.C.; Schwingenschlögl, U. Giant spin-orbit-induced spin splitting in two-dimensional transition-metal dichalcogenide semiconductors. *Phys. Rev. B* **2011**, *84*, 153402.

13. Zibouche, N.; Kuc, A.; Musfeldt, J.; Heine, T. Transition-metal dichalcogenides for spintronic applications. *Ann. der Phys.* **2014**, *526*, 395–401.

14. Nayak, A.P.; Bhattacharyya, S.; Zhu, J.; Liu, J.; Wu, X.; Pandey, T.; Jin, C.; Singh, A.K.; Akinwande, D.; Lin, J.-F. Pressure-induced semiconducting to metallic transition in multilayered molybdenum disulphide. *Nat. Commun.* **2014**, *5*, 3731.

15. Zhao, Z.; Zhang, H.; Yuan, H.; Wang, S.; Lin, Y.; Zeng, Q.; Xu, G.; Liu, Z.; Solanki, G.K.; Patel, K.D.; *et al.* Pressure induced metallization with absence of structural transition in layered molybdenum diselenide. *Nat. Commun.* **2015**, *6*, 7312.

16. Yadgarov, L.; Rosentsveig, R.; Leitus, G.; Albu-Yaron, A.; Moshkovich, A.; Perfilyev, V.; Vasic, R.; Frenkel, A.I.; Enyashin, A.N.; Seifert, G.; *et al.* Controlled Doping of MS_2 (M=W, Mo) Nanotubes and Fullerene-like Nanoparticles. *Angew. Chem. Int. Edit.* **2012**, *51*, 1148–1151.

17. Ivanovskaya, V.V.; Heine, T.; Gemming, S.; Seifert, G. Structure, stability and electronic properties of composite $Mo_{1x}Nb_xS_2$ nanotubes. *Phys. Stat. Solidi B* **2006**, *243*, 1757–1764.

18. Ivanovskaya, V.V.; Seifert, G.; Ivanovskii, A.L. Electronic structure of niobium-doped molybdenum disulfide nanotubes. *Russ. J. Inorg. Chem.* **2006**, *51*, 320–324.

19. Ivanovskaya, V.V.; Zobelli, A.; Gloter, A.; Brun, N.; Serin, V.; Colliex, C. *Ab initio* study of bilateral doping within the MoS_2-NbS_2 system. *Phys. Rev. B* **2008**, *78*, 134104.

20. Tan, C.-K.; Tansu, N. Auger recombination rates in dilute-As GaNAs semiconductor. *AIP Adv.* **2015**, *5*, 057135.

21. Kimura, A.; Paulson, C.A.; Tang, H.F.; Kuech, T.F. Epitaxial $GaN_{1y}As_y$ layers with high As content grown by metalorganic vapor phase epitaxy and their band gap energy. *Appl. Phys. Let.* **2004**, *84*, 1489–1491.

22. Tansu, N.; Yeh, J.-Y.; Mawst, L. High-Performance 1200-nm InGaAs and 1300-nm InGaAsN Quantum Well Lasers by Metalorganic Chemical Vapor Deposition. *Sel. Top. Quantum Electron. IEEE* **2003**, *9*, 1220–1227.

23. Bank, S.; Goddard, L.; Wistey, M.A.; Yuen, H.B.; Harris, J.S. On the temperature sensitivity of 1.5-μm GaInNAsSb lasers. *Sel. Top. Quantum Electron. IEEE* **2005**, *11*, 1089–1098.

24. Gan, L.-Y.; Zhang, Q.; Zhao, Y.-J.; Cheng, Y.; Schwingenschlögl, U. Order-disorder phase transitions in the two-dimensional semiconducting transition metal dichalcogenide alloys $Mo_{1x}W_xX_2$ (X = S, Se, and Te). *Sci. Rep.* **2014**, *4*, 6691.

25. Kutana, A.; Penev, E.S.; Yakobson, B.I. Engineering electronic properties of layered transition-metal dichalcogenide compounds through alloying. *Nanoscale* **2014**, *6*, 5820–5825.

26. Wang, G.; Robert, C.; Suslu, A.; Chen, B.; Yang, S.; Alamdari, S.; Gerber, I.C.; Amand, T.; Marie, X.; Tongay, S.; *et al.* Spin-orbit engineering in transition metal dichalcogenide alloy monolayers. *Nat. Commun.* **2015**.

27. Liu, H.; Antwi, K.K.A.; Chua, S.; Chi, D. Vapor-phase growth and characterization of $Mo_{1x}W_xS_2$ ($0 \leqslant x \leqslant 1$) atomic layers on 2-inch sapphire substrates. *Nanoscale* **2014**, *6*, 624–629.

28. Chen, Y.; Dumcenco, D.O.; Zhu, Y.; Zhang, X.; Mao, N.; Feng, Q.; Zhang, M.; Zhang, J.; Tan, P.-H.; Huang, Y.-S.; *et al.* Composition-dependent Raman modes of $Mo_{1x}W_xS_2$ monolayer alloys. *Nanoscale* **2014**, *6*, 2833–2839.

29. Ren, X.; Ma, Q.; Fan, H.; Pang, L.; Zhang, Y.; Yao, Y.; Ren, X.; Liu, S.F. A Se-doped MoS_2 nanosheet for improved hydrogen evolution reaction. *Chem. Commun.* **2015**, *51*, 15997–16000.

30. Faraji, M.; Sabzali, M.; Yousefzadeh, S.; Sarikhani, N.; Ziashahabi, A.; Zirak, M.; Moshfegh, A.Z. Band engineering and charge separation in the $Mo_{1x}W_xS_2/TiO_2$ heterostructure by alloying: First principle prediction. *RSC Adv.* **2015**, *5*, 28460–28466.

31. Dovesi, R.; Saunders, V.R.; Roetti, R.; Orlando, R.; Zicovich-Wilson, C.M.; Pascale, F.; Civalleri, B.; Doll, K.; Harrison, N.M.; Bush, I.J.; *et al. CRYSTAL09 User's Manual*; University of Torino: Torino, Canada, 2009.

32. Perdew, J.P.; Burke, K.; Ernzerhof, M. Generalized Gradient Approximation Made Simple. *Phys. Rev. Lett.* **1996**, *77*, 3865.

33. Cora, F.; Patel, A.; Harrison, N.M.; Roetti, C.; Catlow, C.R.A. An ab-initio Hartree-Fock study of alpha-MoO_3. *J. Mater. Chem.* **1997**, *7*, 959.

34. Cora, F.; Patel, A.; Harrison, N.M.; Dovesi, R.; Catlow, C.R.A. An ab-initio Hartree-Fock study of the cubic and tetragonal phases of Bulk Tungsten Trioxide. *J. Am. Chem. Soc.* **1996**, *118*, 12174.

35. Dall'Olio, S.; Dovesi, R.; Resta, R. Spontaneous polarization as a Berry phase of the Hartree-Fock wave function: The case of $KNbO_3$. *Phys. Rev. B* **1997**, *56*, 10105–10114.

36. Te Velde, G.; Baerends, E. J. Precise density-functional method for periodic structures. *Phys. Rev. B* **1991**, *44*, 7888–7903.

37. Philipsen, P.; te Velde, G.; Bearends, E.; Berger, J.; de Boeij, P.; Groenveld, J.; Kadantsev, E.; Klooster, R.; Kootstra, F.; Romaniello, P.; *et al. BAND2012, SCM, Theoretical Chemistry*; Vrije Universiteit: Amsterdam, Netherlands, 2012.

38. Monkhorst, H.J.; Pack, J.D. Special points for Brillouin-zone integrations. *Phys. Rev. B* **1976**, *13*, 5188.

39. Wilson, J.A.; Yoffe, A.D. The transition metal dichalcogenides discussion and interpretation of the observed optical, electrical and structural properties. *Adv. Phys.* **1969**, *18*, 193–335.

40. Mattheis, L.F. Band Structures of Transition-Metal-Dichalcogenide Layer Compounds. *Phys. Rev. B* **1973**, *8*, 3719.

41. Coehoorn, R.; Haas, C.; Dijkstra, J.; Flipse, C.J.F.; Degroot, R.A.; Wold, A. Electronic structure of $MoSe_2$, MoS_2, and WSe_2. I. Band-structure calculations and photoelectron spectroscopy. *Phys. Rev. B* **1987**, *35*, 6195–6202.

42. Tongay, S.; Zhou, J.; Ataca, C.; Lo, K.; Matthews, T.S.; Li, J.; Grossman, J.C.; Wu, J. Thermally Driven Crossover from Indirect toward Direct Bandgap in 2D Semiconductors: $MoSe_2$ versus MoS_2. *Nano Lett.* **2012**, *12*, 5576–5580.

43. Kang, J.; Tongay, S.; Li, J.; Wu, J. Monolayer semiconducting transition metal dichalcogenide alloys: Stability and band bowing. *J. Appl. Phys.* **2013**, *113*, 143703.

Equilibrium Molecular Dynamics (MD) Simulation Study of Thermal Conductivity of Graphene Nanoribbon: A Comparative Study on MD Potentials

Asir Intisar Khan, Ishtiaque Ahmed Navid , Maliha Noshin, H. M. Ahsan Uddin, Fahim Ferdous Hossain and Samia Subrina

Abstract: The thermal conductivity of graphene nanoribbons (GNRs) has been investigated using equilibrium molecular dynamics (EMD) simulation based on Green-Kubo (GK) method to compare two interatomic potentials namely optimized Tersoff and 2nd generation Reactive Empirical Bond Order (REBO). Our comparative study includes the estimation of thermal conductivity as a function of temperature, length and width of GNR for both the potentials. The thermal conductivity of graphene nanoribbon decreases with the increase of temperature. Quantum correction has been introduced for thermal conductivity as a function of temperature to include quantum effect below Debye temperature. Our results show that for temperatures up to Debye temperature, thermal conductivity increases, attains its peak and then falls off monotonically. Thermal conductivity is found to decrease with the increasing length for optimized Tersoff potential. However, thermal conductivity has been reported to increase with length using 2nd generation REBO potential for the GNRs of same size. Thermal conductivity, for the specified range of width, demonstrates an increasing trend with the increase of width for both the concerned potentials. In comparison with 2nd generation REBO potential, optimized Tersoff potential demonstrates a better modeling of thermal conductivity as well as provides a more appropriate description of phonon thermal transport in graphene nanoribbon. Such comparative study would provide a good insight for the optimization of the thermal conductivity of graphene nanoribbons under diverse conditions.

Reprinted from *Electronics*. Cite as: Khan, A.I.; Navid, I.A.; Noshin, M.; Uddin, H.M.A.; Hossain, F.F.; Subrina, S. Equilibrium Molecular Dynamics (MD) Simulation Study of Thermal Conductivity of Graphene Nanoribbon: A Comparative Study on MD Potentials. *Electronics* **2015**, *4*, 1109–1124.

1. Introduction

In recent years, high density integration and size minimization have taken a tremendous turn in device technology. As a result, further development of the silicon-based micro-electronic device urges for the search of a new type of high thermal conductivity material. In this context, graphene has drawn a lot of attention as it has a number of unique properties which make it interesting for both fundamental studies and future applications. Graphene nanoribbons (GNRs), the narrow strips of graphene with few nanometers width, are particularly considered as a significant element for future nano-electronics. GNRs exhibit several intriguing electronic [1], thermal [2] and mechanical [3] properties dominated by their geometry *i.e.*, width or edge structure [4–6]. The ultra-thin characteristics of graphene and GNR are extremely beneficial for the aforementioned high-density integration. For a device to perform reliably, it is very important to manage the device heat efficiently. Therefore, it is necessary to investigate the thermal transport properties of graphene as well as graphene nanoribbons deeply [7]. Phonon vibration is the prime thermal energy transport mechanism for graphene or GNRs [2,8,9]. The contribution of phonons to the thermal conductivity is about 50–100 times greater than that of electrons [10]. Particularly in recent experiments, it has been found using micro-Raman spectroscopy that single layer graphene sheets show extremely large values of thermal conductivity [11] . With the use of micro-Raman spectroscopy, the room temperature thermal conductivity for a single layer graphene suspended across a trench was found in the range of 3080–5150 W/m-K by Balandin [12]. Furthermore, significant thermal rectification for triangular-shaped GNR has been reported [13]. The thermal conductivity of graphene nanoribbons found in molecular dynamics simulation is of the same order of magnitude (2000 W/m-K) [13] as that of the experimental value (5300 W/m-K) [8], but the magnitude changes considerably with graphene dimension. For isolated 6 nm × 6 nm single layer graphene sheet, Chen *et al.* [14] reported 1780 W/m-K thermal conductivity while a value of 500 W/m-K for a graphene flake of 200 nm × 2.1 nm has been found in [15]. Furthermore, for 27 nm × 15 nm graphene nanoribbon, the study of [7] obtained a thermal conductivity value of 5200 W/m-k whereas Yu *et al.* [16] observed 550 W/m-K thermal conductivity for a 600 nm × 10.4 nm graphene nanoribbon. Because of the extraordinary high value of thermal conductivity of graphene or GNR, deeper investigation on their thermal transport is necessary for the enhancement of energy-efficient nano-electronics. For the purpose of analyzing the thermal conduction property of graphene or GNR, molecular dynamics (MD) simulation can be employed. In our study, we have performed equilibrium molecular dynamics (EMD) simulation to compute thermal conductivity of GNR. EMD is based on Green-Kubo (GK) method derived from linear response theory [17]. The thermal conductivities extracted from non-equilibrium MD (NEMD) method are generally one order smaller in magnitude than the experimentally obtained results following the similar trend [7]. This is due to the fact that NEMD approach might demonstrate non-linear effects because of applied temperature gradient. This non linear effect may be subjected to the strong scattering caused by the heat source or heat sink used in the NEMD approach [18]. Furthermore, size effects are more prominent in NEMD simulation in comparison with EMD simulation [19]. EMD gives provision of computing the thermal conductivity along all directions in one simulation while NEMD can calculate thermal conductivity in one direction only as it uses thermal gradient in a particular direction. As a result, EMD is highly applicable for the geometries involving periodic boundary conditions. On the other hand, EMD is computationally more expensive and simulation results are more susceptible to parameters. However, the EMD with GK method is advantageous over NEMD because of inclusion of the entire thermal conductivity tensor from one simulation along with some additional parameters like heat current autocorrelation function (HCACF) [20]. This is why, in order to systematically observe the thermal conductivity of GNRs, EMD simulation has been carried out in this paper. In a recent study, Mahdizadeh *et al.* [20] showed the variation of thermal conductivity of graphene nanoribbon (10 nm × 2 nm) with temperature and discussed the size dependence of thermal conductivity only with varying lengths (2 nm to 30 nm) using optimized Tersoff potential. However, the reliability of MD simulations highly depends on the use of appropriate interatomic energies and forces that can be advantageously interpreted by interatomic classical potentials. So, in this study, we have taken into account two of the MD potential fields: optimized Tersoff potential and 2nd generation REBO potential with a view to interpreting the heat transport mechanism in the graphene nanoribbon and thereby discussing the comparative reliability of the two potentials to provide the more appropriate description of the thermal conductivity of graphene nanoribbon. Using these two potential fields, we have performed a comparative analysis for the size dependence, *i.e.* varying lengths (8 nm to 14 nm) and widths (1 nm to 2.5 nm) and temperature dependence of thermal conductivity of graphene nanoribbons.

2. Theory and Simulation

2.1. Interatomic MD Potentials

Theoretical and simulation based approaches of phonon thermal transport in graphene, nanotubes and nanoribbons employing MD simulation require accurate representation of interactions between atoms. Interatomic potential, one of the most dictating fundamentals that influence the accuracy of thermal conductivity [11] can conveniently represent these interactions in a realizable form. Literature reported that the original Tersoff [21] potential tends to overestimate the thermal conductivity of graphene structures [22,23] while original REBO potential [24], adaptive intermolecular REBO or AIREBO [25] are found to underestimate the experimental data on the same [23,26]. 2nd generation REBO potential produces a more reliable function for simultaneously predicting bond lengths, energies, and force constants than the original version of the REBO potential [27]. On the other hand, optimized Tersoff and Brenner potential [23] parameters predict acoustic phonon (majority heat carriers in a material) velocity values to be in better agreement with the experimental data. In our work, we have done comparative computations using 2nd generation reactive empirical bond order potential (REBO) [27] and optimized Tersoff potential [23] because of their better accuracy in describing bond order characteristics and anharmonicity of both sp^2 and sp^3 carbon systems including different carbon structures like graphene, carbon nanotube and graphene nanoribbons.

2.2. Equilibrium Molecular Dynamics Simulation: Green-Kubo Method

The computation of dynamic properties such as thermal conductivity in EMD is based on the fluctuation dissipation and linear response theorem [17]. In equilibrium, the heat flow in a system of particles fluctuates around zero and this fact is directly used in this method. The calculation of the heat flux vectors and their correlations is carried out throughout the simulation [11,19]. In this method, thermal conductivity (K_x) is related to HCACF by the following equation:

$$K_x = \frac{1}{VK_BT^2} \int_0^\tau \langle J_x(t) \cdot J_x(0) \rangle dt \tag{1}$$

where V stands for the system volume defined as the product of ribbon planar area and the nominal graphite interplanar separation, i.e., van der Waals thickness (3.4 Å), K_B is the Boltzmann constant, T is the system temperature and τ is the correlation time needed for reasonable decay of HCACF. J_x stands for the x component of heat flux. The ensemble average of the heat flux autocorrelation function is presented by the $\langle J_x(t) \cdot J_x(0) \rangle$ term. To use the above equation in EMD simulation, the integration of the equation is represented in the discrete form as the following equation including the time averaging:

$$K_x = \frac{\Delta t}{VK_BT^2} \sum_{m=1}^{M} \frac{1}{N-m} \sum_{n=1}^{N-m} J_x(m+n)J_x(n) \tag{2}$$

where Δt is the MD simulation time step, $M\Delta t$ corresponds to the correlation time τ of Equation (1), $J_x(m+n)$ and $J_x(n)$ are the xth components of the heat current in x direction at MD time-steps $(m+n)$ and n respectively. N represents the total number of simulation steps while the number of time steps required for the heat flux correlation vector is denoted by M. Therefore, for obtaining better average from statistical point of view, M should be less than N. The heat current J used in

Equation (2) for pair potential is defined as the time derivative of the sum of the moments of the site energies as

$$J = \frac{d}{dt} \sum_i r_i(t) \varepsilon_i(t) \qquad (3)$$

Here, $r_i(t)$ is the time-dependent position of atom i and $\varepsilon_i(t)$ is the site energy which is the summation of kinetic energy and potential energy of the ith atom. For two body interactions, the total potential energy content of a particle is defined in terms of the pair-wise interactions $u_2(r)$ as

$$E_{pot} = \frac{1}{2} \sum_{ij,i\neq j} u_2(r_{ij}) \qquad (4)$$

and the total site energy is expressed by

$$\varepsilon_i = \frac{1}{2} m_i v_i^2 + \frac{1}{2} \sum_j u_2(r_{ij}) \qquad (5)$$

and subsequently the corresponding heat current can be calculated through the following expression:

$$J(t) = \sum_i \varepsilon_i v_i + \frac{1}{2} \sum_{ij,i\neq j} (F_{ij} \cdot v_i) r_{ij}, \qquad (6)$$

where v_i is the velocity of atom i, $r_{ij} = r_i - r_j$ and F_{ij} is the interatomic force exerted by atom j on atom i [18].

2.3. Quantum Correction

Quantum correction to the classical MD calculations of temperature and thermal conductivity is necessary because of the fact that quantum effects in classical MD approach are neglected below Debye temperature (T_d). As a result, thermal conductivity calculated by MD approach gives erroneous results for temperatures below T_d. Classical energy per carbon atom should be equal to phonon energy per carbon atom at quantum temperature, T_q. Therefore,

$$\int_0^{\omega_d} D(\omega) \left[n(\omega, T_q) + \frac{1}{2} \right] h\omega d\omega = 3 K_b T_{MD} \qquad (7)$$

where T_{MD} is the temperature in MD simulation and $n(\omega, T_q)$ is the occupation number of phonons given by

$$n(\omega, T_q) = \frac{1}{e^{\frac{\hbar\omega}{k_b T_q}} - 1} \tag{8}$$

and the term $\frac{1}{2}$ in Equation (7) represents the effect of zero point energy. Here, phonon density of states (DOS), $D(\omega) = \frac{3\omega}{2\pi N v^2}$, $\omega_d = (4\pi N)^{1/2} v$, N = number of carbon atoms in unit area of GNR and v is the effective phonon sound velocity satisfying

$$\frac{3}{v^2} = \frac{1}{v_{LA}^2} + \frac{1}{v_{TA}^2} + \frac{1}{v_{ZA}^2} \tag{9}$$

where LA, TA, and ZA stand for in-plane longitudinal, in-plane transverse and out-of-plane acoustic phonons respectively. Phonons of three acoustic branches are considered with phonon sound velocities v_{LA} = 19.5 km/s , v_{TA} = 12.2 km/s and v_{ZA} = 1.59 km/s [28].

This process aims at mapping classically calculated results onto their quantum analogs at the same energy level. Quantum corrected temperatures can be obtained from the following equation:

$$T_{MD} = 2\frac{T_q^3}{T_D^2} \int\limits_0^{T_D/T_q} \frac{x^2}{e^x - 1} dx + \frac{1}{3} T_D, \tag{10}$$

where $T_D = \frac{\hbar\omega_d}{k_b}$. T_D is the Debye temperature which is 322 Kelvin. So, 2nd term of the equation becomes 107 K which represents that if MD simulation temperature is less than 107 K, there will be no quantum corrected temperature.

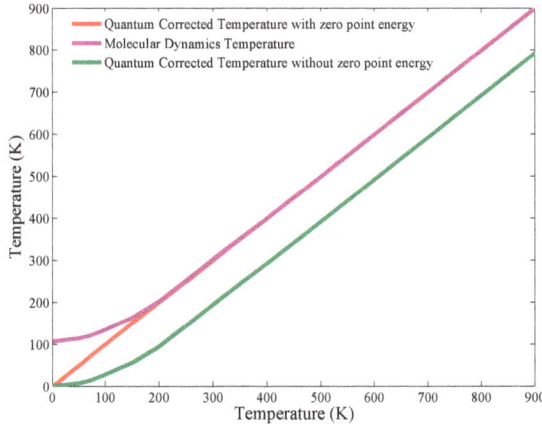

Figure 1. Molecular Dynamics (MD) temperature *versus* Quantum Corrected temperature for graphene nanoribbon (GNR).

The quantum corrected thermal conductivity K_{QC} can be calculated by multiplying the uncorrected thermal conductivity K_{MD} by $\frac{dT_{MD}}{dT_q}$,

$$K_{QC} = \frac{dT_{MD}}{dT_q} \times K_{MD} \tag{11}$$

From Figure 2, we can clearly see that quantum correction is dominant at low temperatures while at higher temperatures it is almost negligible.

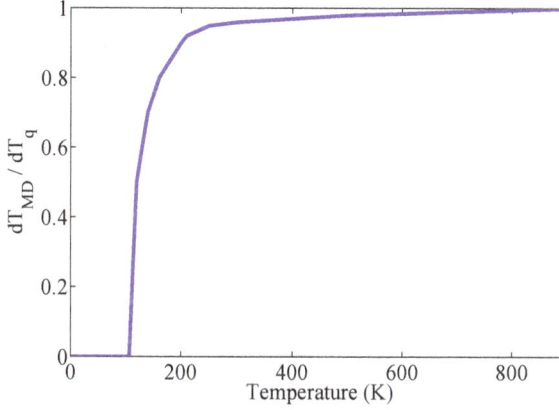

Figure 2. Rate of change of MD temperature with respect to Quantum Corrected temperature versus MD temperature for GNR.

2.4. Simulation Details

In this work, equilibrium molecular dynamics simulations were carried out using LAMMPS [29] with proper periodic boundary conditions applied in the plane, i.e., in the length and width directions of graphene nanoribbon to avoid the effect of fixed walls. We have considered a zigzag graphene nanoribbon (GNR) of 10 nm × 1 nm (length × width) at room temperature which is shown in Figure 3. Two potential fields, namely optimized Tersoff potential and 2nd generation REBO potential were applied to study their effect on thermal transport. The system was thermalized using NoseHoover thermostat for 3×10^5 time steps with a time step of 0.5 fs followed by a switching to NVE ensemble for 2×10^7 time steps. During MD simulations the equations of motion were integrated with a velocity-Verlet integrator. Energy minimization was done using steepest descent algorithm due to its robustness. The heat current data were recorded every 5 steps in the micro canonical ensemble average state. Heat flux autocorrelation values were calculated by averaging five obtained HCACFs. Variable autocorrelation lengths were used for different sizes of GNRs to ensure the reasonable decay of normalized HCACF values.

Figure 3. Schematic atomic structure of 10 nm×1 nm zigzag GNR.

3. Results and Discussion

3.1. Potential Dependence of Thermal Conductivity

Our results show that 2nd generation REBO potential underestimates the thermal conductivity of graphene nanoribbon by a considerable margin in comparison with that of optimized Tersoff potential. In fact, REBO potential measures lower thermal conductivity than even the original Tersoff parameters [23,30,31]. Lattice thermal conductivity depends strongly on the phonon dispersion energies and near-zone centre velocities. With the 2nd generation REBO potential parameters, the velocities of the transverse acoustic (TA) branch and longitudinal acoustic (LA) branch are found to be very low while dispersion of the out of plane branch is also underestimated. In fact, 2nd generation REBO potential does not appropriately measure the zone centre velocities for all the acoustic modes. Strong anharmonicity of 2nd generation REBO potential resulting in high phonon-phonon scattering rates may also contribute to lower thermal conductivity [23]. On the other hand, in our study, thermal conductivity of GNR using optimized Tersoff potential gives much better estimation which is in accordance with the expectation of Lindsay *et al.* [23]. Using optimized Tersoff potential, our extracted value of thermal conductivity for 10 nm × 1 nm GNR at room temperature is 3207 W/m-K while using 2nd generation REBO potential measured thermal conductivity is 1650 W/m-K at 300 K. Evans *et al.* [22] also reported a very high thermal conductivity (~3000 W/m-K at room temperature) of 10 nm × 1 nm sized graphene nanoribbon using Tersoff potential which might be due to the limited number of phonon-phonon scattering in the small systems [32]. Overall improved estimation of ZA, TA and LA phonon branches along with a better fit of near-zone-centre velocities by the optimized Tersoff parameters can be demonstrated to provide a better modeling of thermal conductivity of graphene nanoribbons in contrast to 2nd generation REBO potential.

3.2. Temperature Dependence of Thermal Conductivity

Figure 4 suggests that thermal conductivity of GNR decreases as temperature increases for both optimized Tersoff and 2nd generation REBO potential fields. This phenomena is a consequence of thermal conductivity reduction by phonon-phonon scattering in pure (without defect) lattice structures [33]. For optimized Tersoff potential, the thermal conductivities obtained in our study at 300 K and 400 K are ~3000 W/m-K and ~1650 W/m-K respectively. As temperature increases, the number of high frequency phonon increases. Hence phonon-phonon anharmonic interaction (Umklapp scattering) increases that makes the decay of HCACF profile faster resulting in a decrease of thermal conductivity in EMD method. Our results follow the similar trend of magnitude with the variation of temperature as reported by Mahdizadeh *et al.* [20] (~2500 W/m-K and ~1700 W/m-K at 300 K and 400 K, respectively) where 10 nm × 2 nm sized GNR is considered.

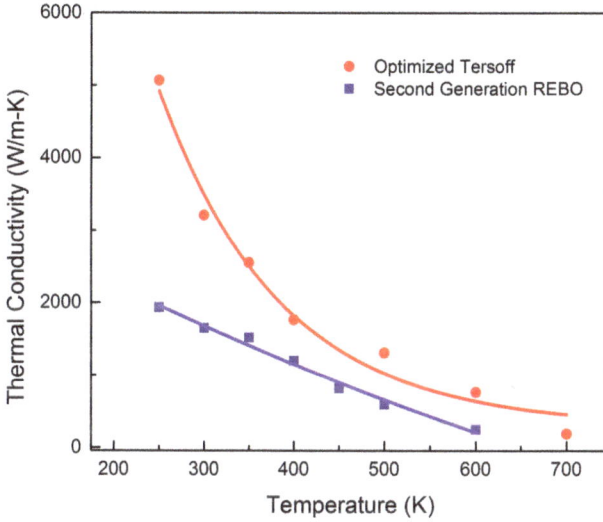

Figure 4. Thermal Conductivity (uncorrected) of GNR (10 nm × 1 nm) as a function of temperature using two different interatomic potentials.

Our results show that the estimated thermal conductivity using optimized Tersoff potential is higher approximately by a factor of 2 compared to that with 2nd generation REBO potential near room temperature and this factor decreases with the increase of temperature. At lower temperature, thermal conductivity for optimized Tersoff potential varies with T^{-1} but it deviates at higher temperature [32]. This might be due to an increased non-linear thermal resistivity considering three phonon-phonon interactions and comparatively weaker but appreciable four phonon-phonon process at an elevated temperature [34]. In addition, anharmonic interactions of two acoustic modes with an optical phonon mode might lead to the observed variation of thermal conductivity at sufficiently high temperatures [35].

Figure 5a,b shows the quantum corrected and uncorrected thermal conductivity of GNR as a function of temperature for optimized Tersoff potential and 2nd generation REBO potential, respectively. At low temperature (upto Debye temperature), quantum corrected thermal conductivity increases almost linearly as temperature increases and reaches a maximum and then drops again. The Umklapp process is supposed to be inactive at low temperature and therefore not available to provide thermal resistance [32] which has been considered through quantum correction. At room temperature and above, Umklapp scattering becomes highly significant [36] and thermally excited high energy phonons dominate the thermal conductivity. As a result, thermal conductivity decreases with the increase of temperature.

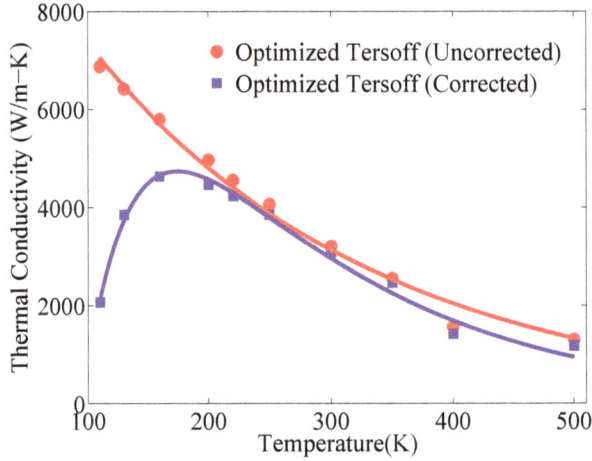

(a) Optimized Tersoff

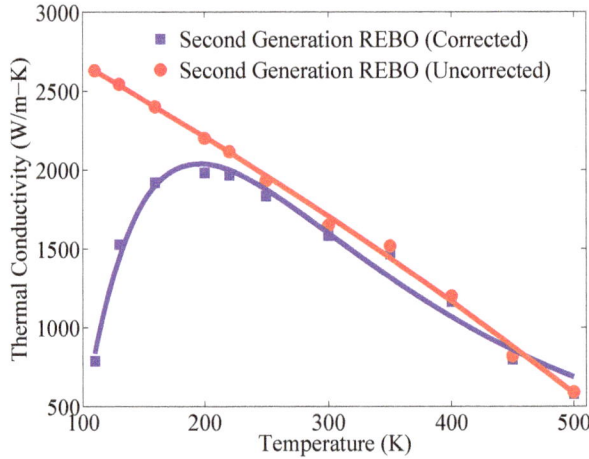

(b) 2nd Generation REBO

Figure 5. Quantum corrected and uncorrected thermal conductivity of GNR (10nm×1nm) as a function of temperature using (**a**) optimized Tersoff and (**b**) 2nd generation Reactive Empirical Bond Order (REBO) potentials.

Figure 6 shows the variation of quantum corrected thermal conductivity as a function of temperature for both optimized Tersoff potential and 2nd generation REBO potentials. The figure shows that the peak thermal conductivity for optimized Tersoff potential is higher than that of 2nd generation REBO potential.

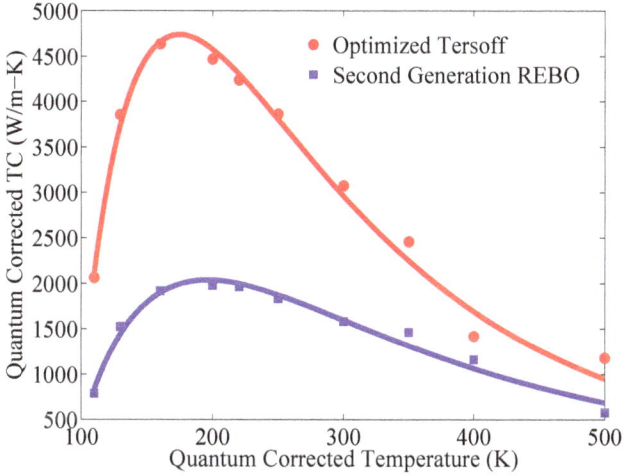

Figure 6. Quantum corrected thermal conductivity (TC) of GNR (10 nm length, 1 nm width) as a function of temperature using optimized Tersoff and 2nd generation REBO potentials.

3.3. Length Dependence of Thermal Conductivity:

In Figure 7, the thermal conductivity of graphene nanoribbon as a function of length for both optimized Tersoff and 2nd generation REBO potential fields has been depicted. The figure shows that the length dependence of thermal conductivity for these two fields is opposite in nature. The thermal conductivity decays with the increase in length for optimized Tersoff potential while it shows an increasing trend with the increase of length for 2nd generation REBO potential field.

According to Figure 7, the thermal conductivity of GNR decreases with the increase of length in a monotonic manner for optimized Tersoff potential field which can be firstly interpreted from the HCACF profile plotted in Figure 8.

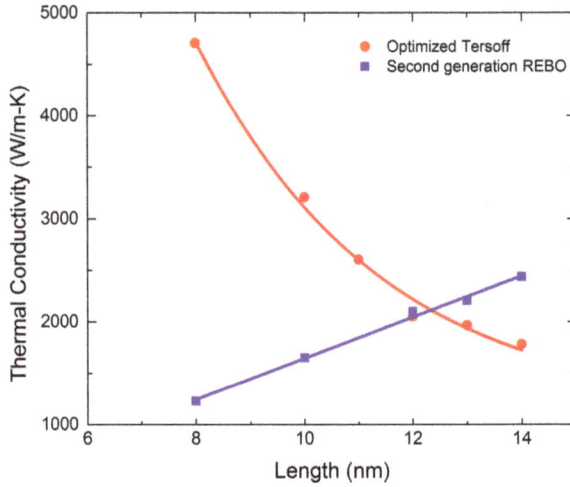

Figure 7. Thermal conductivity variation of GNR with length using optimized Tersoff and 2nd generation REBO potentials at 300 K temperature. The width of GNR is 1 nm.

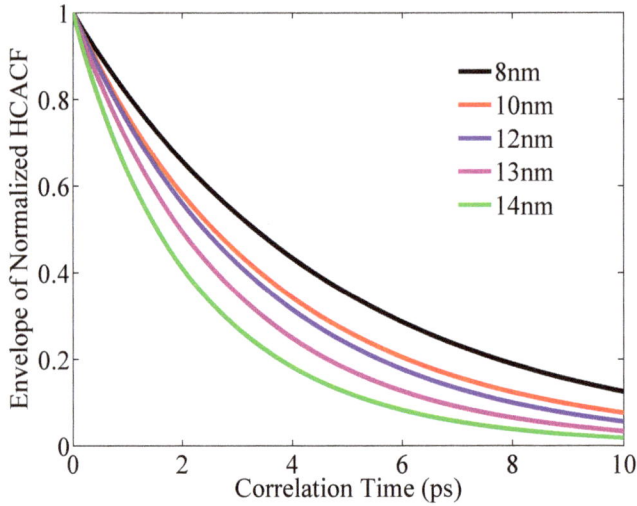

Figure 8. Envelopes of decaying heat current autocorrelation function (HCACF) profiles as a function of correlation length for varying lengths of GNRs (1nm width) using optimized Tersoff potential.

With the increase in GNR length, the number of phonon increases resulting in the rise of phonon-phonon interactions. This causes a faster decay rate for HCACF, *i.e.*, HCACF decays to zero more quickly. Secondly, the contribution of flexural phonons (ZA) to thermal transport of graphene has to be taken into consideration.

144

Using optimized Tersoff potential, Lindsay *et al.* [23] showed the significance of ZA modes as the majority heat carriers in describing thermal transport in contrary to the established hypothesis of ignoring the out of plane phonon contribution. In this context, dispersion characteristics of ZA mode phonons in graphene are accountable for the convergence of thermal conductivity [33]. ZA modes play the dual role of providing a significant source of heat carriers and at the same time preventing the divergence of thermal conductivity through heat carrier scattering. Their group velocity approaches zero near Γ point and hence, the scattering of other heat carriers becomes their prime role in heat transport [23,33]. Thus, ZA modes in the low frequency region of power spectrum lower the in-plane thermal conductivity of GNR as demonstrated by the convergence trends using optimized Tersoff potential. However, converged sampling of the low frequency acoustic flexural modes is required to accurately calculate the thermal conductivity of GNR in molecular dynamics simulation. In this case, our findings disagree with some literature specified in the micrometer range of length [37,38]. This might be due to poor sampling of phase space and poor ergodicity issues, *i.e.*, lack of correspondence between the system's statistical average with the ensemble average. Here, the term ensemble average represents the average taken over all the states present in a system whereas the system statistical average implies the time average of a particular sequence of events rather than all the events or states present [17]. However, Nika *et al.* [38] considered that the out of plane acoustic phonons (ZA) have insignificant contribution to thermal transport. However, recent studies by Nissimagoudar *et al.* [39] found that, ZA branch has significant influence in the thermal properties of graphene nanoribbons due to its quadratic dispersion. At the same time, Sonvane *et al.* [37] believes that flexural phonons dominate the thermal conductivity of graphene with length and width smaller than one micron which is in agreement with our study.

On the other hand, the thermal conductivity rises with the increase in length when the 2nd generation REBO potential is involved. 2nd generation REBO potential underestimates the dispersion of ZA branches in contrast to the optimized Tersoff potential [23]. Moreover, 2nd generation REBO potential includes a dihedral bonding term which poorly fits the phonon frequencies for the ZA branch. In this context, the length dependence of thermal conductivity using 2nd generation REBO potential is mainly governed by the phonon boundary scattering mechanism instead of the dispersion of ZA modes. The length of the GNR controls the phonon mean free path (MFP). With the increase in length, the acoustic phonons with longer wavelengths become available for heat transfer [37]. The phonon transport in graphene, for phonon cutoff frequencies including zero frequency, is two dimensional (2D). In graphene, the long-wavelength phonons weakly scatter in the three-phonon Umklapp processes. This is calculated in the first order and yields the divergent nature of the thermal conductivity. Klemens [40,41], while working with the thermal conductivity of graphite, assumed negligible contribution of long wavelength *i.e.*, low frequency phonons to the in plane thermal conductivity. To avert the problem of long-wavelength phonons in case of 2D lattices, Klemens [40] explained the phenomena using the size-dependent low-bound cut-off frequency $\omega_{min,s}$. $\omega_{min,s}$ is related to finite in plane size by equation $L = \tau(\omega_{min,s})v_s$ where L is the in-plane size, τ is the phonon relaxation time and v_s is the phonon group velocity [42]. Thus he related his study with other literature showing a logarithmic divergence of thermal conductivity as a function of length for 2D lattices. In fact, with the small GNR length compared to phonon MFP (775nm for graphene), Umklapp scattering among phonons becomes negligible and the phonon collision at the edge becomes the significant scattering mechanism. Therefore, the shorter the GNR is, the stronger the edge scattering becomes which abates the thermal conductivity. Conversely, longer GNR results in the weakening of edge scattering leading to the increase of the thermal conductivity [7].

3.4. Width Dependence of Thermal Conductivity

Figure 9 shows the dependence of thermal conductivity on GNR width changing from 1 nm to 2.5 nm at a fixed length of 10 nm. The thermal conductivity increases monotonically in the mentioned width range. The impacts of edge localized phonon effect or boundary scattering effect and the phonon's Umklapp effect, both having negative effects on the thermal conductivity, might be considered. The impact of boundary scattering is weakened with the increase in GNR width. But as the GNR width increases, the probability of Umklapp scattering rises as a result of increased number of phonons and the reduced energy separation between the phonon modes. The dominance between these two mechanisms dictates the nature of thermal conductivity variation. In case of narrower GNRs, the reduced edge-localized phonon effect is dominant with the increase of width leading to the increased thermal conductivity which is reflected in our study range. However, for a large enough width, more and more phonons will be activated with remarkably significant phonon's Umklapp effect and as a consequence, the thermal conductivity decreases with the increase of width of graphene nanoribbon [22,37,43].

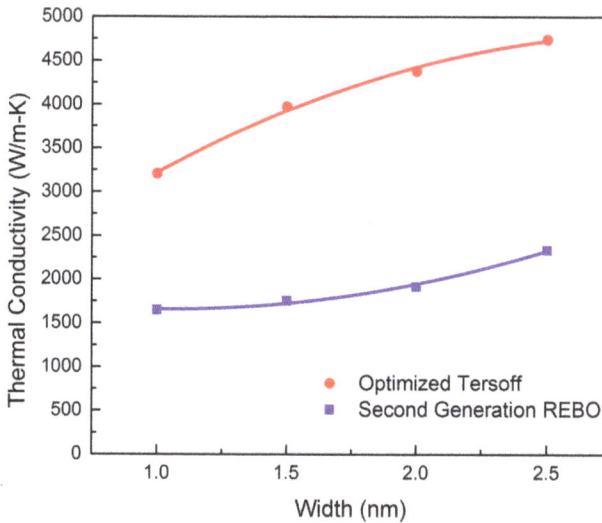

Figure 9. Thermal conductivity variation of GNR with width using optimized Tersoff and 2nd generation REBO potentials at 300 K temperature. The length of GNR is 10 nm.

4. Conclusions

In this study, using EMD simulation based on GK method, we have compared two interatomic potentials, optimized Tersoff and 2nd generation REBO in terms of thermal conductivity of graphene nanoribbons as a function of temperature, length and width for periodic boundary conditions. Our results show that thermal conductivity decreases with the increase of temperature for both optimized Tersoff and 2nd generation REBO potentials because of increasing high frequency phonon-phonon scattering and faster decaying of HCACF values. Furthermore, by introducing quantum correction to include the quantum effect, thermal conductivity increases with temperature up to Debye temperature, achieves its peak value and then falls off monotonically with temperature for both the potentials. Our study of thermal conductivity as a function of length demonstrates that, thermal conductivity decreases with increasing length for optimized Tersoff potential while an increasing trend is observed for 2nd generation REBO potential. Significant contribution of the dispersion of flexural mode phonons which are dominant in optimized Tersoff potential is responsible for convergence of thermal conductivity with increasing length. On the contrary, 2nd generation REBO potential highly underestimates the dispersion and phonon frequencies for the ZA branch as well as TA and LA velocities. As a result, the divergence nature of thermal conductivity of 2nd generation REBO potential is supposed to be dictated by phonon boundary scattering mechanism. It is found that thermal conductivity increases with the increasing width for both

147

the potentials due to the dominant boundary scattering effect which decreases with increasing width. Finally, the estimated thermal conductivity using optimized Tersoff potential is higher approximately by a factor of 2 compared to that of 2nd generation REBO potential near room temperature. Optimized Tersoff potential parameters provide a better fit with various phonon branches, particularly with near zone centre velocities of these branches without altering structural data for graphene nanoribbon in comparison with the 2nd generation REBO potential. Therefore, optimized Terosoff potential has resulted in a better estimation of thermal conductivity and better description of phonon thermal transport of graphene nanoribbons compared to those of 2nd generation REBO potential.

Author Contributions: Asir Intisar Khan; Ishtiaque Ahmed Navid; Maliha Noshin; H. M. Ahsan Uddin and Fahim Ferdous Hossain jointly designed and performed the simulation under the supervision of Samia Subrina All authors discussed and analyzed the data and interpretation, and contributed during the writing of the manuscript. All authors have given approval to the final version of the manuscript.

Conflicts of Interest: The authors declare no conflict of interest.

References

1. Novoselov, K.; Geim, A.; Morozov, S.; Jian, D.; Katsnelson, M.; Grigorieva, I.; Dubonos, S.; Firsov, A. Two-dimensional gas of massless Dirac fermions in graphene. *Nature* **2005**, *438*, 197–200.

2. Seol, J.H.; Jo, I.; Moore, A.L.; Lindsay, L.; Aitken, Z.H.; Pettes, M.T.; Li, X.; Yao, Z.; Huang, R.; Broido, D.; *et al.* Two-Dimensional Phonon Transport in Supported Graphene. *Science* **2010**, *328*, 213–216.

3. Lee, C.; Wei, X.; Kysar, J.W.; Hone, J. Measurement of the Elastic Properties and Intrinsic Strength of Monolayer Graphene. *Science* **2008**, *321*, 385–388.

4. Nakada, K.; Fujita, M. Edge state in graphene ribbons: Nanometer size effect and edge shape dependence. *Phys. Rev. B* **1996**, *54*, 954–960.

5. Son, Y.W.; Cohen, M.L.; Louie, S.G. Energy Gaps in Graphene Nanoribbons. *Phys. Rev. Lett.* **2006**, *97*, 216803.

6. Han, M.Y.; Ozyilmaz, B.; Zhang, Y.; Kim, P. Energy Band-Gap Engineering of Graphene Nanoribbons. *Phys. Rev. Lett.* **2007**, *98*, 206805.

7. Yang, D.; Ma, F.; Sun, Y.; Hu, T.; Xu, K. Influence of typical defects on thermal conductivity of graphene nanoribbons: An equilibrium molecular dynamics simulation. *App. Surf. Sci.* **2012**, *258*, 9926–9931.

8. Balandin, A.A.; Ghosh, S.; Bao, W.; Calizo, I.; Teweldebrhan, D.; Miao, F.; Lau, C.N. Superior Thermal Conductivity of Single-Layer Graphene. *Nano Lett.* **2008**, *8*, 902–907.

9. Singh, V.; Joung, D.; Zhai, L.; Das, S.; Khondaker, S.I.; Seal, S. Graphene based materials: Past, present and future. *Prog. Mater. Sci.* **2011**, *56*, 1178–1271.

10. Hone, J.; Whitney, M.; Piskoti, C.; Zettl, A. Thermal conductivity of single-walled carbon nanotubes. *Phys. Rev. B* **1999**, *59*, 2514–2516.

11. Cao, A. Molecular dynamics simulation study on heat transport in monolayer graphene sheet with various geometries. *App. Phys. Lett.* **2012**, *111*, 083528.

12. Ghosh, S.; Calizo, I.; Teweldebrhan, D.; Pokatilov, E.P.; Nika, D.L.; Balandin, A.A.; Bao, W.; Miao, F.; Lau, C.N. Extremely high thermal conductivity of graphene: Prospects for thermal management applications in nanoelectronic circuits. *App. Phys. Lett.* **2008**, *92*, 151911.

13. Hu, J.; Ruan, X.; Chen, Y.P. Thermal Conductivity and Thermal Rectification in Graphene Nanoribbons: A Molecular Dynamics Study. *Nano Lett.* **2009**, *9*, 2730–2735.

14. Chen, L.; Kumar, S. Thermal transport in graphene supported on copper. *J. Appl. Phys.* **2012**, *112*, 043502.

15. Zhang, Y.; Cheng, Y.; Pei, Q.; Wang, C.; Xiang, Y. Thermal conductivity of defective graphene. *Phys. Lett. A* **2012**, *376*, 3668–3672.

16. Yu, C.; Zhang, G. Impacts of length and geometry deformation on thermal conductivity of graphene nanoribbons. *J. App. Phys.* **2013**, *113*, doi:10.1063/1.4788813.

17. Frenkel, D.; Smit, B. *Understanding Molecular Simulation: From Algorithms to Applications*; Academic Press: San Diego, CA, USA, 2002.

18. Schelling, P.K.; Phillpot, S.R.; Keblinski, P. Comparison of atomic-level simulation methods for computing thermal conductivity. *Phys. Rev. B* **2001**, *65*, 144306.

19. Khadem, M.H.; Wemhoff, A.P. Comparison of Green-Kubo and NEMD heat flux formulations for thermal conductivity prediction using the Tersoff potential. *Comp. Mater. Sci.* **2013**, *69*, 428–434.

20. Mahdizadeh, S.J.; Goharshadi, E.K. Thermal conductivity and heat transport properties of graphene nanoribbons. *Springer* **2014**, *16*, 1–12.

21. Tersoff, J. Empirical Interatomic Potential for Carbon, with Applications to Amorphous Carbon. *Phys. Rev. Lett.* **1988**, *61*, 2879–2882.

22. Evans, W.J.; Hu, L.; Keblinski, P. Thermal conductivity of graphene ribbons from equilibrium molecular dynamics: Effect of ribbon width, edge roughness, and hydrogen termination. *Appl. Phys. Lett.* **2010**, *96*, 203112.

23. Lindsay, L.; Broido, D.A. Optimized Tersoff and Brenner empirical potential parameters for lattice dynamics and phonon thermal transport in carbon nanotubes and graphene. *Phys. Rev. B* **2010**, *81*, 205441.

24. Brenner, D.W. Empirical potential for hydrocarbons for use in simulating the chemical vapor deposition of diamond films. *Phys. Rev. B* **1990**, *42*, 9458–9471.

25. Stuart, S.J.; Tutein, A.B.; Harrison, J.A. A reactive potential for hydrocarbons with intermolecular interactions. *J. Chem. Phys.* **2000**, *112*, 6472–6486.

26. Zhang, H.; Lee, G.; Fonseca, A.F.; Borders, T.L.; Cho, K. Isotope Effect on the Thermal Conductivity of Graphene. *J. Nanomater* **2010**, *2010*, 537657.

27. Brenner, D.W.; Shenderova, O.A.; Harrison, J.A.; Stuart, S.J.; Ni, B.; Sinnott, S.B. A second-generation reactive empirical bond order (REBO) potential energy expression for hydrocarbons. *J. Phys. Condens. Matter* **2002**, *14*, 783–802.

28. Lee, Y.H.; Biswas, R.; Soukoulis, C.M.; Wang, C.Z.; Chan, C.T.; Ho, K.M. Molecular-Dynamics Simulation of Thermal Conductivity in Amorphous Silicon. *Phys. Rev. B* **1991**, *43*, 6573–6580.

29. Plimpton, S. Fast Parallel Algorithms for Short-Range Molecular Dynamics. *J. Comput. Phys.* **1995**, *117*, 1–19.

30. Thomas, J.A.; Iutzi, R.M.; McGaughey, A.J.H. Thermal conductivity and phonon transport in empty and water-filled carbon nanotubes. *Phys. Rev. B* **2010**, *81*, 045413.

31. Yu, C.; Shi, L.; Yao, Z.; Li, D.; Majumdar, A. Thermal Conductance and Thermopower of an Individual Single-Wall Carbon Nanotube. *Nano Lett.* **2005**, *5*, 1842–1846.

32. ZIMAN, J.M. *Electrons and Phonons: The Theory of Transport Phenomena in Solids*; Oxford University Press: Amen House, London, UK, 1960.

33. Pereira, L.F.C.; Donadio, D. Divergence of the thermal conductivity in uniaxially strained graphene. *Phys. Rev. B* **2013**, *87*, 125424.

34. Ecsedy, D.J.; Klemens, P.G. Thermal resistivity of dielectric crystals due to four-phonon processes and optical modes. *Phys. Rev. B* **1977**, *15*, 5957–5962.

35. Steigmeier, E.F.; Kudman, I. Acoustical-Optical Phonon Scattering in Ge, Si, and III-V Compounds. *J. Appl. Phys.* **1966**, *141*, 767–774.

36. Pichanusakorn, P.; Bandaru, P. Nanostructured thermoelectrics. *Mater. Sci. Eng. R* **2010**, *67*, 19–63.

37. Sonvane, Y.; Gupta, S.K.; Raval, P.; Lukacevic, I.; Thakor, P. Length, Width and Roughness Dependent Thermal Conductivity of Graphene Nanoribbons. *Chem. Phys. Lett.* **2015**, *634*, doi:10.1016/j.cplett.2015.05.036.

38. Nika, D.; Ghosh, S.; Pokatilov, E.; Balandin, A. Lattice Thermal conductivity of graphene flakes: Comparison with bulk graphite. *App. Phys. Lett.* **2009**, *94*, 203103.

39. Nissimagoudar, A.S.; Sankeshwar, N.S. Significant reduction of lattice thermal conductivity due to phonon confinement in graphene nanoribbons. *Phys. Rev. B* **2014**, *89*, 235422.

40. Klemens, P.G. Theory of the A-Plane Thermal Conductivity of Graphite. *J. Wide Bandgap Mater.* **2000**, *7*, 332–339.

41. Klemens, P.G.; Pedraza, D.F. Thermal Conductivity of Graphite in the Basal Plane. *Carbon* **1994**, *32*, 735–741.

42. Nika, D.L.; Askerov, A.S.; Balandin, A.A. Anomalous Size Dependence of the Thermal Conductivity of Graphene Ribbons. *Nano Lett.* **2012**, *12*, 3238–3244.

43. Cao, H.Y.; Guo, Z.X.; Xiang, H.; Gong, X.G. Layer and size dependence of thermal conductivity in multilayer graphene nanoribbons. *Phys. Lett. A* **2012**, *376*, 525–528.

Enhanced Visibility of MoS$_2$, MoSe$_2$, WSe$_2$ and Black-Phosphorus: Making Optical Identification of 2D Semiconductors Easier

Gabino Rubio-Bollinger, Ruben Guerrero, David Perez de Lara, Jorge Quereda, Nicolas Agraït, Rudolf Bratschitsch and Andres Castellanos-Gomez

Abstract: We explore the use of Si$_3$N$_4$/Si substrates as a substitute of the standard SiO$_2$/Si substrates employed nowadays to fabricate nanodevices based on 2D materials. We systematically study the visibility of several 2D semiconducting materials that are attracting a great deal of interest in nanoelectronics and optoelectronics: MoS$_2$, MoSe$_2$, WSe$_2$ and black-phosphorus. We find that the use of Si$_3$N$_4$/Si substrates provides an increase of the optical contrast up to a 50%–100% and also the maximum contrast shifts towards wavelength values optimal for human eye detection, making optical identification of 2D semiconductors easier.

Reprinted from *Electronics*. Cite as: Rubio-Bollinger, G.; Guerrero, R.; de Lara, D.P.; Quereda, J.; Vaquero-Garzon, L.; Agraït, N.; Bratschitsch, R.; Castellanos-Gomez, A. Enhanced Visibility of MoS$_2$, MoSe$_2$, WSe$_2$ and Black-Phosphorus: Making Optical Identification of 2D Semiconductors Easier. *Electronics* **2015**, *4*, 847–855.

1. Introduction

Mechanical exfoliation has proven to be a very powerful tool to isolate two-dimensional material out of bulk layered crystal [1]. The produced atomically thin layers, however, are randomly deposited on the substrate surface and are typically surrounded by thick crystal which hampers the identification of the thinner material. Optical microscopy is a perfect complement to mechanical exfoliation as is a reliable and non-destructive method and it allows distinguishing the atomically thick layers from their bulk-like counterparts. This technique is valid despite the reduced thickness of 2D materials. They can be seen through an optical microscope with the naked eye, because of the wavelength dependent reflectivity of the dielectric/2D material system [2–6]. This dependence can be exploited to easily identify and isolate 2D material single layer flakes by modifying the substrate dielectric thickness and permittivity. In addition to increasing the visibility, the use of different substrate materials may improve the performance of the produced devices if the chosen substrate has good dielectric properties.

In this work we systematically study the visibility of several 2D materials with potential applications in electronics and optoelectronics, such as MoS$_2$, MoSe$_2$, WSe$_2$ and black-phosphorus (BP) [7–11]. The performed experiments and analysis are general, and can be applied to any kind of 2D materials. Here, we explore the use of silicon nitride (Si$_3$N$_4$), a high k dielectric material (κ ~7) commonly used in the semiconductor industry, as a substitute of the silicon oxide layer, which is almost exclusively used nowadays to fabricate nanodevices based on 2D materials. We show how the use of silicon nitride strongly enhances the optical contrast of 2D semiconductors, making the identification of ultrathin sheets easier. Moreover, by using a Si$_3$N$_4$ spacer layer of 75 nm in thickness, the optical contrast reaches its maximum value at a wavelength of 550 nm (which is the optimal wavelength detection of the human eye) [12], while 285 nm of SiO$_2$ spacer layer (the standard in graphene and MoS$_2$ research nowadays) has its maximum contrast value at 460 nm, in the deep-blue/violet part of the visible spectrum.

Apart from the enhanced visibility, the use of Si$_3$N$_4$ as spacer layer also has the potential to improve the electrical performance of nanoelectronic devices due to its high dielectric constant (almost twice that of SiO$_2$) that can help to screen Coulomb scatterers and, thus, to improve the mobility [13]. Additionally, the use of Si$_3$N$_4$ do not present any disadvantage with respect to SiO$_2$ layers in terms of processing as Si$_3$N$_4$ is compatible with most common semiconductor industry processes. Moreover, Si$_3$N$_4$ substrates can be used for other fabrication processes different than mechanical exfoliation, such as CVD [14–16] (because of its high thermal stability) and inkjet printing [17–19] as its surface chemistry has the potential to be tuned using similar recipes to those used for SiO$_2$ substrates. In summary due to the good dielectric performance of Si$_3$N$_4$ and its deposition compatibility with other semiconductor industry processes, we believe that the use of Si$_3$N$_4$ as spacer layer for 2D semiconductor applications will become popular in the near future.

2. Experimental Section

Two-dimensional semiconductor samples are prepared using a recently developed deterministic transfer process [20]. First, we mechanically exfoliate bulk MoS$_2$, MoSe$_2$, WSe$_2$ or black phosphorous using clear Nitto tape (SPV 224) (Osaka, Japan). Bulk crystals were synthetic (grown by vapor transport method) in all cases except the MoS$_2$ crystal that was obtained from naturally occurring molybdenite (Moly Hill mine, Quebec, QC, Canada). The freshly cleaved flakes are then deposited onto a viscoelastic poly-dimethylsiloxane (PDMS) substrate. Subsequently, the flakes are transferred onto two different silicon substrates: one with a 285 nm thick SiO$_2$ oxide layer on top and another one with a 75 nm thick Si$_3$N$_4$ layer. The latter thickness was chosen after the theoretical analysis explained in Section 3 in order to maximize the contrast at a wavelength of 550 nm.

Few-layer flakes are identified under an optical microscope (Nikon Eclipse LV100) (Nikon Corporation, Tokyo, Japan) and the number of layers is determined by a combination of quantitative optical microscopy and contact mode atomic force microscopy (used instead of tapping mode to avoid artifacts in the thickness determination). The optical properties of the nanosheets have been studied with a modification of a home-built hyperspectral imaging setup, described in detail in Reference [21].

3. Results and Discussion

3.1. Optical Contrast Calculation

In order to evaluate the potential of Si_3N_4 to enhance the optical visibility of 2D semiconductors we have first calculated the optical contrast of monolayer MoS_2, $MoSe_2$ and WSe_2 as function of the illumination wavelength for substrates with Si_3N_4 and SiO_2 layers of different thickness. The model is based on the Fresnel law and more details can be found in the literature [2,3,22–25]. Briefly, the optical contrast of atomically thin materials is due to a combination of: (1) interference between the reflection paths that originate from the interfaces between the different media and (2) thickness dependent transparency of the 2D material that strongly modulates the relative amplitude of the different reflection paths. These two effects combined lead to color shifts (dependent on the thickness of the 2D material) that can be appreciated by eye. Figure 1 displays colormaps that represent the optical contrast (defined as $C = (I_{\text{flake}} - I_{\text{substrate}})/(I_{\text{flake}} + I_{\text{substrate}})$) as a function of the illumination wavelength (vertical axis) and the thickness of the dielectric layer (horizontal axis). The references employed to extract the refractive indexes for the different materials employed in the calculation of the optical contrast are summarized in Table 1. One can clearly see how the optical contrast for Si_3N_4 substrates is much higher than for SiO2. Moreover, substrates with a 75 nm Si_3N_4 layer have a maximum contrast at a wavelength around 550 nm, which is optimal for human eye detection. The strong optical contrast enhancement observed for 75 nm thick Si_3N_4 layers on Si substrates can be easily understood as the combination of thickness and refractive index of this layer makes it an almost perfect anti-reflective coating. An optimal anti-reflective coating should have a refractive index $n_{ar} = \sqrt{n_{air} \cdot n_{Si}} \approx 2$ (very similar to the n_{Si3N4} in the visible part of the spectrum) and a thickness $d_{ar} = \frac{\lambda}{4 n_{ar}} \approx 65 - 75$ nm (to minimize the reflection within the visible part of the spectrum). In fact, the reflectivity of a 75 nm thick Si_3N_4 layer on Si is almost zero in the visible range of spectrum. Therefore, the contrast of the 2D materials is enhanced, because their surrounding substrate almost does not reflect any light.

Table 1. Summary of the references with the refractive index values necessary for the calculation of the optical contrast.

Material	Reference
SiO_2	[26]
Si_3N_4	[27]
MoS_2	[28]
$MoSe_2$	[28]
WSe_2	[29]

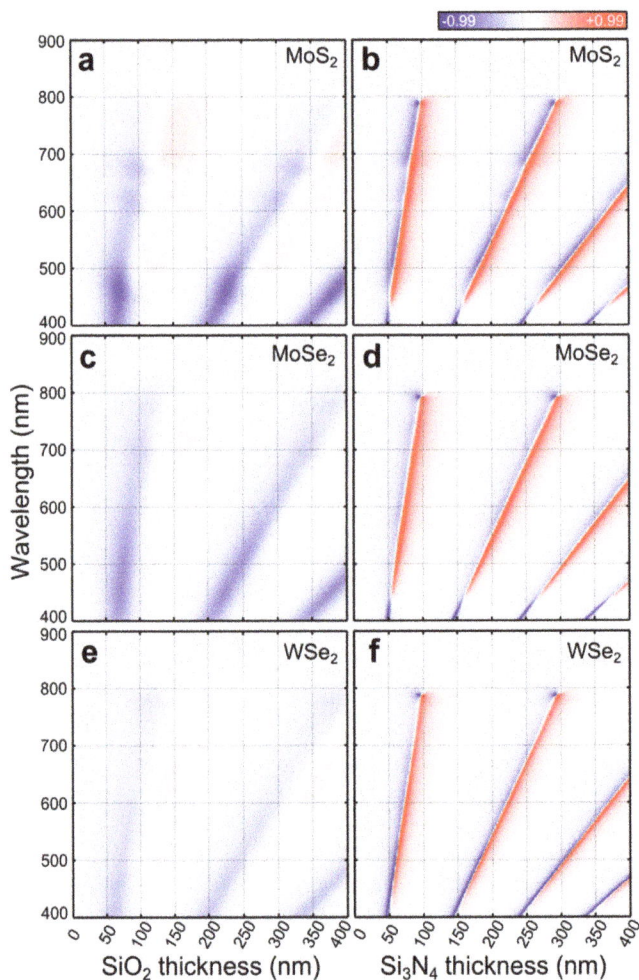

Figure 1. Calculated optical contrast as a function of the illumination wavelength and spacer layer thickness for monolayer MoS_2 (**a,b**), $MoSe_2$ (**c,d**) and WSe_2 (**e,f**) for substrates with SiO_2 (left panel) and Si_3N_4 (right panel) spacer layers.

As one can see from Figure 1, the optical contrast on Si_3N_4/Si substrates is more sensitive to small variations of the spacing layer than the SiO_2/Si substrates. While for SiO_2/Si substrates can present the SiO_2 thickness variations of $\sim \pm 10$ nm, the Si_3N_4 layer thickness should present variations within $\sim \pm 3$ nm to avoid substantial contrast variations within the sample. We address the readers to the Supporting Information to see a comparison between two horizontal linecuts along 550 nm wavelength in panel (a) and (b).

The result of the calculations displayed in Figure 1 illustrates the potential of Si_3N_4 spacer layers with a thickness of 50 nm–100 nm to enhance the optical contrast significantly with respect to conventionally used SiO_2 spacer layers. Therefore, we have explored the potential of Si_3N_4 in 2D semiconductor research by experimentally studying the optical contrast of several 2D semiconductors (MoS_2, $MoSe_2$, WSe_2 and black phosphorus) on silicon substrates with a 75 nm thick Si_3N_4 layer (IDB Technologies Ltd, Whitley, Wiltshire, UK). We also fabricated samples on Si/SiO_2 substrates with 285 nm SiO_2 in order to compare the measured optical contrast with the most extended dielectric layer used in 2D materials research nowadays.

3.2. Hyperspectral Imaging

The optical contrast is measured at different illumination wavelengths with a modified hyperspectral imaging setup described in Reference [21]. The sample is illuminated with monochromatic light with the help of a monochromator. The measurement is carried out by sweeping the illumination wavelength from 450 nm to 900 nm in 5 nm steps, and acquiring an image with a monochrome camera at each wavelength step. The thickness of the studied flakes has been determined by atomic force microscopy prior to the hyperspectral imaging measurements (see Figure 2). Raman spectroscopy or photoluminescence can be also used to characterize and to determine the thickness of the exfoliated flakes on Si_3N_4 surfaces [30], see Supporting Information for Raman spectra acquired for MoS_2 flakes on a 75 nm Si_3N_4/Si substrate and a comparison with the spectra reported for flakes on 285 nm SiO_2/Si substrates [31,32].

Figure 3 shows the obtained optical contrast maps of MoS_2 flakes with a single-layer region (highlighted in the Figure with "1 L") on a 285 nm SiO_2/Si substrate (a) and on a 75 nm Si_3N_4/Si substrate (b) under illumination with different wavelengths: 500 nm, 550 nm, 600 nm, 650 nm, 700 nm and 750 nm. The comparison between the results obtained for the SiO_2 and Si_3N_4 layers illustrates how the optical contrast of monolayer MoS_2 on SiO_2 is weaker within the visible part of the spectrum, whereas for Si_3N_4 around 500–600 nm the monolayer contrast reaches the highest value.

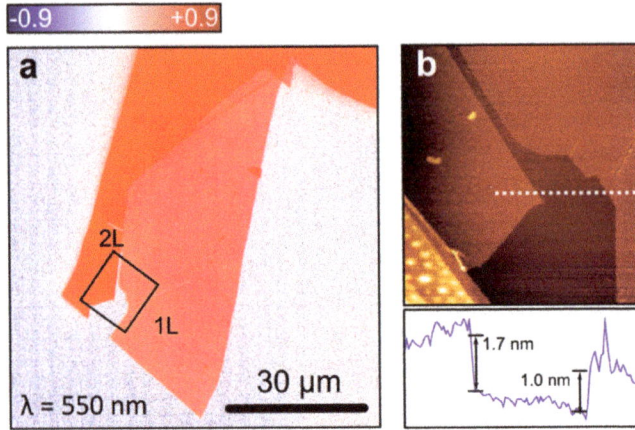

Figure 2. (a) Contrast map of a MoS$_2$ flake deposited onto a 75 nm Si$_3$N$_4$/Si substrate under illumination with 550 nm wavelength; (b) Topographic atomic force microscopy image acquired on the region highlighted with the square in (a), a topographic line profile is also shown below.

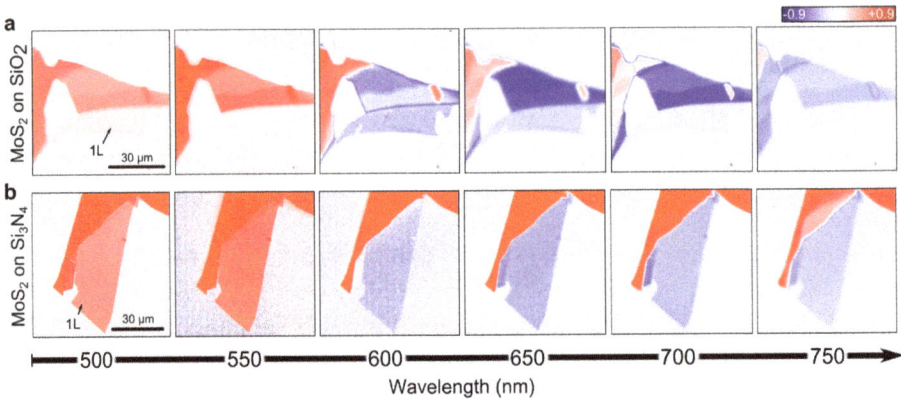

Figure 3. Contrast maps obtained by hyperspectral imaging of MoS$_2$ flakes on Si substrate with spacer layers of, (a) 285 nm of SiO$_2$ or (b) 75 nm of Si$_3$N$_4$, at different illuminating wavelengths. The contrast in the single layer case is maximum at 550 nm for Si$_3$N$_4$, whereas in the SiO$_2$ case is at 600 nm. It allows for the direct identification of single layer flakes.

3.3. Wavelength Dependent Optical Contrast

From the contrast maps at different wavelengths one can extract the wavelength dependence of the optical contrast for flakes with different thicknesses. Figure 4 summarizes the contrast *vs.* wavelength dependence measured for MoS$_2$, MoSe$_2$, WSe$_2$ and black phosphorus on both substrates. For all the studied materials the optical contrast is enhanced on substrates with Si_3N_4 by a 50%–100%. The wavelength with the maximum optical contrast is also shifted: while on SiO_2/Si substrates it is ~650 nm, on Si_3N_4 the maximum contrast is at ~550 nm.

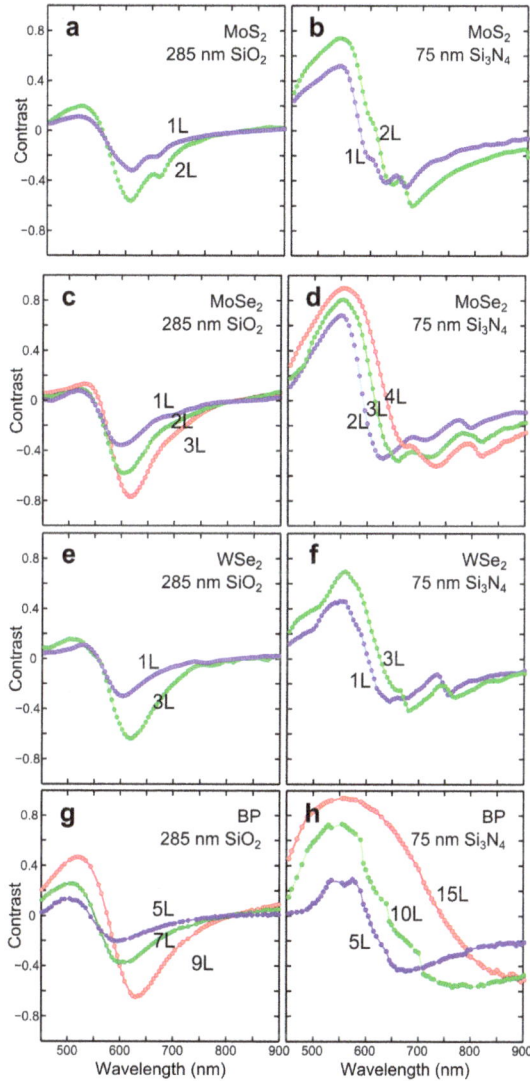

Figure 4. Wavelength dependence of the optical contrast measured for MoS_2, $MoSe_2$, WSe_2 and black phosphorus flakes of different thickness. (**a,b**) optical contrast of MoS_2 on 285 nm SiO_2/Si and 75 nm Si_3N_4/Si, respectively; (**c,d**) optical contrast of $MoSe_2$ on 285 nm SiO_2/Si and 75 nm Si_3N_4/Si, respectively; (**e,f**) optical contrast of WSe_2 on 285 nm SiO_2/Si and 75 nm Si_3N_4/Si, respectively; (**g,h**) optical contrast of black phosphorus on 285 nm SiO_2/Si and 75 nm Si_3N_4/Si, respectively.

4. Conclusions

In summary, we have explored the use of Si_3N_4 as dielectric layer for 2D semiconductor research. We systematically studied the optical contrast of several 2D semiconductors (MoS_2, $MoSe_2$, WSe_2 and black phosphorus) on silicon substrates with 75 nm of Si_3N_4 spacer layers, which according to our calculations should substantially enhance the optical contrast. We experimentally demonstrated the optical contrast enhancement due to 75 nm Si_3N_4/Si substrates by measuring the optical contrast in the range of 450 nm to 900 nm by hyperspectral imaging. We compared the measured contrast to that acquired for samples fabricated on the standard 285 nm SiO_2/Si substrates, finding an increase of the optical contrast up to a 50%–100%. The maximum contrast also shifts in wavelength towards wavelength values optimal for human eye detection. The obtained results provide a way of improving optical identification of single layers of 2D materials.

Acknowledgments: A.C.-G. acknowledges financial support from the BBVA Foundation through the fellowship "I Convocatoria de Ayudas Fundacion BBVA a Investigadores, Innovadores y Creadores Culturales", from the MINECO (Ramón y Cajal 2014 program, RYC-2014-01406) and from the MICINN (MAT2014-58399-JIN). R.G. acknowledges financial support from the AMAROUT-Marie Curie program. We also acknowledge funding from the projects MAT2014-57915-R (MINECO) and MAD2D project S2013/MIT-3007 (Comunidad Autónoma de Madrid).

Author Contributions: G.R.-B. developed the experimental setup. R.G., D.P.D.L., L.V.-G. and A.C.-G. fabricated the samples. J.Q. performed the AFM characterization. G.R.-B., R.G., D.P.D.L., L.V.-G. carried out the hyperspectral measurements. A.C.-G. directed the research project and wrote the manuscript. All authors discussed the data and interpretation, and contributed during the writing of the manuscript. All authors have given approval to the final version of the manuscript.

Conflicts of Interest: The authors declare no conflict of interest.

References

1. Novoselov, K.S.; Jiang, D.; Schedin, F.; Booth, T.J.; Khotkevich, V.V.; Morozov, S.V.; Geim, A.K. Two-dimensional atomic crystals. *Proc. Natl. Acad. Sci. USA* **2005**, *102*, 10451–10453.
2. Blake, P.; Hill, E.W.; Castro Neto, A.H.; Novoselov, K.S.; Jiang, D.; Yang, R.; Booth, T.J.; Geim, A.K. Making graphene visible. *Appl. Phys. Lett.* **2007**, *91*.
3. Abergel, D.S.L.; Russell, A.; Fal'ko, V.I. Visibility of graphene flakes on a dielectric substrate. *Appl. Phys. Lett.* **2007**, *91*, 063125.
4. Casiraghi, C.; Hartschuh, A.; Lidorikis, E.; Qian, H.; Harutyunyan, H.; Gokus, T.; Novoselov, K.S.; Ferrari, A.C. Rayleigh imaging of graphene and graphene layers. *Nano Lett.* **2007**, *7*, 2711–2717.
5. Ni, Z.H.; Wang, H.M.; Kasim, J.; Fan, H.M.; Yu, T.; Wu, Y.H.; Feng, Y.P.; Shen, Z.X. Graphene thickness determination using reflection and contrast spectroscopy. *Nano Lett.* **2007**, *7*, 2758–2763.

6. Roddaro, S.; Pingue, P.; Piazza, V.; Pellegrini, V.; Beltram, F. The optical visibility of graphene: Interference colors of ultrathin graphite on SiO_2. *Nano Lett.* **2007**, *7*, 2707–2710.

7. Wang, Q.H.; Kalantar-Zadeh, K.; Kis, A.; Coleman, J.N.; Strano, M.S. Electronics and optoelectronics of two-dimensional transition metal dichalcogenides. *Nat. Nanotechnol.* **2012**, *7*, 699–712.

8. Butler, S.Z.; Hollen, S.M.; Cao, L.; Cui, Y.; Gupta, J.A.; Gutiérrez, H.R.; Heinz, T.F.; Hong, S.S.; Huang, J.; Ismach, A.F.; *et al.* Progress, challenges, and opportunities in two-dimensional materials beyond graphene. *ACS Nano* **2013**, *7*, 2898–2926.

9. Koppens, F.H.L.; Mueller, T.; Avouris, P.; Ferrari, A.C.; Vitiello, M.S.; Polini, M. Photodetectors based on graphene, other two-dimensional materials and hybrid systems. *Nat. Nanotechnol.* **2014**, *9*, 780–793.

10. Buscema, M.; Island, J.O.; Groenendijk, D.J.; Blanter, S.I.; Steele, G.A.; van der Zant, H.S.J.; Castellanos-Gomez, A. Photocurrent generation with two-dimensional van der Waals semiconductors. *Chem. Soc. Rev.* **2015**, *44*, 3691–3718.

11. Castellanos-Gomez, A. Black phosphorus: narrow gap, wide applications. *J. Phys. Chem. Lett.* **2015**, *6*, 4280–4291.

12. Wald, G. Human Vision and the Spectrum. *Science* **1945**, *101*, 653–658.

13. Ponomarenko, L.A.; Yang, R.; Mohiuddin, T.M.; Katsnelson, M.I.; Novoselov, K.S.; Morozov, S.V.; Zhukov, A.A.; Schedin, F.; Hill, E.W.; Geim, A.K. Effect of a High-κ Environment on Charge Carrier Mobility in Graphene. *Phys. Rev. Lett.* **2009**, *102*, 206603.

14. Perea-López, N.; Lin, Z.; Pradhan, N.R.; Iñiguez-Rábago, A.; Laura Elías, A.; McCreary, A.; Lou, J.; Ajayan, P.M.; Terrones, H.; Balicas, L.; *et al.* CVD-grown monolayered MoS 2 as an effective photosensor operating at low-voltage. *2D Mater.* **2014**, *1*, 011004.

15. Xia, J.; Huang, X.; Liu, L.-Z.; Wang, M.; Wang, L.; Huang, B.; Zhu, D.-D.; Li, J.-J.; Gu, C.-Z.; Meng, X.-M. CVD synthesis of large-area, highly crystalline MoSe2 atomic layers on diverse substrates and application to photodetectors. *Nanoscale* **2014**, *6*, 8949–8955.

16. Van der Zande, A.M.; Huang, P.Y.; Chenet, D.A.; Berkelbach, T.C.; You, Y.; Lee, G.-H.; Heinz, T.F.; Reichman, D.R.; Muller, D.A.; Hone, J.C. Grains and grain boundaries in highly crystalline monolayer molybdenum disulphide. *Nat. Mater.* **2013**, *12*, 554–561.

17. Torrisi, F.; Hasan, T.; Wu, W.; Sun, Z.; Lombardo, A.; Kulmala, T.S.; Hsieh, G.-W.; Jung, S.; Bonaccorso, F.; Paul, P.J.; *et al.* Inkjet-printed graphene electronics. *ACS Nano* **2012**, *6*, 2992–3006.

18. Li, J.; Naiini, M.M.; Vaziri, S.; Lemme, M.C.; Östling, M. Inkjet Printing of MoS 2. *Adv. Funct. Mater.* **2014**, *24*, 6524–6531.

19. Withers, F.; Yang, H.; Britnell, L.; Rooney, A.P.; Lewis, E.; Felten, A.; Woods, C.R.; Sanchez Romaguera, V.; Georgiou, T.; Eckmann, A.; *et al.* Heterostructures produced from nanosheet-based inks. *Nano Lett.* **2014**, *14*, 3987–3992.

20. Castellanos-Gomez, A.; Buscema, M.; Molenaar, R.; Singh, V.; Janssen, L.; van der Zant, H.S.J.; Steele, G.A. Deterministic transfer of two-dimensional materials by all-dry viscoelastic stamping. *2D Mater.* **2014**, *1*, 011002.

21. Castellanos-Gomez, A.; Quereda, J.; van der Meulen, H.P.; Agraït, N.; Rubio-Bollinger, G. Spatially resolved optical absorption spectroscopy of single- and few-layer MoS2 by hyperspectral imaging. *arXiv* **2015**, *1507.00869*. Available online: http://arxiv.org/abs/1507.00869 (accessed on 3 July 2015).

22. Castellanos-Gomez, A.; Agraït, N.; Rubio-Bollinger, G. Optical identification of atomically thin dichalcogenide crystals. *Appl. Phys. Lett.* **2010**, *96*, 213116.

23. Li, H.; Lu, G.; Yin, Z.; He, Q.; Li, H.; Zhang, Q.; Zhang, H. Optical Identification of Single- and Few-Layer MoS2 Sheets. *Small* **2012**, *8*, 682–686.

24. Li, H.; Wu, J.; Huang, X.; Lu, G.; Yang, J.; Lu, X.; Xiong, Q.; Zhang, H. Rapid and reliable thickness identification of two-dimensional nanosheets using optical microscopy. *ACS Nano* **2013**, *7*, 10344–10353.

25. Dols-Perez, A.; Sisquella, X.; Fumagalli, L.; Gomila, G. Optical visualization of ultrathin mica flakes on semitransparent gold substrates. *Nanoscale Res. Lett.* **2013**, *8*, 305.

26. Jung, I.; Pelton, M.; Piner, R.; Dikin, D.A.; Stankovich, S.; Watcharotone, S.; Hausner, M.; Ruoff, R.S. Simple Approach for High-Contrast Optical Imaging and Characterization of Graphene-Based Sheets. *Nano Lett.* **2007**, *7*, 3569–3575.

27. Refractive index of Si_3N_4. Available online: www.filmetrics.com (accessed on 1 August 2015).

28. Beal, A.R.; Hughes, H.P. Kramers-Kronig analysis of the reflectivity spectra of 2H-MoS 2, 2H-MoSe 2 and 2H-MoTe 2. *J. Phys. C Solid State Phys.* **1979**, *12*, 881–890.

29. Beal, A.R.; Liang, W.Y.; Hughes, H.P. Kramers-Kronig analysis of the reflectivity spectra of 3R-WS 2 and 2H-WSe 2. *J. Phys. C Solid State Phys.* **1976**, *9*, 2449–2457.

30. Tóvári, E.; Csontos, M.; Kriváchy, T.; Fürjes, P.; Csonka, S. Characterization of SiO_2/SiNx gate insulators for graphene based nanoelectromechanical systems. *Appl. Phys. Lett.* **2014**, *105*, 123114.

31. Lee, C.; Yan, H.; Brus, L.E.; Heinz, T.F.; Hone, Ḱ.J.; Ryu, S. Anomalous Lattice Vibrations of Single-and Few-Layer MoS2. *ACS Nano* **2010**, *4*, 2695–2700.

32. Buscema, M.; Steele, G.A.; van der Zant, H.S.J.; Castellanos-Gomez, A. The effect of the substrate on the Raman and photoluminescence emission of single-layer MoS2. *Nano Res.* **2014**, *7*, 561–571.

Effect of Edge Roughness on Static Characteristics of Graphene Nanoribbon Field Effect Transistor

Yaser M. Banadaki and Ashok Srivastava

Abstract: In this paper, we present a physics-based analytical model of GNR FET, which allows for the evaluation of GNR FET performance including the effects of line-edge roughness as its practical specific non-ideality. The line-edge roughness is modeled in edge-enhanced band-to-band-tunneling and localization regimes, and then verified for various roughness amplitudes. Corresponding to these two regimes, the off-current is initially increased, then decreased; while, on the other hand, the on-current is continuously decreased by increasing the roughness amplitude.

Reprinted from *Electronics*. Cite as: Banadaki, Y.M.; Srivastava, A. Effect of Edge Roughness on Static Characteristics of Graphene Nanoribbon Field Effect Transistor. *Electronics* **2016**, *5*, 11.

1. Introduction

The exponential trend in scaling MOSFET has satisfied Moore's law for decades, leading to denser chips with more functionality, a lower price per chip, faster switching and lower power consumption. However, the International Technology Roadmap for Semiconductors (ITRS) [1] has predicted the demise of silicon-CMOS technology due to the fundamental limits of CMOS FETs, and has put forward alternate channel materials to silicon such as carbon nanotubes and newly discovered graphene [2]. Graphene is one atomic layer of carbon sheet in a honeycomb lattice, which can outperform state-of-the-art silicon in many applications [2,3] due to its excellent electronic properties. The carrier transport in graphene is similar to the transport of massless particles since 2D electron gas in graphene [4] provides both high carrier velocity and high carrier concentration, resulting in large carrier mobility and, consequently, faster switching capability [5]. Despite the fascinating properties of graphene, it is a semimetal with an overlapping zero bandgap and is not satisfactory for digital applications [6]. The quantum confinement of graphene sheet in the form of one-dimensional (1D) strips with a very narrow width known as graphene nanoribbon (GNR) provides the energy gap of several hundred meV required for FET operations in digital applications [7,8]. As the fabrication technology of GNR FETs in this structure is still in an early stage, performance evaluation of futuristic graphene-based circuits requires a SPICE-compatible model. The state-of-the-art patterning technique is far from achieving atomic-scale precision, and

GNRs with perfect smooth edges cannot be fabricated; such that, line-edge roughness may play an important role in the production of narrow GNRs for channel material of GNR FETs. The edge roughness enhances the edge scattering and generates edge states in the bandgap, which can significantly enhance the leakage current and reduce the drive current. Thus, modeling edge roughness is very useful to examine the effect of process variation on circuit performance of GNR FET. The dispersion of the electrical characteristics due to random edge defects in realistic nanoribbons can be precisely evaluated by statistical analysis at the device-level, based on the atomistic quantum transport simulations of large ensembles of randomly-generated GNRs [9]. However, the device-level analysis requires extensive computational time; therefore, the same statistical approach cannot be used for circuit-level simulations.

The ideal smooth-edge GNR FETs give an estimation of the upper bound performance, however, the line-edge roughness needs to be considered for practical GNR FETs which deteriorate their performance. A semi-analytical model for GNR FET with perfectly smooth edges was developed in [10], which involved numerical integrations; thereby, it cannot be used for circuit simulation. In [11], a circuit model was implemented based on lookup table techniques to use the results of device-level quantum transport simulations for circuit simulations. However, with a single change in a design parameter, the intensive device-level simulations need to be repeated to rebuild the model accordingly, which makes it inappropriate for evaluating the optimized design parameters of GNR FET circuits. A SPICE-compatible model of GNR FET including the edge roughness is presented in [12]. In this model, the effect of rough edges on the increasing leakage current of GNR (N,0) is considered by effective bandgap due to the bandgap of GNR (N-1,0), while the real GNRs with rough edges are composed of all neighboring GNRs. Also, the effect of rough edges on decreasing on-current was modeled by a fitting equation regardless of the physical scattering mechanisms in a GNR channel. In addition, this model cannot capture the effect of large line-edge roughness on localization of carriers, which tends to reduce both off- and on-currents of GNR FETs. It has been shown both experimentally [13] and theoretically [14] that strong localization can appear in single layer GNRs for high line-edge roughness.

In this work, we develop a physics-based analytical model for circuit simulation of GNR FETs. The band-to-band-tunneling (BTBT) from drain to channel regions can be important for small bandgap GNRs, which has been modeled by a current source in parallel with another current source for the thermionic current. The line-edge roughness in GNRs is modeled using an exponential autocorrelation function. The model incorporates the effect of edge states on the initial increase of BTBT and high edge scattering of carriers in a localization regime. The device-level simulation is performed to evaluate the static performance of GNR FETs in edge-enhanced BTBT and localization regimes. The results of our analytical model are verified by numerical

results from accurate quantum transport simulations based on non-equilibrium Green function (NEGF) formalism. The organization of this paper is as follows: In Section 2, we describe the structure of GNR FETs in all-graphene architecture; Section 3 presents the model equations and equivalent circuit model of GNR FET for the circuit simulation in SPICE. Model validation is described in Sections 4 and 5; finally, the last section draws summarizing conclusions.

2. GNR FET Structure

Figure 1 shows the 3D view of a GNR FET, where the ribbon of the armchair chirality GNR is the channel material in a MOSFET-like structure. This structure is expected to demonstrate a higher I_{ON}/I_{OFF} ratio, outperforming the GNR FET with Schottky barriers in logic applications [15]. The intrinsic GNR channel (L_{CH}) has the same length underneath as the gate contact (L_G), while its width (W_G) is extended equally from each side of the GNR channel. The width of the intrinsic GNR is $W_{GNR} = (N+1)\sqrt{3}a_{cc}/2$, where a_{cc} is the carbon-carbon bonding length and N is the number of dimer lines for the armchair GNR (N,0). The symmetric regions of the GNR channel between the gate and contacts with the length of L_{RES} are doped with the n-type dopants concentration of f_{dop} per carbon atom as the source and drain reservoirs. The metallic source and drain electrodes are omitted in the model as the two doped regions of GNRs can be directly connected to the GNR interconnect in all-graphene architecture [16], avoiding the series resistance of metal-to-graphene contacts. Aluminum nitride (AlN) insulator layers with a relative dielectric permittivity of $\kappa = 9$ are assumed. The large-scale and cost-effective production of thin AlN dielectric layers with good reproducibility and uniformity [17,18] can result in small equivalent oxide thickness (EOT) while reducing phonon scattering in epitaxial graphene, enabling near ballistic carrier transport in a short channel GNR FET [19].

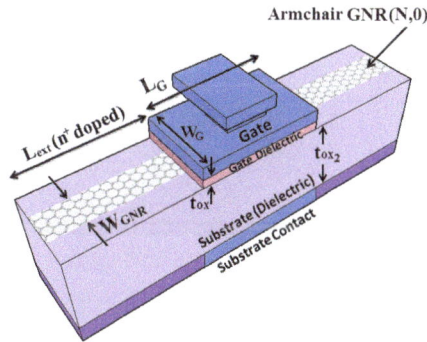

Figure 1. 3D view of a graphene nanoribbon field effect transistor (GNR FET) with armchair GNR (N,0) as channel material together with the device geometries.

3. GNR FET Model

Figure 2a shows the energy band diagram and the corresponding components in the equivalent circuit model of the GNR FET. The model contains four capacitors, $C_{G,CH}$, $C_{S,CH}$, $C_{B,CH}$, $C_{D,CH}$, to account for the electrostatic coupling of the channel to the potentials at gate, source, substrate and drain electrodes, respectively. Two current sources model the thermal current flowing through the channel and band-to-band-tunneling (BTBT) current from drain to channel regions. These current sources account for the DC behavior while a voltage-controlled voltage source (V_{CH}) in the model accounts for the charging and discharging the GNR channel, thereby the transient AC behavior of GNR FETs.

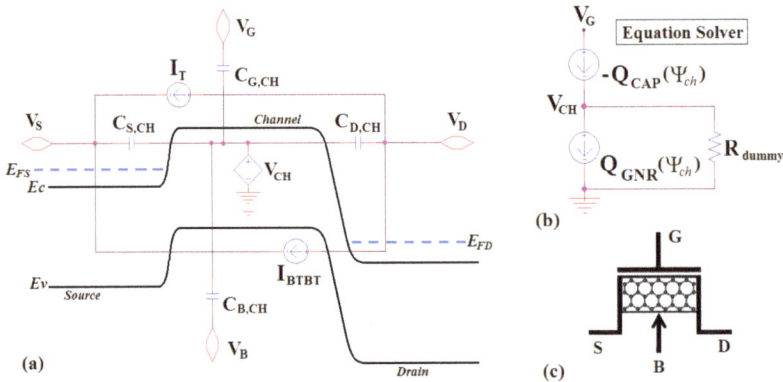

Figure 2. (a) Energy band diagram and the corresponding components in the equivalent circuit model of a GNR FET; (b) Series implementation of two voltage-controlled current sources in SPICE to obtain the channel surface potential; (c) GNR FET circuit symbol.

3.1. Computing GNR Subbands

The minimum energy (E_b) and effective mass (m_b^*) of subbands for different armchair GNRs need to be obtained for transport equations. Tight-binding (TB) calculation can be employed based on the nearest neighbor orthogonal p_z orbitals as basis functions equal to the number of atoms in a desired unit cell in the transverse direction [20]. The nearest neighbor hopping energy for the atoms not located at the edge is $t = -2.7$ eV, while it is assumed 1.12 t for the pairs of carbon atoms along the edges of the GNR, to take into account the edge bond relaxation due to the lattice termination and occupation of hydrogen atoms at the edges [21]. Further detail about the TB calculation and the effective mass extraction using non-parabolic effective mass model can be followed in [22].

The bandgap is increased by decreasing the GNR width due to the quantum confinement of carriers in one dimension while it increases the effective mass due to the degradation of the band linearity near the Dirac point. For narrow armchair GNRs, removing or adding one edge atom along the nanoribbon can significantly change the bandgap energy and effective mass of armchair GNRs. Two thirds of armchair GNRs, GNR (3p+1,0) and GNR (3p,0) have proper bandgaps, large enough for replacing silicon as a channel material. The third subclass, GNR (3p+2,0), has a very small bandgap. Although, the bandgap of GNR (3p+1,0) is slightly larger than GNR (3p,0), both GNR families follow the same trends of decreasing bandgaps by increasing the GNR widths. Thus, we have omitted one semiconducting family in our studies to prevent the confusion resulted from chirality dependence of bandgap. Also, we have omitted the incorporation of upper subbands in the model as the first three subbands can accurately describe the carrier transport of GNR FETs, considering the bias voltages and the position of minimum conduction subbands in energy.

3.2. Finding Channel Surface Potential

The key parameter for evaluating GNR FET current is to find the variation of channel surface potential (Ψ_{ch}) in response to the variation of gate and source/drain voltages. For a channel material with infinite density of states (DOS), the channel potential can be obtained by geometrical transient capacitance network, such that the channel potential is dominantly controlled by the voltage bias at electrodes, especially gate voltage for long channel devices. For a semiconducting channel with a finite DOS, the channel surface potential changes with the gate bias at a rate $\Delta\Psi_{ch}/\Delta V_G < 1$. This promotes the device operation close to quantum capacitance limit (QCL) [23]. GNR has very small DOS due to the atomically thin channel in vertical direction and quantum mechanical confinement in the transverse direction. Thus, evaluating the channel surface potential is very important, which can be modeled using the charge conservation equations by series implementation of two voltage-controlled current sources in SPICE simulation as shown in Figure 2b [24]. This forces the two currents to be equal in magnitude. In other words, the charge induced by the capacitances networks connected to the contacts (Q_{CAP}) has to be equal to the charge capacity of the GNR channel limited by its DOS. This implementation results in the automatic calculation of the channel voltage (V_{CH}) and the corresponding channel surface potential (Ψ_{ch}).

3.2.1. Computing Channel Charge

In n-type GNR FETs, the hole concentration in the channel is suppressed due to the n-doped drain and source reservoirs, thereby, the electron density of bth subband in GNR channel (n_b) can be obtained considering DOS of GNR ($D_b(E)$) from the carrier density relationship as follows [10]:

$$n_b = \int_0^\infty f(E)\, D_b(E)\, dE \tag{1}$$

$$D_b(E) = \frac{2(E_b + E)}{\pi \hbar} \sqrt{\frac{m_b^*}{E_b E\,(E + 2E_b)}} \tag{2}$$

where $\hbar$ is the reduced Planck constant and $f(E)$ is the Fermi-Dirac distribution function. In order to solve the integral in Equation (1) and obtain an analytical equation, the Fermi-Dirac distribution function can be approximated by Boltzmann distribution $f(E) = \exp((E_F - E)/kT)$ when the Fermi level is more than 3 kT away from the subband energy [10]. As the bandgap of GNRs can be very small, the assumption is inaccurate in many bias conditions of the GNR FETs. In order to accurately evaluate the electron density in GNRs, $f(E)$ needs to be approximated depending on the relative location of Fermi levels at the terminals to conduction band energy ($E_{FC}^b = E_{F,i} - E_C^b$). Equation (3) provides a smooth transition between two approximations: (1) exponential carrier concentration ($n_b^{\exp}$) when the Fermi level is near the conduction band (high DOS, $E_{FC} \cong 0$); and (2) step carrier concentration (n_b^{step}) when the Fermi level is 3 kT away from the subband energy ($E_{FC} > 3$ KT) [12].

$$n_b(E_{FC}) = w \times n_b^{\exp}(E_{FC}^b) + (1 - w) \times n_b^{step}(E_{FC}^b) \tag{3}$$

$$n_b^{\exp}(E_{FC}) = \frac{\sqrt{m_b^*\, \alpha^3}\,(1 + 2E_b/\alpha)}{2\pi \hbar E_b}\exp(E_{FC}/\alpha) \tag{4}$$

$$n_b^{step}(E_{FC}) = (2\sqrt{m_b^*/\pi \hbar}) \times \sqrt{\max((E_{FC}\,(E_{FC} + 2E_b)/E_b), 0)} \tag{5}$$

where $w = 1/[1 + \exp(3(E_{FC} - kT)/kT)]$ is the relative weight of the two approximations and $\alpha = 3kT/\ln[f(E_{FC}) \times (1 + \exp((3kT - E_{FC})/kT))]$. Thus, the total electron density in the GNR channel can be obtained by summation over the carrier density of subbands as follows:

$$Q_{GNR}^n = -\frac{q\,L_{CH}}{2}\sum_b [n_b(E - (E_{FS} - E_C^b)) + n_b(E - (E_{FD} - E_C^b))] \tag{6}$$

where q is an electron charge; $E_C^b = E_b - \Psi_{ch}$ is the conduction band energy; $E_{FS} = E_F - qV_S$ and $E_{FD} = E_F - qV_D$ are the Fermi levels corresponding to the voltages at source and drain electrodes, respectively. The equilibrium Fermi level of doped reservoirs (E_F) sets both E_{FS} and E_{FD} above the conduction band of source and drain regions. We only consider n-type GNR FET throughout this paper, though similar analysis can be applied to p-type GNR FET. In the same scenario, the energy difference between the valence band and the Fermi-level at the terminals, $E_{VF}^b = E_V^b - E_{F,i}$ and the polarity of the terminal voltage needs only to be changed due to the symmetry of the conduction and valence subbands in GNRs.

3.2.2. Computing Transient Capacitance Charge

The induced charge by capacitance network can be calculated as follows [25]:

$$Q_{CAP}^n = \sum_{i=G,B} C_{i,CH} \times (V_i - V_{FB,i} - q\Psi_{ch}) \tag{7}$$

where V_{FB} is the flat-band voltage due to the work function difference between metal and graphene. The geometrical capacitances, $C_{G,CH}$ and $C_{B,CH}$ model the electrostatic coupling between the GNR channel and two electrodes of the gate and the substrate. As the gate width is larger than the GNR width and the oxide thickness, these capacitances can be modeled by the analytical equation of micro-strip lines as follows [26]:

$$C_{i,CH} = \beta L_G \frac{5.55 \times 10^{-11} \, \varepsilon_r}{\ln\left[5.98 \, t_{ox}/(0.8 W_{GNR} + t_{GNR})\right]} \tag{8}$$

where $t_{GNR} \cong 0$ is the GNR thickness, t_{ox} is the dielectric thickness, and $\beta = (1 + 1.5 \, t_{ox}/W_G)^{-1}$ is a correction term for a case when the gate width is not much larger than the oxide thickness [27].

3.3. Current Modeling

Given the surface potential (Ψ_{ch}), both the DC and AC behaviors of GNR FETs can be incorporated in the current calculation. Figure 3a shows the energy position resolved local density of states (LDOS) of a typical GNR FET, which is numerically simulated by NEGF formalism [22]. The bandgap with quite low local density of states, potential barrier of the channel together with the source and drain regions, can be easily identified. The quantum interference pattern due to incident and reflected electron waves in the generated quantum well in the valence band of the channel is also apparent. As can be seen from the LDOS of GNR FET, the carrier transport can be associated with three mechanisms: (1) thermionic current (I_T) for electrons with energies above the channel potential barrier; (2) direct source-to-drain tunneling current through channel potential barrier and (3) band-to-band tunneling, (I_{BTBT}) between hole states in the source and electron states in the drain. While the study on the direct source-to-drain tunneling [28] shows that its contribution in leakage current can be dominant well below 10 nm channel length, the band-to-band tunneling can be comparable to thermionic current at subthreshold regions of GNR

FETs, depending on the bias condition and the bandgap of GNR channel. We will show that incorporating the two mechanisms of the thermionic current and the band-to-band tunneling current using two current sources in the equivalent model of GNR FET can result in sufficiently accurate I-V characteristics of GNR FET for the channel length larger than 10 nm.

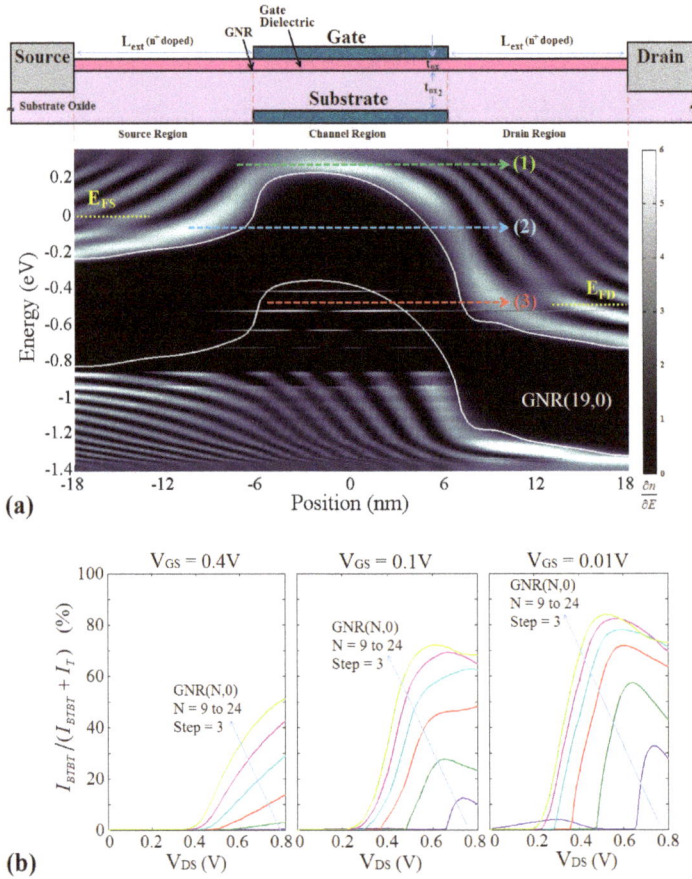

Figure 3. (a) Energy-position-resolved local density of states of a typical GNR FET simulated with NEGF formalism [22], which shows three possible regions for carrier transport in GNR FETs; (b) Contribution of the BTBT current in total current of GNR FETs as a function of drain voltages for different gate voltages and various GNR widths.

3.3.1. Computing Thermionic Current

The thermionic current can be computed using the Landauer-Buttiker formalism [29], in which the probability of the electrons being injected onto the conduction band from the source side is subtracted from the probability of the electrons being injected onto the conduction band from the drain side as follows:

$$I_T = \frac{2q}{h} \sum_b \int_0^\infty T(E)[f(E - (E_{FS} - E_C^b)) - f(E - (E_{FD} - E_C^b))] \, dE \qquad (9)$$

where h is Planck constant. The integral in the above expression can be evaluated analytically considering the Fermi-Dirac integral of order 0, which results in the current at the ballistic limits as follows:

$$I_T = \frac{2q}{h} k_B T \sum_b [\ln(1 + \exp((E_{FS} - E_C^b)/kT)) - \ln(1 + \exp((E_{FD} - E_C^b)/kT))] \qquad (10)$$

3.3.2. Evaluating Band-to-Band-Tunneling Current and Charge

While the thermionic current strongly dominates the carriers transport at very high drain voltages, the band-to-band-tunneling (BTBT) can be comparable to the thermionic current in subthreshold region (small V_{GS}) and sufficiently high V_{DS} as shown in Figure 3b. As the wider GNRs can have very small bandgaps and high effective masses, the BTBT can significantly increase leakage current of GNR FETs. Band-to-band-tunneling occurs when the confined states in the valence band of the channel align with the occupied states in the drain. This can happen when the conduction band at the drain side is below the valance band at channel side ($V_{CH,D} > 2E_b$) and empty states were sufficiently available at the drain side for the electrons that tunnel from the channel region. Assuming ballistic transport for the tunneling process, the BTBT current can be approximated by the maximum possible tunneling current integrating from the conduction band at the drain side up to the valance band at the source side times the BTBT tunneling probability as follows [24]:

$$I_{BTBT} = \frac{2q}{h} k_B T \sum_b [T_{BTBT} \ln(\frac{1 + \exp((qV_{CH,D} - E_b - E_F)/k_BT)}{1 + \exp((E_b - E_F)/k_BT)}) \times \frac{\max(qV_{CH,D} - 2E_b, 0)}{qV_{CH,D} - 2E_b}] \qquad (11)$$

where E_F is the Fermi level of the doped regions at the drain side of GNR FET. T_{BTBT} is the Wentzel–Kramers–Brillouin-like transmission coefficient that can be calculated following the work of Kane [30] as follows:

$$T_{BTBT} \approx \frac{\pi^2}{9} \exp\left(-\frac{\pi m_b^{*(1/2)}(\eta_b 2E_b)^{3/2}}{2^{3/2} q \hbar F}\right) \qquad (12)$$

In Equation (12), $F = (V_{CH,D} + (E_F - \Psi_{ch})/q)/l_{relax}$ is the electrical field triggering the tunneling process through the junction at the drain side of the GNR channel when the potential across the drain-channel junction is $V_{CH,D}$.

169

η_b models the bandgap narrowing effect under a high electrical field [31], which is set to 0.5 corresponding to that of carbon nanotube in [24]. Basically, a graphene nanoribbon channel is an unfolded lattice structure of carbon nanotube with the same one-dimensional channel. As such, it is an appropriate assumption that both have the same bandgap narrowing effects under a high electrical field. l_{relax} is the relaxation length of potential drop, which has been extracted with the procedure explained in Section 3.5.

Although the thermionic emission of holes into the channel is negligible for n-type GNR FET, the band-to-band-tunneling significantly increases the accumulation of holes in the channel, especially at a high V_{DS}. As such, both the charges of the GNR channel (Equation (6)) and the charge induced by the capacitance network (Equation (7)) need to be corrected corresponding to the tunneling coefficient (Tr) as follows:

$$Q_{GNR} = Q_{GNR}^n + Tr. \, p_b(E_V^b - E_{FD}) \tag{13}$$

$$Q_{CAP} = Q_{CAP}^n + Tr. \, \beta . \, C_{i,CH} \times ((E_V^b - E_{FD})/q) \tag{14}$$

$$Tr = 1 - [1 + \exp(\frac{qV_{CH,D} - \eta_b \, E_b - E_F}{\delta})]^{-1} \tag{15}$$

where $E_V^b = -E_b - \Psi_{ch}$ is the valence band energy and $\beta = l_{relax}/L_{CH}$. δ is a fitting parameter, which controls how fast Tr increases by increasing the band bending between the channel and drain, corresponding to the value of $V_{CH,D}$. The transient capacitance network in Figure 2b can be computed by introducing the intrinsic capacitors as the derivatives of the channel charge with respect to drain and source voltages, $C_{S,CH} = \partial Q_{GNR}/\partial V_S$ and $C_{D,CH} = \partial Q_{GNR}/\partial V_D$, which can be implemented in SPICE by voltage-controlled capacitors.

3.4. Non-Ballistic Transport

The experimental results show that the carrier mobility in graphene can be as high as 200,000 cm^2/Vs [32]. However, different scattering mechanisms due to the intrinsic acoustic phonons (AP) and optical phonons (OP) of graphene [33], the interaction of carriers with optical phonons of the substrate [34] and the line-edge roughness (LER) in narrow GNRs [35] can limit its mobility to orders of magnitude lower values. While the transmission coefficient of the carriers can be assumed to be at unity for developing the compact model based on ballistic assumptions [25], these scattering mechanisms must be incorporated in the model as they have been shown to play an important role in the performance of GNR FETs [36]. The transmission of carriers is decreased by scattering in the channel, which can cause a carrier to return to the source region under a low drain bias. The back-scattering of carriers to the source continues under a high drain bias within an approximate critical length of $l = (\hbar \omega_{op}/qV_D)L_{CH}$ near the source end of the channel while the scattered carriers in the channel will be absorbed by the drain beyond this critical distance without having a direct effect on the source-drain current [36]. In the absence of scattering mechanisms, the carrier transport is in the ballistic regime and the conductance is independent of the device length. However, carrier scattering in the channel results in the diffusive transport of carriers, thereby making the conductance inversely proportional to the channel length. The channel transmission coefficient provides a simple way to describe the device in the presence of scattering mechanisms [37] as follows:

$$
T = \begin{cases} \lambda^{eff}/(\lambda^{eff} + L_{CH}) & if \ qV_D < \hbar\omega_{op} \\ \dfrac{\lambda^{eff}}{(\lambda^{eff} + (\hbar\omega_{op}/qV_D)L_{CH})} & if \ qV_D > \hbar\omega_{op} \end{cases} \tag{16}
$$

where $L_{CH}(=L_G)$ is the channel length and $\hbar\omega_{op} \cong 0.18\,\mathrm{eV}$ [38] is the OP energy and λ^{eff} is the effective mean free path (MFP) of GNR, which can be obtained using Mattheissen's rule as follows [39],

$$
\frac{1}{\lambda^{eff}} = \frac{1}{\lambda^{sub}} + \frac{1}{\lambda^{ac}} + \frac{1}{\lambda^{LER}} \tag{17}
$$

where λ^{sub} is the substrate-limited MFP which is reported close to 100 nm and 300 nm for the GNR on top of SiO$_2$ and h-BN dielectrics [39], respectively. λ^{ac} is the acoustic phonons-limited MFP [34] as follows:

$$
\lambda_{ap} = \frac{h^2 \rho_s v_s^2 v_f^2 W_{GNR}}{\pi^2 D_A^2 k_B T} \tag{18}
$$

where $v_s = 2.1 \times 10^4$ m/s is the sound velocity in graphene, $D_A = 17 \pm 1\,\mathrm{eV}$ is the acoustic deformation potential, and $\rho_s = 6.5 \times 10^{-7}$ kg/m^2 is the 2D mass density of graphene. λ^{LER} is the line-edge roughness (LER) scattering-limited MFP which can be as small as a few tens of nanometers [40], which can exhibit the dominant scattering mechanism in narrow GNRs as it has been predicted in both experimental data [35] and theoretical studies [41,42]. The edge disorder has been analytically modeled by Anderson distribution in [43] assuming that atoms at the edges are randomly removed with uniform probability, such that the correlation between edge disorders has been neglected. As line-edge roughness is a statistical phenomenon, a more realistic model needs an autocorrelation function as have been already used for modeling Si/SiO$_2$ interface roughness [44] and the line-edge roughness in GNRs [45,46]. In this paper, we consider an exponential spatial autocorrelation function as follows:

$$
R(x) = \Delta W^2 \exp(-\frac{|x|}{\Delta L}) \tag{19}
$$

where ΔW is the root mean square of the width fluctuation amplitude or roughness amplitude and ΔL is the roughness correlation. Increasing ΔW or decreasing ΔL initially makes the carrier transport more diffusive. The line-edge roughness causes fluctuations in the edge potential and bandgap modulation due to the localized edge states as shown in Figure 4a. The decrease in the conductivity of narrow GNRs as a result of line-edge roughness can be incorporated by introducing the effective bandgap in the transport calculation corresponding to the LER scattering-limited MFP as follows [46]:

$$\frac{1}{\lambda^{LER}} = \sum_b \frac{1}{\lambda_b^{LER}} = \sum_b \frac{1}{A\{(E - E^b) + B\,(E - E^b)^2\}} \qquad (20)$$

$$\Delta E_g = 2(\frac{2k_B T\,L_{CH}}{A\,B})^{1/3} \qquad (21)$$

$$A = (\frac{W_{GNR}}{\Delta W})^2 \frac{\hbar^2}{8\,m_b^*\,\Delta L\,E_b^2} \quad , \quad B = \frac{8\,m_b^*\,\Delta L^2}{\hbar^2} \qquad (22)$$

Using Equation (21), the effective subband energy of GNR $(N,0)$, $E_{b,eff}^N$ can be modeled as follows:

$$E_{b,eff}^N = E_b^N + \gamma\,(\Delta E_g/2)\,(E_b^N/E_1^N) \qquad (23)$$

where γ is a fitting parameter, which weights the increase in the subbands energy corresponding to the calculated effective bandgap due to LER scattering mechanism. While the increase in subbands' energy of GNRs in Equation (23) can model the decrease in carrier transport, the line-edge roughness can also contribute to the formation of some localized states in the band gap, which enhances the band-to-band-tunneling of carriers, leading to the initial increase in off-current of GNR FETs at small roughness amplitudes [47]. As GNR channel has a large number of carbon rings and is long enough to provide sufficient averaging, the increase in BTBT current of GNR $(N,0)$ due to the variation in its width can be analytically modeled by summation over the BTBT of its neighbor GNRs (conceptual example of $N = 9$, 10 and 11 is shown in Figure 4b) as follows:

$$I_{BTBT,rough}^N = I_{BTBT}^N + \alpha_r \frac{\sum_{i=1}^m [(I_{BTBT}^{N-i} + I_{BTBT}^{N+i})(\frac{\Delta W}{W_{GNR}} - \frac{i}{N-1})^{-1}]}{\sum_{i=1}^m [\frac{\Delta W}{W_{GNR}} - \frac{i}{N-1}]^{-1}} \exp(-((\frac{\Delta W}{W_{GNR}} - \frac{\Delta W_c}{W_{GNR}})/\beta_r)^2 + \frac{N}{2}) \qquad (24)$$

where α_r and β_r are fitting parameters, which model the dominant effects between carriers localization and carriers tunneling corresponding to the amount of roughness amplitude. Equation (24) models the increase in BTBT due to the edge states in the bandgap by summing and normalizing the neighboring GNRs. The integer m has the value of the ratio $\Delta W/\sqrt{3}a_{cc}$. $\Delta W_c = 0.04$ is the critical width fluctuation amplitude.

The BTBT through edge states in the bandgap leads to the increase in net transport of carriers from source to drain; consequently, the increase in the conduction of GNR FET for $\Delta W_c > \Delta W$. By increasing the roughness amplitude larger than ΔW_c, however, the tunneling of carriers occurs mostly between the localized states without a net transport of carriers from source to drain regions. The exponential decrease in device conductivity in the localization regime has been modeled with both the increase in the subband energies in Equation (23) and in the exponential term in Equation (24).

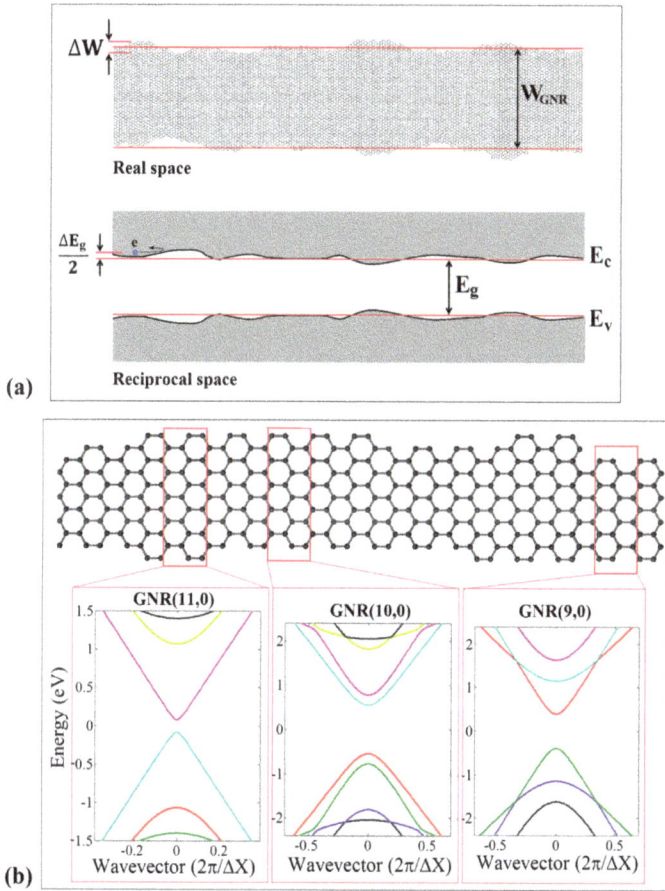

(a)

(b)

Figure 4. (a) Line-edge roughness scattering of a graphene nanoribbon in real space and reciprocal space. Note: The variation of the GNR width in real space causes the generation of edge states and potential variation of its bandgap in reciprocal space; (b) Removing or adding carbon atoms at the edge (e.g., GNR (10,0)) can significantly change the local bandgap of GNR and the corresponding change in the local states can contribute in the enhancement of the BTBT current at small roughness amplitude.

3.5. Extracting Fitting Parameters

The fitting parameters in our developed GNR FET model have been extracted by matching its transfer characteristics with numerical data in regards to the following procedure:

(1) Obtain the numerical data from NEGF simulation for bias conditions and the device geometries related to BTBT phenomena.
(2) Obtain the analytical results using the developed model for the same bias conditions and device geometries for a given fitting parameter.
(3) Change a fitting parameter according to its broad range to determine the best value in which the root mean square error (RMSE) between the numerical and analytical data is minimized.

The above procedure is relatively quick since numerical data, which can be computationally intensive, needs to be calculated only once for a device setting. Then, the analytical model provides prompt results that need to be repeated to search for the best value of the fitting parameters. In addition, all the fitting parameters in this work are used for modeling BTBT from drain to channel with only dominant effects on the total current of FET structure for narrow-bandgap GNRs and under bias conditions of high V_{DS} and low V_{GS} (see Figure 3b). As such, the fitting procedure can be limited to those bias conditions (subthreshold region) and wider GNR channels, which corresponds to higher RMSE between numerical and analytical results.

The dependence of fitting parameters on the GNR width can be eliminated by including another fitting dimension for semiconducting GNR chiralities such as GNR(N,0) shown in Figure 5. The figure shows an example of a two-dimensional fitting procedure, in which all the analytical results *versus* gate voltage and GNR widths are repeated for different values of a fitting parameter, e.g., δ, approaching the best fit with smallest average RMSE in two dimensions.

The fitting procedure is repeated until the RMSE reaches $\pm10\%$ and $\pm20\%$ of the numerical data for fitting parameters related to ideal GNRs and non-ideal GNRs, respectively. The maximum error values are our assigned acceptable errors for obtaining general fitting parameters independent of device dimensions, which correspond to the GNR FETs with 16 nm channel length of GNR(24,0) at $V_{DS} = 0.8$ V and $V_{GS} = 0.05$ (significant BTBT phenomena). Two fitting parameters, l_{relax} and δ in Equation (15) are associated with ballistic transport in ideal GNRs while γ, α_r and β_r in Equations (23) and (24) are related to the non-ballistic transport modeling of carriers in rough-edge GNRs. Table 1 shows the searching ranges and obtained values of fitting parameters, as well as the maximum errors with regard to the numerical data. For example, by altering l_{relax} in a range from 10 nm to 100 nm, the analytical results match the numerical data in the subthreshold region within the acceptable RMSE at $l_{relax} = 40$ nm. A similar method was used to obtain l_{relax} of Stanford CNT

FET model [24]. As the fitting parameters in our GNR FET model only need to be obtained just once and their values are valid for all the device geometries and line-edge roughness in this study, as indicated in Table 2.

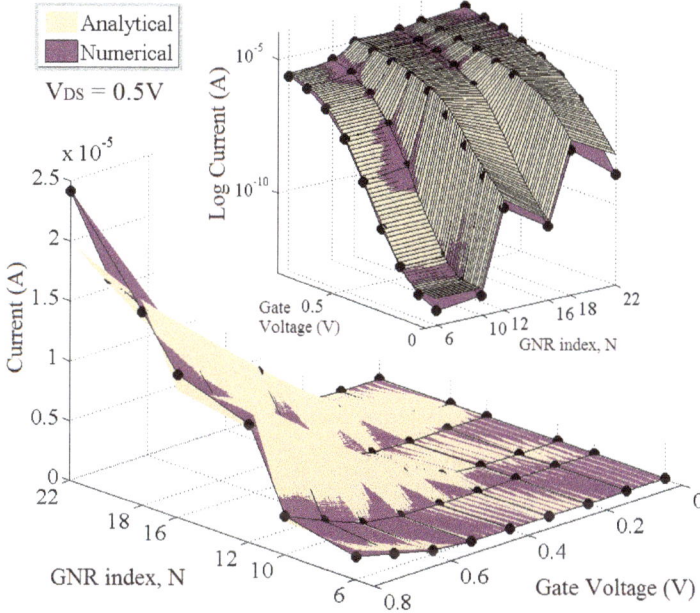

Figure 5. An example of two-dimensional fitting of continuous analytical results to the variety of discrete numerical data from the NEGF simulation, which has been used to obtain the fitting parameter $\delta = 0.05$.

Table 1. Obtained values of fitting parameters, the searching ranges and step, along with the corresponding errors.

Fitting Parameter	Obtained Value	Searching Range	Searching Step	% RMSE
l_{relax}	40 nm	10 nm–100 nm	5 nm	<10%
δ	0.05	0.01–0.1	0.025	
γ	2.9	0–3	0.5, then 0.1	
α_r	6×10^{-6}	From 0	Different (Nonlinear)	<20%
β_r	0.0145	From 0	0.01, 0.001 and 0.0001	

Table 2. Scope of the device dimension and line-edge roughness in GNR FET Model.

Device Dimension	**GNR Channel Length (L_{CH})**	$10\,\text{nm} - 45\,\text{nm}$
	GNR Channel Width (W_{GNR}, GNR(N,0))	$\sim 0.87\,\text{nm}\,(N = 6) - \sim 3.18\,\text{nm}\,(N = 25)$
	Oxide Thickness (t_{OX})	$0.5\,\text{nm} - 2.0\,\text{nm}$
Line-edge Roughness	**Roughness Amplitude (ΔW)**	$0.0 - 0.2$
	Roughness Correlation (ΔL)	$3\,\text{nm} - 10\,\text{nm}$

4. Model Validation

The dimension and bias conditions of most experimental works are much larger than the target ranges of GNR FET for emerging technology. For example, the GNR width (W_{GNR}) and gate voltage, (V_G) are mostly studied in the range of 10–150 nm and 10–50 V, respectively [48–50]. Therefore, we have validated the accuracy of our developed analytical model against the device-level atomistic numerical simulation based on NEGF formalism as described for ideal GNR FET in [22] and for GNR FET with line-edge roughness in [46]. Figure 6a,b shows the I_{DS}–V_{GS} characteristic of GNR FETs with three different GNR indices of $N = 6$, 12 and 18 for drain voltages of 0.1 V and 0.5 V, respectively. It is shown that our analytical model agrees well with numerical simulations. The device dimensions in the simulation are, $L_G = 16$ nm, $W_G = 2$ nm and $t_{ox} = 1$ nm. For the comparison with device-level simulation, the Fermi level due to the work function difference between metallic gate and graphene is set to $E_F = 0$, while its nominal value is kept equal to $E_F = 0.4$ eV for the other simulations. It can be observed that increasing GNR width or drain voltage leads to the increase in off-current as both can provide more subbands to incorporate in BTBT of carriers from drain to channel. Figure 6c shows the I_{DS}–V_{DS} characteristics of GNR FET at different gate voltages, demonstrating that the results of the analytical model in this work are in correlation with those of numerical simulations.

Figure 6d shows the effect of line-edge roughness on the off-state and on-state currents of GNR FET with the channel of GNR (15,0). It can be seen that the analytical model agrees very well with the numerical simulations, which can be obtained at the expense of long computational time by statistical averaging many GNR samples with the same roughness parameters [14]. Furthermore, the GNR FET with perfectly smooth edges can have more than three orders of magnitude difference between on- and off-currents. The off-current increases by increasing the roughness amplitude, as the formation of some localized states in the band gap significantly increases the band-to-band-tunneling (BTBT) of carriers. Conversely, the on-current decreases by increasing the roughness amplitude due to the increase in carrier scattering associated with line-edge roughness (LER). By increasing the line-edge roughness, the carrier transport becomes more diffusive and the trend continues up to the critical roughness amplitude (ΔW_c). For LER amplitude larger than ΔW_c, the BTBT of carriers through localized edge states is dominated by the localization of carriers in those states; while, the back-scattering of carriers occur due to the localized potential barriers of edge disorders. The critical roughness amplitude corresponds with the maximum off-current and the minimum value of I_{ON}/I_{OFF} ratio. By increasing the roughness amplitude, the off-current is decreased as the carriers transport takes place by tunneling between localized edge states, leading to the reduction in the effective net transport of carriers from drain to channel. It is worth mentioning that the uniform edge roughness has been assumed in some works [51,52] by defining roughness percentage, p, which neglects the correlation between edge disorders. This assumption underestimates the effect of line-edge roughness; thereby, only the initial increase in off-current due to BTBT phenomena has been exhibited while larger line-edge roughness reduces off-current similar to on-state current.

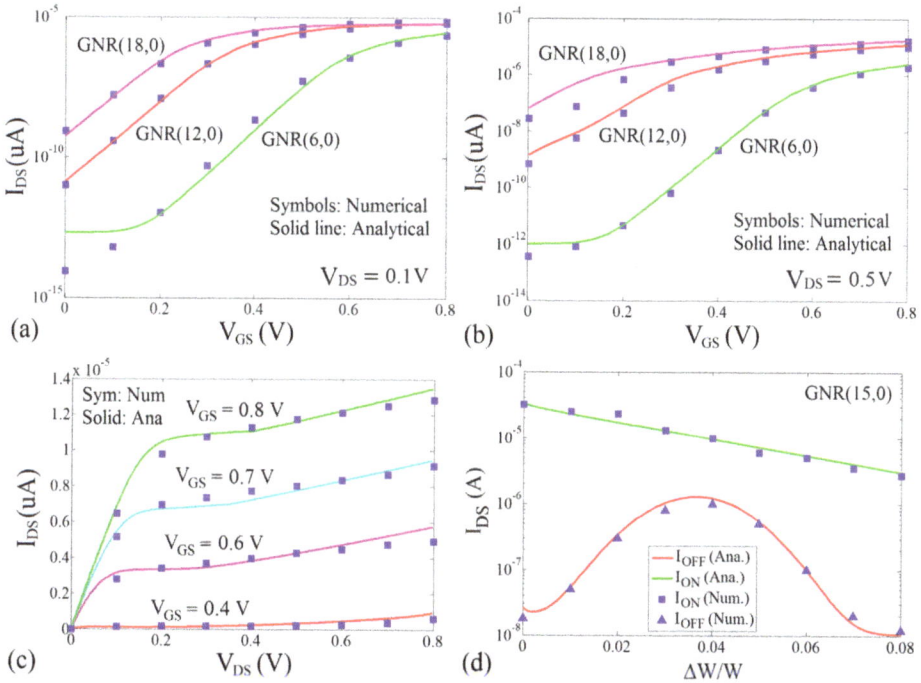

Figure 6. I_{DS}–V_{GS} characteristic of GNR FETs with three different GNR indices of N = 6, 12 and 18 for (**a**) V_{DS} = 0.1 V and (**b**) V_{DS} = 0.5 V, respectively; (**c**) I_{DS}–V_{DS} characteristics of GNR FET at different gate voltages. Note: The numerical results are shown with dotted symbols and the analytical results with solid lines; (**d**) I_D (ON) and I_D (OFF) *vs.* $\Delta W/W$ for L_{CH} = 16 nm.

Figure 7 shows the I_{DS}–V_{GS} characteristic of three GNR FETs with the channels of GNR (9,0), GNR (15,0) and GNR (24,0) at V_{DS} = 0.25 V for various LER amplitudes and constant roughness correlation of ΔL = 10 nm. It can be seen that the narrower ribbon, *i.e.*, GNR (9,0), has lower off-current than GNR (15,0) and GNR (24,0) for all LER amplitudes. At V_{DS} = 0.25 V, the BTBT of GNR FET with larger bandgap, *i.e.*, GNR (9,0), is very small, thus the edge states introduced in the bandgap by increasing LER amplitude cannot lead to a significant increase in off-state current (V_{GS} = 0). This increasing LER amplitude only increases the scattering of carriers, consequently, reducing the currents. For GNR(24,0), however, the bandgap is small and the BTBT can be significant at V_{DS} = 0.25 V, thus the off-current initially increases by increasing the roughness amplitude up to the critical roughness amplitude (ΔW_c = 0.04), then decreases at larger roughness amplitude due to the back-scattering of carriers from localized edge states. However, the edge roughness continuously increases the scattering of thermionic carriers at on-state; thereby the reduction of on-current is continuous, correspondingly. Both increasing off-current and decreasing on-current can deteriorate the device performance at a critical roughness amplitude. For large LER amplitude, however, the main source of performance degradation is the significant decrease in on-current of GNR FET. The same effect of roughness amplitude can be observed in narrower GNRs (larger bandgaps) and at higher V_{DS}.

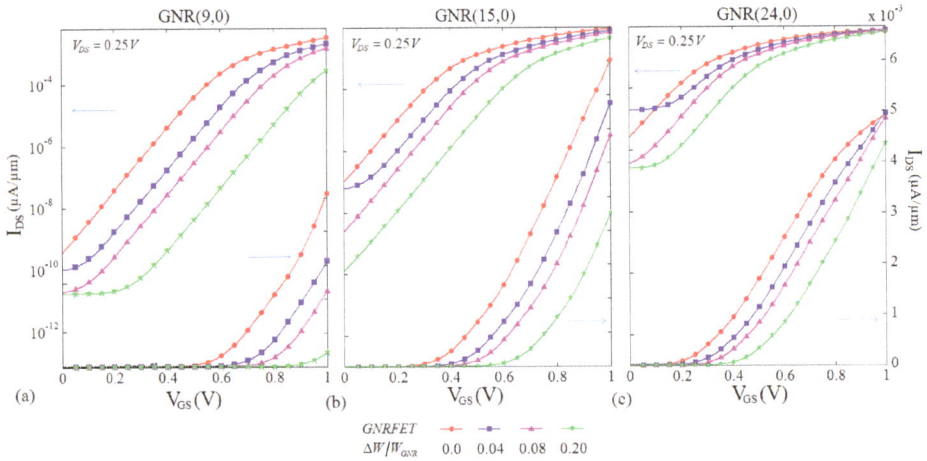

Figure 7. I_{DS}–V_{GS} characteristic of three GNR FETs with the channel of (**a**) GNR (9,0); (**b**) GNR (15,0) and (**c**) GNR (24,0) at V_{DS} = 0.25 V. The ideal edge GNR FETs and non-ideal GNR FETs with edge roughness variations of 0.04, 0.08 and 0.2 are shown.

Figure 8 shows the off-current, on-current and I_{ON}/I_{OFF} ratio as a function of GNR width (Semiconducting GNR index N = 3p, where p is an integer). It can be seen that both on- and off-currents of ideal GNR FETs increases by widening GNR width, as larger numbers of subbands can contribute to carrier transport by decreasing bandgap at a given supply voltage. The I_{ON}/I_{OFF} ratio is increased by decreasing the GNR width, as the leakage current associated with BTBT phenomena is significantly decreased by increasing the bandgap. At the scaled supply voltage ($V_{DD} = 0.5$ V), GNR(6,0) with larger bandgap is not fully at the on-state resulting in smaller I_{ON}/I_{OFF} ratio. GNR FETs demonstrate larger off-current at the critical LER amplitude of $\Delta W/W_{GNR} = 0.04$ due to the maximum number of BTBT from drain to channel occurring through the generated edge states in the bandgap. While I_{OFF} is larger and I_{ON} is smaller than ideal narrow GNR FETs due to the decrease in carrier scattering at on-state, which leads to significant deterioration of the I_{ON}/I_{OFF} ratio for $\Delta W/W_{GNR} = 0.04$. Larger line-edge roughness amplitudes, e.g., $\Delta W/W_{GNR} = 0.1$, can set the carrier transport in localization regime such that both off- and on- currents are reduced resulting from the decrease in the MFP of edge scattering and the increase in the back-scattering of carriers occurred by a larger number of localized edge states.

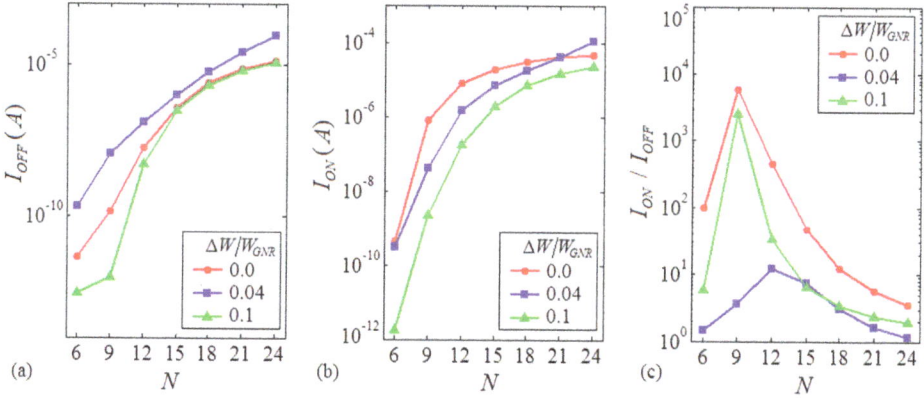

Figure 8. (a) Off-current; (b) on-current and (c) I_{ON}/I_{OFF} ratio as a function of armchair GNR index N. Note: $V_{DD} = 0.5$ V.

5. Validation of Single-Particle Calculations

It has been shown both experimentally [13,53] and theoretically [14] that strong localization can appear in a single-layer GNR for large line-edge roughness, known as Anderson-type localization [54]. Edge-disorder increases the number of localized carriers and their corresponding density of states at the edges, which blocks the conductive paths of carrier transport through the graphene nanoribbon. This can potentially increase the importance of electron–electron interaction for carrier transport, such that the validity of single-particle calculations for this problem needs to be justified. The effects of electron–electron interactions on carrier transport need to be investigated using numerical simulations, then the corresponding results can be incorporated in the analytical calculations. Many samples need to be created with given roughness parameters and each has to be simulated by a self-consistent iteration algorithm between the carrier transport problem using the non-equilibrium Green's function formalism and the electrostatic problem using the Poisson equation. Incorporating the electron–electron interaction in the problem will add another iteration algorithm between the Green's function and the electron-electron self-energy [55]. From the obtained numerical data and using equation: $< T(E) > = M(E)/(1 + L_{CH}/\lambda(E))$, where $M(E)$ is the number of active conduction channels and $< T(E) >$ is the average transmission probability, the mean free path of the carriers can be extracted to investigate the role of line-edge roughness scattering and the electron–electron interaction in the transport problem [55].

Figure 9 shows the mean free paths associated with line-edge roughness and the electron–electron interaction. In the simulation, the edge roughness is set to be highly correlated, $\Delta L = 3$ nm, which reduces the mean free path below the nanometer range and the electron–electron scattering rate is assumed 50 ps for a momentum conserving interaction [56]. It can be seen that the analytical results of Equation (20) for line-edge roughness are in agreement with the numerical simulations and are in correspondence with Fermi's golden rule ($\lambda^{LER} \propto 1/\Delta W^2$) [45]. By incorporating the electron–electron interaction, there is a deviation in total mean free path of carriers at small roughness amplitudes. The total MFP is dominated by the effects of the electron–electron interaction at small roughness while the line-edge roughness scattering is the dominant scattering mechanism at larger roughness amplitudes. Electron–electron interactions are known to have significant effects on the carrier transport in materials with one-dimensional (1D) confinement, e.g., GNR interconnects. As the GNRs in this application need to be long, the 1D problem is very strong and carrier transport is deep in the localization regime [57]. The carriers transport can only take place by tunneling between localized edge states [58] which results in significant reduction of transmission probability [54]. However, the GNR length as a channel material is 16 nm in our simulation, which is a typical channel length for emerging transistors and the GNR width is ~1.6 nm; thus, the strength of 1D problem corresponding to the ratio of the GNR channel length to GNR width ($W/L \approx 1/10$) is not large enough to make electron–electron interaction a considerable portion for most roughness amplitudes. However, the soundness of single-particle calculations with regard to width-to-length ratio of GNR is still controversial in the literature, such that charge localization due to strong edge roughness has been reported in some works [59,60] as a responsible mechanism for Coulomb blockade and the formation of bottlenecks for carrier transport in GNR channels. Further research is required to understand the exact mechanism for charge localization along the edge. Particularly, the manner in which the edge disorder changes the potential landscape in a graphene nanoribbon so that electrons prefer to be located along the edge rather than transported in the bulk needs to be investigated.

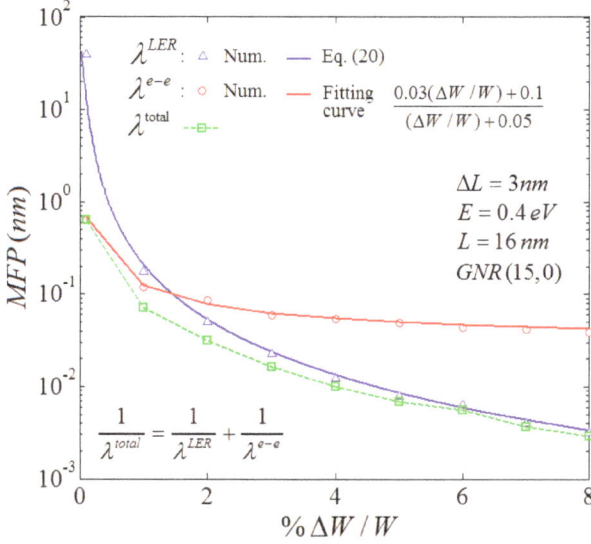

Figure 9. Mean free path (MFP) of line-edge roughness and electron–electron interaction. Note: Symbol shows the numerical results and solid curves show the analytical results from Equation (20) for the mean free path of line-edge roughness together with the fitting equation for the extracted mean free path of the electron–electron interaction.

6. Conclusions

We present a physics-based analytical model compatible with SPICE for the circuit simulation of GNR FETs. The carrier charge and current have been analytically calculated and compared with the numerical data obtained by NEGF formalism. We presented the device-level performance of non-ideal GNR FETs in edge-enhanced band-to-band-tunneling and localization regimes for various roughness amplitudes. Corresponding to these two regimes, I_{OFF} is initially increased, then decreased; while, on the other hand, I_{ON} is continuously decreased by increasing roughness amplitude. The band-to-band-tunneling and the corresponding increase in off-current can be reduced by accurately choosing the supply voltage for a given GNR width. The line-edge roughness increases the threshold voltage and limits the advantage of bandgap engineering of GNR FETs for scaling down the supply voltages. Our results show that the line-edge roughness in the GNR channels associated with the lack of atomic-scale precision in current patterning technique significantly degrades the performance of GNR FETs.

181

Author Contributions: Both authors have contributed equally to the paper.

Conflicts of Interest: The authors declare no conflict of interest.

References

1. Wilson, L. *International Technology Roadmap for Semiconductors (ITRS)*; Semiconductor Industry Association: Washington, DC, USA, 2013.
2. Novoselov, K.S.; Fal, V.; Colombo, L.; Gellert, P.; Schwab, M.; Kim, K. A roadmap for graphene. *Nature* **2012**, *490*, 192–200.
3. Obeng, Y.; De Gendt, S.; SRinivasan, P.; Misra, D.; Iwai, H.; Karim, Z.; Hess, D.; Grebel, H. *Graphene and Emerging Materials for Post-CMOS Applications*; Electrochemical Society: Pennington, NJ, USA, 2009.
4. Novoselov, K.; Geim, A.K.; Morozov, S.; Jiang, D.; Katsnelson, M.; Grigorieva, I.; Dubonos, S.; Firsov, A. Two-dimensional gas of massless dirac fermions in graphene. *Nature* **2005**, *438*, 197–200.
5. Cooper, D.R.; D'Anjou, B.; Ghattamaneni, N.; Harack, B.; Hilke, M.; Horth, A.; Majlis, N.; Massicotte, M.; Vandsburger, L.; Whiteway, E. Experimental review of graphene. *Int. Sch. Res. Not.* **2012**, *2012*.
6. Schwierz, F. Graphene transistors: Status, prospects, and problems. *Proc. IEEE* **2013**, *101*, 1567–1584.
7. Harada, N.; Sato, S.; Yokoyama, N. Theoretical investigation of graphene nanoribbon field-effect transistors designed for digital applications. *Jpn. J. Appl. Phys.* **2013**, *52*, 094301.
8. Johari, Z.; Hamid, F.; Tan, M.L.P.; Ahmadi, M.T.; Harun, F.; Ismail, R. Graphene nanoribbon field effect transistor logic gates performance projection. *J. Comput. Theor. Nanosci.* **2013**, *10*, 1164–1170.
9. Poljak, M.; Wang, K.L.; Suligoj, T. Variability of bandgap and carrier mobility caused by edge defects in ultra-narrow graphene nanoribbons. *Solid-State Electron.* **2015**, *108*, 67–74.
10. Michetti, P.; Iannaccone, G. Analytical model of one-dimensional carbon-based schottky-barrier transistors. *IEEE Trans. Electron Dev.* **2010**, *57*, 1616–1625.
11. Choudhury, M.; Yoon, Y.; Guo, J.; Mohanram, K. Technology Exploration for Graphene Nanoribbon F s. In Proceedings of the 45th Annual Design Automation Conference, Anaheim, CA, USA, 8–13 June 2008; pp. 272–277.
12. Chen, Y.-Y.; Rogachev, A.; Sangai, A.; Iannaccone, G.; Fiori, G.; Chen, D. A SPICE-Compatible Model of Graphene nano-Ribbon Field-Effect Transistors Enabling Circuit-Level Delay and Power Analysis under Process Variation. In Proceedings of the IEEE Design, Automation & Test in Europe Conference & Exhibition (DATE), Grenoble, France, 18-22 March 2013; pp. 1789–1794.
13. Xu, G.; Torres Jr, C.M.; Tang, J.; Bai, J.; Song, E.B.; Huang, Y.; Duan, X.; Zhang, Y.; Wang, K.L. Edge effect on resistance scaling rules in graphene nanostructures. *Nano Lett.* **2011**, *11*, 1082–1086.

14. Yazdanpanah, A.; Pourfath, M.; Fathipour, M.; Kosina, H.; Selberherr, S. A numerical study of line-edge roughness scattering in graphene nanoribbons. *IEEE Trans. Electron Dev.* **2012**, *59*, 433–440.

15. Yoon, Y.; Fiori, G.; Hong, S.; Iannaccone, G.; Guo, J. Performance comparison of graphene nanoribbon fets with schottky contacts and doped reservoirs. *IEEE Trans. Electron Dev.* **2008**, *55*, 2314–2323.

16. Kang, J.; Sarkar, D.; Khatami, Y.; Banerjee, K. Proposal for all-graphene monolithic logic circuits. *Appl. Phys. Lett.* **2013**, *103*, 083113.

17. Owlia, H.; Keshavarzi, P. Investigation of the novel attributes of a double-gate graphene nanoribbon fet with aln high-κ dielectrics. *Superlattices Microstruct.* **2014**, *75*, 613–620.

18. Oh, J.G.; Hong, S.K.; Kim, C.-K.; Bong, J.H.; Shin, J.; Choi, S.-Y.; Cho, B.J. High performance graphene field effect transistors on an aluminum nitride substrate with high surface phonon energy. *Appl. Phys. Lett.* **2014**, *104*, 193112.

19. Konar, A.; Fang, T.; Jena, D. Effect of high-k dielectrics on charge transport in graphene. 2009, arXiv:0902.0819.

20. Banadaki, Y.; Srivastava, A. Scaling effects on static metrics and switching attributes of graphene nanoribbon fet for emerging technology. *IEEE Trans. Emerg. Top. Comput.* **2015**, *4*, 458–469.

21. Grassi, R.; Poli, S.; Gnani, E.; Gnudi, A.; Reggiani, S.; Baccarani, G. Tight-binding and effective mass modeling of armchair graphene nanoribbon fets. *Solid-State Electron.* **2009**, *53*, 462–467.

22. Banadaki, Y.M.; Srivastava, A. Investigation of the width-dependent static characteristics of graphene nanoribbon field effect transistors using non-parabolic quantum-based model. *Solid-State Electron.* **2015**, *111*, 80–90.

23. Kliros, G.S. Scaling Effects on the Gate Capacitance of Graphene Nanoribbon Transistors. In Proceedings of the IEEE International Semiconductor Conference (CAS), Sinaia, Romania, 15–17 October 2012; pp. 83–86.

24. Deng, J.; Wong, H.P. A compact spice model for carbon-nanotube field-effect transistors including nonidealities and its application-part i: Model of the intrinsic channel region. *IEEE Trans. Electron Dev.* **2007**, *54*, 3186–3194.

25. Gholipour, M.; Chen, Y.-Y.; Sangai, A.; Chen, D. Highly Accurate SPICE-Compatible Modeling for Single-and Double-Gate Gnrfets with Studies on Technology Scaling. In Proceedings of the Conference on Design, Automation & Test in Europe, Dresden, Germany, 24–28 March 2014; p. 120.

26. Capacitor Calculator. Available online: http://www.Technick.Net/public/code/cp_dpage.Php?Aiocp_dp=util_ pcb_imp_microstrip_embed (accessed on 7 January 2015).

27. Garg, R.; Bahl, I.; Bozzi, M. *Microstrip Lines and Slotlines*; Artech house: Norwood, MA, USA, 2013.

28. Wang, J.; Lundstrom, M. Does Source-to-Drain Tunneling Limit the Ultimate Scaling of Mosfets? In Proceedings of the IEEE International Electron Devices Meeting (IEDM'02), San Francisco, CA, USA, 8–11 December 2002; pp. 707–710.

29. Datta, S. *Quantum Transport: Atom to Transistor*; Cambridge University Press: Cambridge, UK, 2005.

30. Kane, E.O. Theory of tunneling. *J. Appl. Phys.* **1961**, *32*, 83–91.

31. Geist, J.; Lowney, J. Effect of band-gap narrowing on the built-in electric field in n-type silicon. *J. Appl. Phys.* **1981**, *52*, 1121–1123.

32. Geim, A.K.; Novoselov, K.S. The rise of graphene. *Nat. Mater.* **2007**, *6*, 183–191.

33. Akturk, A.; Goldsman, N. Electron transport and full-band electron-phonon interactions in graphene. *J. Appl. Phys.* **2008**, *103*, 053702.

34. Chen, J.-H.; Jang, C.; Xiao, S.; Ishigami, M.; Fuhrer, M.S. Intrinsic and extrinsic performance limits of graphene devices on sio2. *Nat. Nanotechnol.* **2008**, *3*, 206–209.

35. Yang, Y.; Murali, R. Impact of size effect on graphene nanoribbon transport. *IEEE Electron Dev. Lett.* **2010**, *31*, 237–239.

36. Zhao, P.; Choudhury, M.; Mohanram, K.; Guo, J. Computational model of edge effects in graphene nanoribbon transistors. *Nano Res.* **2008**, *1*, 395–402.

37. Lundstrom, M.; Guo, J. *Nanoscale Transistors: Device Physics, Modeling and Simulation*; Springer Science & Business Media: New York, NY, USA, 2006.

38. Perebeinos, V.; Tersoff, J.; Avouris, P. Electron-phonon interaction and transport in semiconducting carbon nanotubes. *Phys. Rev. Lett.* **2005**, *94*, 086802.

39. Rakheja, S.; Kumar, V.; Naeemi, A. Evaluation of the potential performance of graphene nanoribbons as on-chip interconnects. *Proc. IEEE* **2013**, *101*, 1740–1765.

40. Wang, X.; Ouyang, Y.; Li, X.; Wang, H.; Guo, J.; Dai, H. Room-temperature all-semiconducting sub-10-nm graphene nanoribbon field-effect transistors. *Phys. Rev. Lett.* **2008**, *100*, 206803.

41. Areshkin, D.A.; Gunlycke, D.; White, C.T. Ballistic transport in graphene nanostrips in the presence of disorder: Importance of edge effects. *Nano Lett.* **2007**, *7*, 204–210.

42. Gunlycke, D.; Areshkin, D.; White, C. Semiconducting graphene nanostrips with edge disorder. *Appl. Phys. Lett.* **2007**, *90*, 142104.

43. Gunlycke, D.; White, C. Scaling of the localization length in armchair-edge graphene nanoribbons. *Phys. Rev. B* **2010**, *81*, 075434.

44. Goodnick, S.; Ferry, D.; Wilmsen, C.; Liliental, Z.; Fathy, D.; Krivanek, O. Surface roughness at the Si(100)-Sio2 interface. *Phys. Rev. B* **1985**, *32*, 8171.

45. Fang, T.; Konar, A.; Xing, H.; Jena, D. Mobility in semiconducting graphene nanoribbons: Phonon, impurity, and edge roughness scattering. *Phys. Rev. B* **2008**, *78*, 205403.

46. Goharrizi, A.Y.; Pourfath, M.; Fathipour, M.; Kosina, H.; Selberherr, S. An analytical model for line-edge roughness limited mobility of graphene nanoribbons. *IEEE Trans. Electron Dev.* **2011**, *58*, 3725–3735.

47. Luisier, M.; Klimeck, G. Performance analysis of statistical samples of graphene nanoribbon tunneling transistors with line edge roughness. *Appl. Phys. Lett.* **2009**, *94*, 223505.

48. Bai, J.; Duan, X.; Huang, Y. Rational fabrication of graphene nanoribbons using a nanowire etch mask. *Nano Lett.* **2009**, *9*, 2083–2087.

49. Lin, M.-W.; Ling, C.; Zhang, Y.; Yoon, H.J.; Cheng, M.M.-C.; Agapito, L.A.; Kioussis, N.; Widjaja, N.; Zhou, Z. Room-temperature high on/off ratio in suspended graphene nanoribbon field-effect transistors. *Nanotechnology* **2011**, *22*, 265201.

50. Huang, C.-H.; Su, C.-Y.; Okada, T.; Li, L.-J.; Ho, K.-I.; Li, P.-W.; Chen, I.-H.; Chou, C.; Lai, C.-S.; Samukawa, S. Ultra-low-edge-defect graphene nanoribbons patterned by neutral beam. *Carbon* **2013**, *61*, 229–235.

51. Yoon, Y.; Guo, J. Effect of edge roughness in graphene nanoribbon transistors. *Appl. Phys. Lett.* **2007**, *91*, 073103.

52. Fiori, G.; Iannaccone, G. Simulation of graphene nanoribbon field-effect transistors. *Electron Dev. Lett. IEEE* **2007**, *28*, 760–762.

53. Sprinkle, M.; Ruan, M.; Hu, Y.; Hankinson, J.; Rubio-Roy, M.; Zhang, B.; Wu, X.; Berger, C.; De Heer, W.A. Scalable templated growth of graphene nanoribbons on SiC. *Nat. Nanotechnol.* **2010**, *5*, 727–731.

54. Evaldsson, M.; Zozoulenko, I.V.; Xu, H.; Heinzel, T. Edge-disorder-induced anderson localization and conduction gap in graphene nanoribbons. *Phys. Rev. B* **2008**, *78*, 161407.

55. Kahnoj, S.S.; Touski, S.B.; Pourfath, M. The effect of electron-electron interaction induced dephasing on electronic transport in graphene nanoribbons. *Appl. Phys. Lett.* **2014**, *105*, 103502.

56. Li, X.; Barry, E.; Zavada, J.; Nardelli, M.B.; Kim, K. Influence of electron-electron scattering on transport characteristics in monolayer graphene. *Appl. Phys. Lett.* **2010**, *97*, 082101.

57. Lherbier, A.; Biel, B.; Niquet, Y.-M.; Roche, S. Transport length scales in disordered graphene-based materials: Strong localization regimes and dimensionality effects. *Phys. Rev. Lett.* **2008**, *100*, 036803.

58. Datta, S. *Electronic Transport in Mesoscopic Systems*; Cambridge University Press: Cambridge, UK, 1997.

59. Bischoff, D.; Varlet, A.; Simonet, P.; Eich, M.; Overweg, H.; Ihn, T.; Ensslin, K. Localized charge carriers in graphene nanodevices. *Appl. Phys. Rev.* **2015**, *2*, 031301.

60. Sols, F.; Guinea, F.; Neto, A.C. Coulomb blockade in graphene nanoribbons. *Phys. Rev. Lett.* **2007**, *99*, 166803.

Simulation of 50-nm Gate Graphene Nanoribbon Transistors

Cedric Nanmeni Bondja, Zhansong Geng, Ralf Granzner, Jörg Pezoldt and Frank Schwierz

Abstract: An approach to simulate the steady-state and small-signal behavior of GNR MOSFETs (graphene nanoribbon metal-semiconductor-oxide field-effect transistor) is presented. GNR material parameters and a method to account for the density of states of one-dimensional systems like GNRs are implemented in a commercial device simulator. This modified tool is used to calculate the current-voltage characteristics as well the cutoff frequency f_T and the maximum frequency of oscillation f_{max} of GNR MOSFETs. Exemplarily, we consider 50-nm gate GNR MOSFETs with $N = 7$ armchair GNR channels and examine two transistor configurations. The first configuration is a simplified MOSFET structure with a single GNR channel as usually studied by other groups. Furthermore, and for the first time in the literature, we study in detail a transistor structure with multiple parallel GNR channels and interribbon gates. It is shown that the calculated f_T of GNR MOSFETs is significantly lower than that of GFETs (FET with gapless large-area graphene channel) with comparable gate length due to the mobility degradation in GNRs. On the other hand, GNR MOSFETs show much higher f_{max} compared to experimental GFETs due the semiconducting nature of the GNR channels and the resulting better saturation of the drain current. Finally, it is shown that the gate control in FETs with multiple parallel GNR channels is improved while the cutoff frequency is degraded compared to single-channel GNR MOSFETs due to parasitic capacitances of the interribbon gates.

Reprinted from *Electronics*. Cite as: Bondja, C.N.; Geng, Z.; Granzner, R.; Pezoldt, J.; Schwierz, F. Simulation of 50-nm Gate Graphene Nanoribbon Transistors. *Electronics* **2016**, *5*, 3.

1. Introduction

The 2D (two-dimensional) carbon-based material graphene has attracted significant attention during the past 10 years [1]. During the early years of graphene research, *i.e.*, in the period 2004–2009, particularly the high carrier mobilities observed in large-area graphene raised expectations that graphene could be an excellent channel material for future MOSFET (metal-oxide-semiconductor field-effect transistor) generations. Unfortunately, large-area graphene as the natural form of appearance of graphene does not possess a bandgap. Thus, MOSFETs with large-area graphene channels (designated as GFETs in the following) do not switch off and cannot be used for digital logic. RF (radio frequency) FETs, on the other hand,

186

do not necessarily need to be switched off. Therefore a lot of work has been done to develop GFETs for RF applications and indeed experimental GFETs with cutoff frequencies f_T in excess of 300 GHz have been reported [2–4]. Due to the missing gap, however, GFETs suffer from an unsatisfying saturation of the drain current causing poor power gain and low maximum frequency of oscillation f_{max} [5,6]. This seriously limits the potential of GFETs for high-performance RF applications. Thus, for both logic and high-performance RF FETs, a semiconducting channel is needed.

There are different options to open a gap in graphene, *i.e.*, to make this material semiconducting. First, it has been shown that by using bilayer graphene instead of single-layer material and applying a perpendicular field, a gap is formed [7,8]. A second option is to create narrow confined graphene structures, such as GNRs (graphene nanoribbon) [9,10] or graphene nanomeshes [11,12], in which a gap opens. In the present work, we focus on GNRs and use these as MOSFET channels. The gap opening in GNRs has been predicted by first-principle calculations [13–15] and confirmed by experiments [9,10]. Meanwhile the International Technology Roadmap for Semiconductors considers GNRs as a viable channel replacement material for future MOSFET generations [16]. Recently back-gate GNR MOSFETs with ribbon widths down to 2 nm showing excellent switch-off and on-off ratios in excess of 10^6 have been demonstrated [17–19] and top-gate GNR MOSFETs with 10 to 20-nm wide channels and on-off ratios around 70 [20] have been reported.

On the theoretical side, GNR MOSFETs have been simulated at different levels of complexity and physics involved. Steady-state quantum simulations based on the NEGF (nonequilibrium Green's function) approach assuming ballistic transport have been performed [21–25] and GNR MOSFET simulations taking edge scattering [26,27] and phonon scattering [25] into account have been conducted. Moreover, the RF performance of GNR MOSFETs has been investigated by numerical simulations [21,22,25] and analytical equations have been developed to calculate the steady-state behavior and the RF properties of GNR MOSFETs [28].

While these simulations have provided valuable insights in the operation and physics of GNR MOSFETs, so far only simplified transistor structures with a single GNR channel and, in many cases, idealized conditions such as ballistic carrier transport have been considered. This has led to overly optimistic performance predictions such as unrealistically high simulated cutoff frequencies [21,22,25,28]. Moreover, most simulations have been performed using in-house tools not accessible by the community. An exception worth mentioning is the open-source multiscale simulation framework for the investigation of nanoscale devices such as GNR MOSFETs presented in [29]. However, commercial device simulators which are very popular in the semiconductor industry so far have not been applied to the investigation of GNR MOSFETs.

In the present work, we develop an approach to describe the steady-state and RF behavior of GNR MOSFETs in the framework of a commercial device simulator. Since so far neither graphene nor GNR models are implemented in commercial tools, in Section 2 the GNR models we have implemented in the device simulator ATLAS [30] are described and an approach to appropriately account for the DOS (density of states) and quantum capacitance of 1D (one-dimensional) systems such as GNRs in commercial simulation tools is presented. Section 3 summarizes the results of our ATLAS simulations, first for a simplified single-channel GNR MOSFET structure with 50 nm gate length and next for GNR MOSFETs with multiple parallel GNR channels and interribbon gates. Such multiple-channel GNR MOSFETs are studied here for the first time in detail. Finally, Section 4 concludes the paper.

2. Simulation Framework and GNR Models

Appropriate models for the material and carrier transport parameters for GNRs and a formalism to correctly account for the DOS in 1D structures are still missing in ATLAS. Therefore, the first steps of our work are (i) the compilation of data for the material and transport parameters needed for the simulation of GNR MOSFETs; (ii) the elaboration of a suitable approach to describe the DOS and the quantum capacitance in 1D GNR MOS structures properly; and (iii) the implementation of these features in ATLAS.

2.1. Models for Bangap and Carrier Effective Mass

It is well established that in narrow GNRs a sizeable bandgap can be opened and that the gap E_G critically depends on the GNR edge configuration, *i.e.*, ac (armchair) or zz (zigzag), and on the ribbon width w. GNRs of the ac configuration constitute the three different families $3p$, $3p + 1$, and $3p + 2$ (p is an integer). For example, for a GNR with $N = 7$ carbon atoms along its width, p equals 2 and the ribbon belongs to the $3p + 1$ family. Every family of ac GNRs obeys a specific gap-width relationship [4,13–15,31]. On the other hand, there is still a controversial debate on whether zz GNRs are metallic or semiconducting. In the present work, we exemplarily consider ac $3p + 1$ GNRs and focus on the simulation of MOSFETs with $N = 7$ GNR channels.

The bandgap data for ac GNRs available in the literature consistently show for each family a decreasing gap for increasing ribbon width. For a given p, the gap obeys the relation $E_G(3p + 1) > E_G(3p) > E_G(3p + 2)$. However, as shown in Figure 1, the published E_G-w data scatter considerably. This makes it difficult to develop a reliable E_G-w model needed for device simulation. Obviously the reported experimental bandgap data points for the narrow $N = 7$ and $N = 13$ GNRs are located in between the results of the *GW* simulations of Yang *et al.* [14] and the predictions of Raza and

Kan [13]. Moreover, most experimental bandgaps for GNR with widths from below 1 nm to about 20 nm are located in the range between these two predictions.

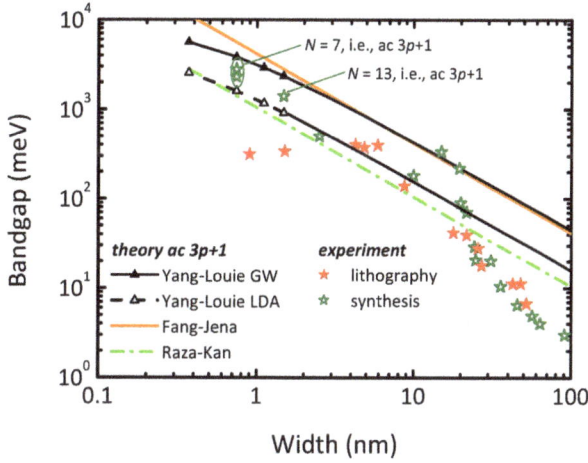

Figure 1. Measured and calculated bandgap of ac GNRs (graphene nanoribbon) *vs.* ribbon width. The experimental data for the $N = 7$ and $N = 13$ ac GNRs are taken from [10,32,33] and the other experimental data are taken from the compilation in [31]. The calculated bandgaps are taken from [13,14,34].

In the present work, we use the E_G-w relation from [13], which for the $3p + 1$ family reads as

$$E_G \quad = \quad 1.04\,eV\,/\,w\;(nm) \tag{1}$$

with

$$w\;(nm) \quad = \quad \frac{0.246}{2}\,(N - 1) \tag{2}$$

Not only the bandgap but also the carrier effective mass m_{eff} in GNRs depends on the GNR edge configuration, family, and width. For ac GNRs of the $3p + 1$ family, Raza and Kan suggested the expression [13]

$$m_{\text{eff}} \quad = \quad \frac{0.16\,m_0}{w} \tag{3}$$

where m_0 is the electron rest mass and w is the ribbon, according to Equation (2), in nm. The $3p + 1$ family has been chosen in the present work since it provides, for a given p, the widest gap of the three families. We note, however, that our approach can be applied to the other two families of ac GNRs as well.

2.2. Transport Model

In the present work, carrier transport is described in the framework of the classical DD (drift-diffusion) model. It is frequently argued that in MOSFETs with sub-100-nm gates the DD model does no longer provide a sufficiently correct description of carrier transport due to the appearance of nonstationary and/or quasi-ballistic transport effects. We have demonstrated, however, that the standard DD model is well suited for the simulation of Si MOSFETs with gate length down to 30 nm, and by assuming a modified gate-length-dependent saturation velocity it provides reasonable results even down to 5 nm gate length [35]. Moreover, it has been shown that carrier transport in 2D MoS$_2$ MOSFETs with 10-nm, and even sub-10-nm, channels is still far from ballistic but dissipative and scattering-dominated instead [36–38]. Finally, the DD approach is considered a reasonable guide even in the presence of nonstationary and quasi-ballistic transport in short-channel MOSFETs since it correctly accounts for both device geometry and electrostatics [39].

For a proper description of carrier transport in the DD model, the v-E (velocity-field) characteristics including the low-field mobility μ_0, the high-field saturation velocity v_{sat}, and the abruptness of the transition between the low-field and high-field regions have to be modeled correctly. The dependence of the effective mass on the ribbon width is already an indication of the fact that the properties of carrier transport in GNRs are related to the ribbon width. This has been confirmed by Monte Carlo transport simulations for 1.12, 2.62, and 4.86 nm wide ac ribbons belonging to the $3p + 1$ family [40]. In the present work, we assume a Caughey–Thomas-type v-E characteristics with soft velocity saturation according to [41]

$$v = \frac{u_0 E}{\left[1 + \left(\dfrac{u_0 E}{v_{sat}} \right)^{\beta} \right]^{1/\beta}} \tag{4}$$

the v-E characteristics of the three GNRs reported in [40], and the parameters μ_0, v_{sat}, and β, see Table 1, are obtained by fitting.

Table 1. Parameters for the Caughey–Thomas fit of the v-E characteristics for three ac $3p + 1$ graphene nanoribbons.

w (nm)	μ_0 (cm^2/Vs)	v_{sat} (10^7 cm/s)	β	Remark
1.12	460	2.2	1.4	This work
2.62	2700	3.2	1.3	[40]
4.86	12,000	3.3	1.3	[40]

Note that the $N = 7$, $w = 0.74$ nm GNR considered in the present work is outside the width range covered by the GNRs in Table 1. Therefore we extrapolate the trends for the parameters from Table 1 towards smaller widths and obtain $\mu_0 = 195$ cm^2/Vs, $v_{sat} = 1.83 \times 10^7$ cm/s, and $\beta = 1.4$ for $N = 7$ GNRs.

2.3. Modeling the Density of States and Quantum Capacitance of 1D Systems

To describe the gate control and electrostatics in low-dimensional MOS systems correctly, it is mandatory to model the DOS (density of states) and the gate capacitance C_G accurately. In general, the gate capacitance of a MOS structure (regardless of the dimensionality of the channel) consists of a combination of the three capacitance components C_{ox}, C_{es}, and C_q connected in series. Note that in the following we do not consider the absolute capacitance but the capacitance per unit area. C_{ox} is the oxide capacitance given by $C_{ox} = \varepsilon_{ox}/t_{ox}$ where ε_{ox} is the dielectric constant and t_{ox} the thickness of the gate oxide; C_{es} is the electrostatic capacitance of the channel related to the average distance of the carriers from the oxide-channel interface; and C_q is the quantum capacitance of the channel related to the finite DOS of the channel. The simulator ATLAS (as most conventional device simulators) *per se* considers all semiconducting device regions (and thus the GNR channel of a GNR MOSFET) as a 3D material. Quantization effects occurring in the 2D system of a conventional MOS inversion channel can be taken into account by using quantum correction models implemented in ATLAS, while appropriate models for 1D systems, such as GNRs, are not yet available in commercial device simulators. On the other hand, the DOS of 1D systems differs significantly from that of 3D and 2D systems. If this is neglected in device simulations, the quantum capacitance may be dramatically underestimated leading to an undervalued gate control in 1D channels, particularly in GNR MOSFETs with thin gate oxide. Therefore, in the following we present an approach to emulate the effects of the 1D DOS of GNRs on the quantum capacitance and on the carrier density in the channel in the framework of ATLAS. In [42], we have elaborated physically correct analytical expressions for the carrier sheet density and the quantum capacitance in 1D systems. The corresponding equations, together with those for 2D and 3D systems, can be found in the Appendix.

Figure 2a shows the quantum capacitance of a $N = 7$ GNR as a function of the relative position of the Fermi level, *i.e.*, E_C–E_F where E_C is the conduction band edge and E_F is the Fermi level, calculated using the expressions from the Appendix for the 1D–3D cases. We have, as in all simulations in the present work, taken the first two subbands with a subband separation of 0.4 eV [43] into account, assumed GNRs with n-type conductivity, modeled the residual electron concentration by assuming a homogeneous n-type doping of the GNR of 2×10^{20} cm^{-3} which corresponds to an electron sheet density of 7×10^{12} cm^{-2}, and used a relative dielectric constant of 1.8 for the GNR. It can be seen from Figure 2a that the quantum capacitance for the 1D case can easily exceed C_q assuming 3D conditions by one order of magnitude due to the huge qualitative and quantitative differences between the 1D and 3D densities of state.

Since the density of states depends on the carrier effective mass m_{eff}, a proper adjustment of m_{eff} in the expressions for the sheet density and quantum capacitance for the 3D case should, at least within a reasonably wide (E_C–E_F) range, lead to a proper reproduction of the 1D quantum capacitance. Figure 2b shows the quantum capacitance for the 1D–3D cases calculated with the expression from the Appendix (now as a function of the electron sheet density), together with a second curve for the 3D case where we have used an adjusted effective mass, namely 24 times as large as the effective mass for the $N = 7$ GNR according to Equation (3). It can be seen that by using this modified effective mass, the quantum capacitance for the 3D case, *i.e.*, the case handled by ATLAS, reproduces the quantum capacitance obtained from the correct equations for the 1D case almost perfectly for sheet densities up to 4×10^{13} cm^{-3}.

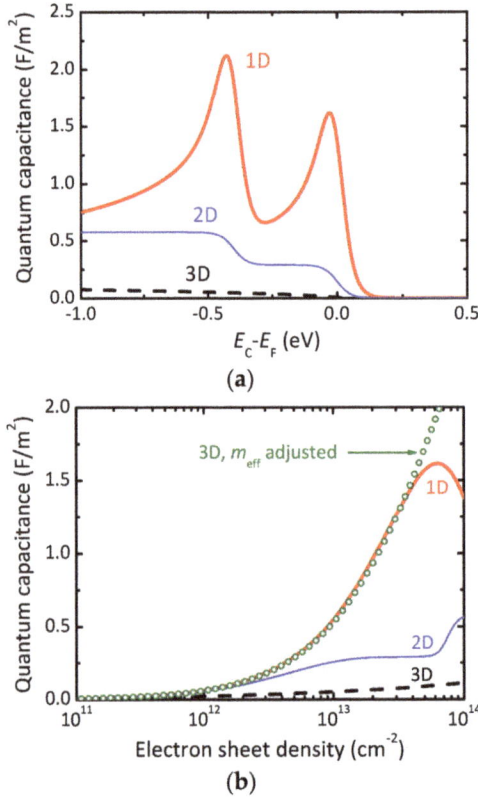

Figure 2. *Cont.*

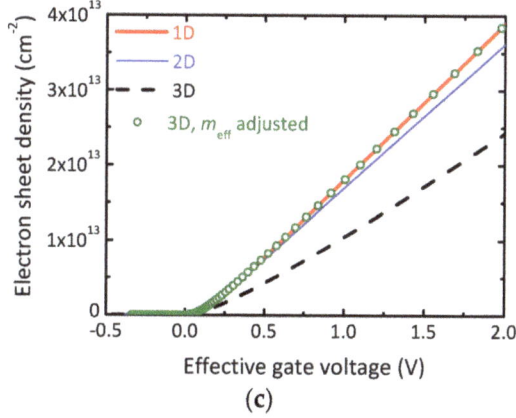

Figure 2. (**a**) Quantum capacitance of a $N = 7$ ac GNR (graphene nanoribbon) channel *vs.* Fermi level position assuming a 1D, 2D, and 3D density of states. (**b**) Quantum capacitance of the corresponding GNR MOS structure with 1 nm equivalent oxide thickness and a gate metal work function equal to the GNR electron affinity. (**c**) Electron sheet density of the GNR MOS structure from (**b**) *vs.* effective gate voltage.

Finally, Figure 2c shows the electron sheet density n_{sh} *versus* the effective gate voltage $V_{G-eff} = V_G - V_{\text{Th}}$, where V_G is the applied gate voltage and V_{Th} is the threshold voltage defined as the gate voltage at which the electron sheet density n_{sh} equals 3×10^{10} cm^{-2}. It can be seen that n_{sh} calculated for the 3D case is almost 40% lower compared to that for the 1D case. On the other hand, considering the 3D case and using the adjusted effective mass $m_{eff-adj} = 24 \times m_{\text{eff}}$, the sheet density for the 1D case is perfectly reproduced over the entire gate voltage range considered. This brings us to the conclusion that using the expressions for the 3D case and assuming an adjusted carrier effective mass is an appropriate means to account for the specifics of the DOS of 1D systems properly. Therefore, in the ATLAS simulations to be discussed in the remainder of the paper, we always use the modified effective mass.

3. Simulated Transistor Structures, Simulation Results, and Discussion

To describe the behavior of GNR MOSFETs properly, actually 3D device simulations should be performed. Note that here the term 3D means that the GNR MOSFET by nature is a 3D device similar as the tri-gate MOSFET or the FinFET and that therefore in the simulation process the semiconductor equations (Poisson, current, and continuity equations) should be solved in three spatial dimensions. This should not be confused with the 3D case mentioned in Section 2.3. Since 3D simulations are computationally more expensive and sometimes more critical regarding convergence, we simulate the 2D GNR MOSFET structure shown in

193

Figure 3 and initially consider only the effect of the top-gate on the channel and the electrostatics. The results of this study are presented in Section 3.1. In Section 3.2, we additionally consider the effect of interribbon gates on the operation of GNR MOSFETs with multiple parallel GNR channels.

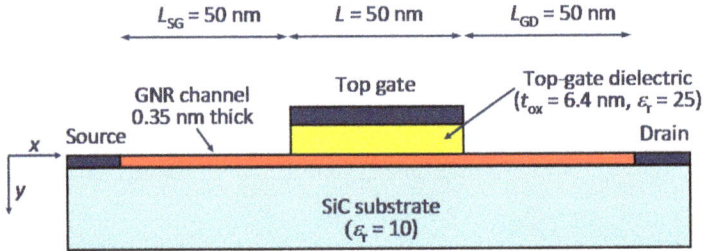

Figure 3. Cross-section (in the x-y plane) of the simulated graphene nanoribbon MOSFET.

3.1. Simulated Transistor Structures

The basic structure of our simulated GNR MOSFETs is shown in Figure 3 in a cross-sectional view. The current flows in the x direction, the width of the GNR extends into the z direction (not shown), and the thicknesses of the substrate, the GNR channel, and the gate dielectric extend into the y direction. The structure consists of a semi-insulating SiC substrate, of which the upper part with a thickness of 1 μm is taken into account in the simulation.

On top of the substrate, an epitaxial GNR channel consisting of a $N = 7$ ac GNR with a thickness of 0.35 nm is located. The GNR is assumed to possess ideal ohmic source and drain contacts at its left and right ends and the top-gate dielectric is formed by a 6.4 nm thick HfO_2 layer with a relative dielectric constant 25. This corresponds to an equivalent oxide thickness EOT of 1 nm. The gate has a length L of 50 nm and a work function equal to the electron affinity of the GNR, and the source-gate and gate-drain separations are 50 nm. Two layout configurations of the GNR MOSFET from Figure 3 are considered in the following. The first one is the single-channel transistor depicted in Figure 4a showing the cross-section in the y-z plane and in Figure 4b showing its top view. This configuration has only a top gate and no interribbon gate and for its investigation 2D device simulations are sufficient. We note that such single-channel devices have been considered in most previous theoretical investigations of GNR MOSFETs [21–27].

As has been shown in Figure 1, GNR channels with sufficiently wide bandgap are very narrow and therefore show only a limited current driving capability. To increase the transistor's drain current and to achieve the required current drivability, structures with multiple parallel GNR channels have to be used. The multiple-channel GNR MOSFET shown in Figure 4c,d constitutes the second configuration investigated in our study. The gate stripe of such a transistor consists of portions acting as the actual gate directly above the GNRs, *i.e.*, the top gate, and of portions called interribbon gate located (on top of the HfO$_2$ dielectric) between the parallel GNR channels. The control effect of the top gate is indicated by the straight black arrows in Figure 4a,c and the effect of the interribbon gates is indicated by the curved red arrows in Figure 4c.

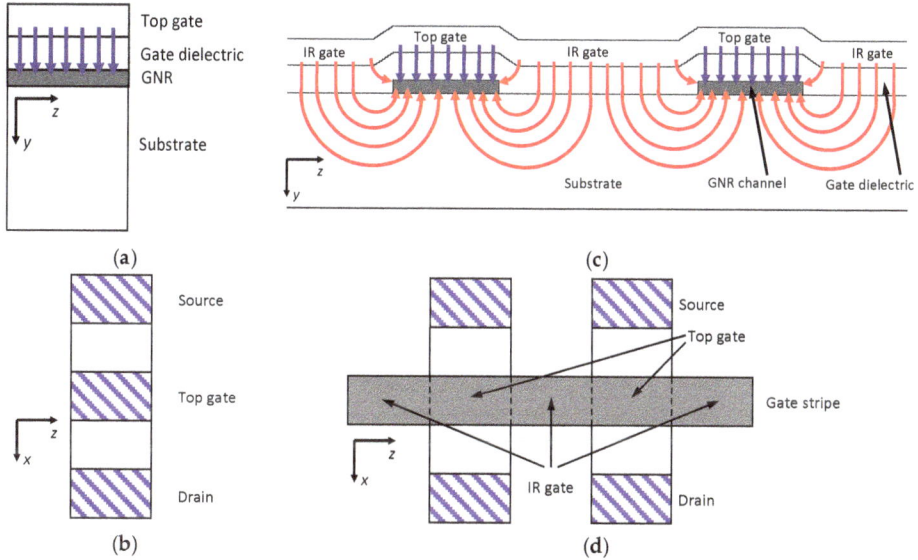

Figure 4. Cross section and top view of a single-channel GNR MOSFET with top gate only and of a multiple-channel GNR MOSFET with interribbon gate. (**a**) Cross section of a single-channel GNR MOSFET in the *y-z* plane. (**b**) Top view of the single-channel GNR MOSFET from (**a**). (**c**) Cross section of a multiple-channel MOSFET with two parallel GNR channels and IR (interribbon) gate in the *y-z* plane. (**d**) Top view of the multiple-channel GNR MOSFET from (**c**). Note that in Figure 4a,c the current flows perpendicular to the paper plane.

3.2. Simulation Results for Single-Channel GNR MOSFETs

Figure 5a shows the simulated transfer characteristics of the 50-nm gate single-channel $N = 7$ ac GNR MOSFET for a drain-source voltage V_{DS} of 1 V. We define transistor's threshold voltage V_{Th} as the gate-source voltage for which at $V_{DS} = 1$ V a drain current of 10^{-7} A $\times w/L$ flows and the effective gate-source voltage V_{GS-eff} is related to the applied gate-source voltage V_{GS} by $V_{GS-eff} = V_{GS} - V_{Th}$. As to be expected from the 1.4 eV bandgap of the $N = 7$ GNR channel, the transistors shows excellent switch-off, an on-off ratio of 1.5×10^6 for a 1 V gate voltage swing (from V_{GS-eff} = −0.25 V to +0.75 V), and a nearly ideal subthreshold swing SS of 64 mV/dec. The transconductance (not shown in the Figure) peaks at an effective gate voltage around 0.68 V reaching 1.25 mS/μm.

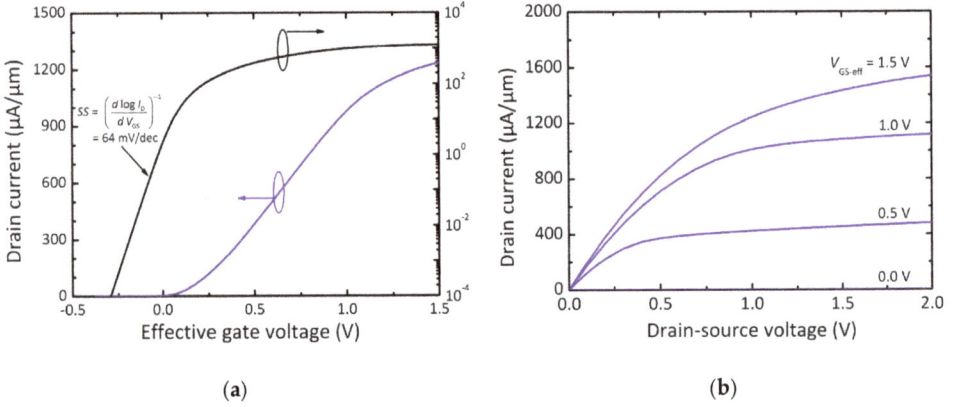

Figure 5. Steady-state characteristics of the simulated 50-nm gate single-channel GNR MOSFET: (**a**) transfer characteristics; and (**b**) output characteristics.

The output characteristics of the same transistor depicted in Figure 5b shows a pronounced saturation of the drain current and a low drain conductance of 76 μS/μm at V_{DS} = 1 V and V_{GS-eff} = 0.5 V. The good current saturation is caused by the semiconducting nature of the GNR channel and marks, in addition to the high on-off ratio, an important improvement compared to GFETs with gapless large-area graphene channels which suffer from a weak saturation and a large drain conductance.

To get an impression on RF potential of GNR FETs, we also perform small-signal analyses for the 50-nm gate single-channel GNR MOSFET and calculate its small-signal current gain h_{21} and unilateral power gain U at a frequency of 10 GHz for V_{DS} = 1 V and varying V_{GS}-V_{Th}. The cutoff frequency f_T and the maximum frequency of oscillation f_{max} are then obtained by extrapolating h_{21} and U with the characteristic slope of −20 dB/dec to zero dB [44]. A peak cutoff frequency of 215 GHz is obtained at V_{GS-eff} around 0.56 V. Figure 6a compares this result with the best experimental f_T data reported for competing RF FETs, *i.e.*, GFETs, Si MOSFETs, and III–V HEMTs (high electron mobility transistor) with comparable gate lengths. As can be seen, in terms of f_T our GNR MOSFET performs worse compared to best GFETs and the other competing RF FETs. This was to be expected because of the relatively low mobility in the GNR channel, particularly compared to the gapless large-area graphene channels of GFET and the InGaAs channels (with high In content) of the III–V HEMTs.

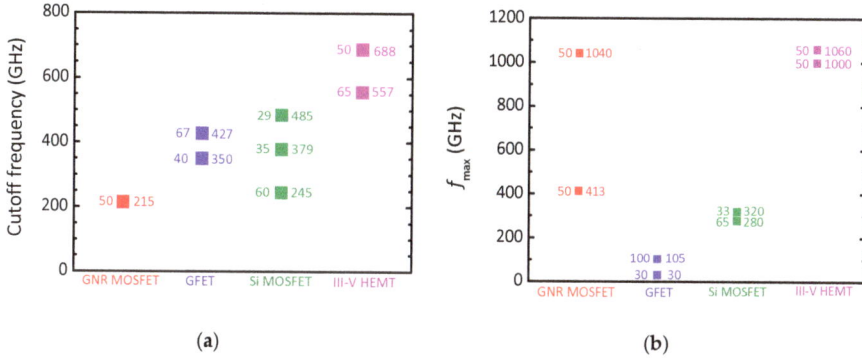

Figure 6. RF performance of the simulated 50-nm gate single-channel GNR MOSFET and of competing RF FETs (experimental data) with comparable gate length in terms of cutoff frequency f_T and maximum frequency of oscillation f_{max}: (a) cutoff frequency; and (b) maximum frequency of oscillation. The numbers at the data points indicate the gate length of the corresponding transistor in nm (at the left) and the frequency f_T or f_{max} in GHz (at the right). The data for the experimental GFETs, Si MOSFETs, and III–V HEMTs are taken from the compilations in [4–6,31,45].

While a high f_T is certainly desirable for a good RF FET, the more important RF figure of merit is the maximum frequency of oscillation f_{max}. Figure 6b compares experimental f_{max} data of competing RF FETs with the simulated f_{max} of our 50-nm gate GNR MOSFETs. Since the gate resistance has a strong impact on f_{max}, Figure 6b contains two simulated f_{max} data points for the GNR MOSFET. The higher f_{max} of 1.04 THz has been simulated assuming the idealized case of zero gate resistance and the second f_{max} of 413 GHz has been calculated for the more realistic case assuming a gate resistance R_G equal to the source access resistance R_S of the GNR MOSFET. As can be seen, the experimental GFETs suffer from poor maximum frequencies of oscillation, mainly due to their unsatisfying current saturation and the resulting large drain conductance causing limited power gain [5]. On the other hand, the simulated f_{max} performance of the GNR MOSFET is better than that of the best Si MOSFETs, even for the case $R_G = R_S$, and only the III–V HEMTs perform noticeably better than the GNR MOSFET.

3.3. Simulation Results for Multiple-Channel GNR MOSFETs with Interribbon Gates

To simulate multiple-channel GNR MOSFETs with interribbon gates as shown in Figure 4c,d correctly and to describe the interribbon gate effect accurately, full 3D device simulations would be necessary. It is possible, however, to get a sufficiently good impression on the behavior of multiple-channel GNR MOSFETs by 2D simulations when applying the approach described in the following.

197

In a first step we perform 2D simulations perpendicular to the direction of current flow, *i.e.*, in the y-z plane, see Figure 4, for zero applied drain-source voltage and calculate the electron sheet density n_{sh} and the gate capacitance C_G given by

$$C_G = q \frac{d\, n_{sh}}{d\, V_{GS}} \qquad (5)$$

This is done twice, first for the simplified structure without interribbon gate shown in Figure 4a and second for structures with interribbon gates, see Figure 4c. Figure 7a shows the calculated electron sheet density as a function of the effective gate voltage for GNR MOS structures (i) with a single GNR channel and top gate only; and (ii) multiple parallel GNR channels, interribbon gates, and varying separations d_{GNR} between adjacent GNRs. Clearly the interribbon gates have a significant effect on the sheet density (n_{sh} is much larger for the structures with interribbon gate compared to the simple structure without interribbon gate) and this effect is getting more pronounced for increasing GNR separation. The corresponding gate capacitance is shown in Figure 7b. A simple way to emulate the effect of the interribbon gates on the channel, even if only the simplified structure from Figure 4a, *i.e.*, without interribbon gate, is simulated, is to modify (increase) the dielectric constants of the gate oxide and of the GNRs by a correction factor. Figure 8 shows the correction factor for the gate capacitance, and thus for the dielectric constants, needed to reproduce the gate capacitance for a multiple-channel structure with interribbon gate.

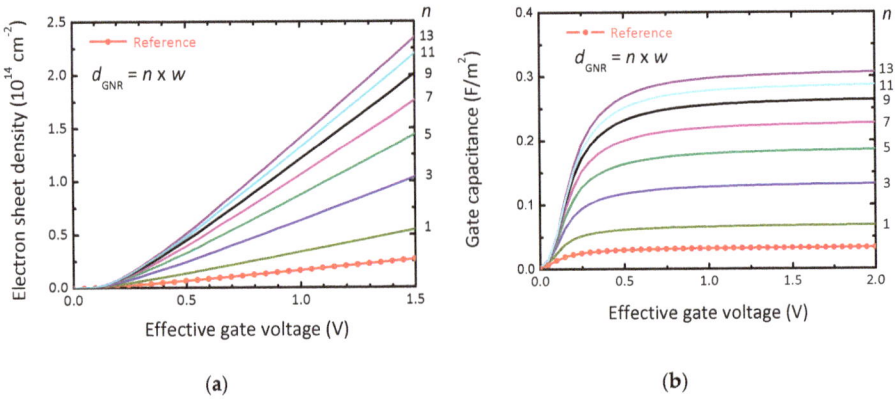

(a)   (b)

Figure 7. (**a**) Electron sheet density and (**b**) gate capacitance of multiple-channel GNR MOS structures with interribbon gates as a function of the effective gate voltage for different separations d_{GNR} between adjacent GNRs (lines). The sheet density and the gate capacitance obtained for the single-channel GNR MOS structure without interribbon gate are also shown (red lines with symbols, designated as Reference).

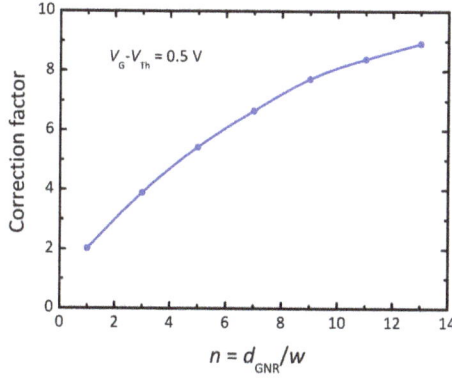

Figure 8. Correction factor for the dielectric constants of the top-gate dielectric and the GNR that reproduces the effect of the interribbon gate. Note that the correction factor has been determined for one single effective gate voltage (0.5 V).

Figure 9 shows that by applying the correction factor approach, the sheet density of multiple-channel GNR MOS structures with interribbon gates can be perfectly reproduced, even if only a single-channel structure without interribbon-gate is simulated. Note that the perfect agreement for effective gate voltages in the range 0–1.5 V has been obtained by multiplying the original dielectric constants for both the top-gate dielectric ($\varepsilon_r = 25$) and the GNR ($\varepsilon_r = 1.8$) by the correction factor from Figure 8, $i.e.$, the correction factor that has been elaborated for one single operating point ($V_{GS-eff} = 0.5$ V).

Having the correction factor approach established, in a second step we investigate how the interribbon gates affect the drain currents and the RF performance (in terms of f_T) of GNR MOSFETs by applying this approach. Figure 10 shows the transfer characteristics of three multiple-channel GNR MOSFETs having different separations between adjacent channels and of a single-channel transistor with top gate only and no interribbon gates. As to be expected from the enhanced carrier sheet density (see Figure 9), the drain currents of the structures with interribbon gates are noticeably larger compared to that of the simplified structure without interribbon gate. Moreover, the slopes of the transfer characteristics for the multiple-channel GNR MOSFETs are larger than that of the single-channel MOSFET, $i.e.$, multiple-channel MOSFETs show a higher transconductance.

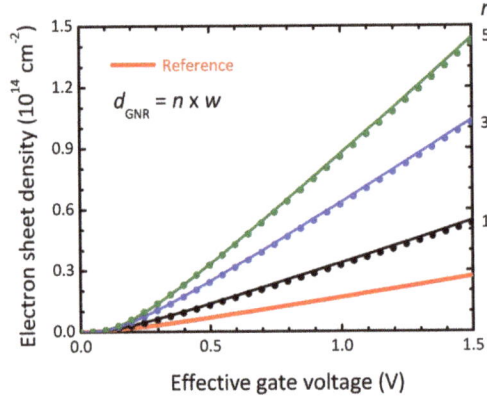

Figure 9. Electron sheet density in multiple-channel GNR MOS structures as a function of effective gate voltage. Green, blue, and black lines: Obtained when simulating the multiple-channel GNR MOS structure from Figure 4c, thereby using the original values for the dielectric constants for the gate dielectric ($\varepsilon_r = 25$) and the GNRs ($\varepsilon_r = 1.8$). Symbols: Obtained by simulating the single-channel GNR MOS structure from Figure 4a applying the correction factor method. Thick red line: Obtained by simulating the single-channel GNR MOS structure from Figure 4a, thereby using the original values for the dielectric constants for the gate dielectric ($\varepsilon_r = 25$) and the GNRs ($\varepsilon_r = 1.8$), designated as Reference.

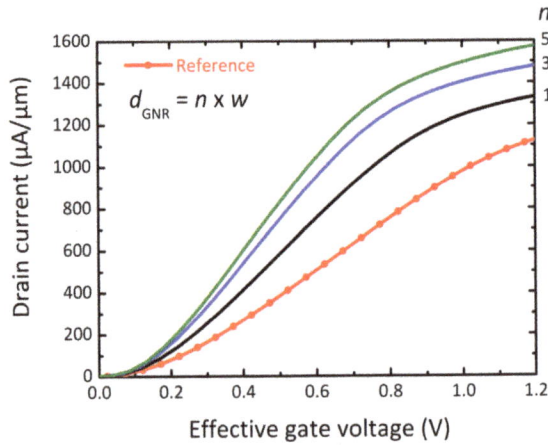

Figure 10. Transfer characteristics of multiple-channel GNR MOSFETs applying the correction factor approach (green, blue, and black lines). The transfer characteristics of the single-channel GNR MOSFET without interribbon gate (red line with symbols, designated as Reference), *i.e.*, the characteristics obtained using the original values of the dielectric constants of the gate dielectric ($\varepsilon_r = 25$) and the GNRs ($\varepsilon_r = 1.8$) are also shown.

200

On the other hand, the interribbon gates deteriorate the RF performance. While, as already shown in Figure 6a, the single-channel GNR MOSFET achieves a peak f_T of 215 GHz, the multiple-channel transistors show lower cutoff frequencies of 184 GHz, 158 GHz, and 145 GHz for GNR separations d_{GNR} of $1 \times w$, $3 \times w$, and $5 \times w$ (w is the GNR width), respectively. This effect looking surprising on first view is closely related to the observed degradation of the RF performance of Si FinFETs and Si tri-gate MOSFETs compared to their planar counterparts [46]. The degraded cutoff frequencies originate from additional capacitance components caused by the interribbon gates. These contribute to the current control less efficiently than the top gate capacitance leading to the situation that the effect of the increased gate capacitance cannot be fully compensated by the enhanced transconductance, *i.e.*, in multiple-channel GNR MOSFETs with interribbon gates the transconductance increases to a lesser extent than the gate capacitance. We note, however, that the interribbon gates will lead to a better suppression of short-channel effects and improve the scaling behavior of GNR MOSFETs. Such a combination of an improved scaling behavior and simultaneously degraded RF performance is not specific for GNR MOSFETs but has also been observed in Si FinFETs and Si tri-gate MOSFETs since the interribbon gate of multiple-channel GNR MOSFETs resembles the sidewall gates of FinFETs and tri-gate FETs. The maximum frequency of oscillation f_{max} of multiple-channel GNR MOSFETs will be affected by the additional capacitance of the interribbon gates to a similar extent as the cutoff frequency f_T since both f_T and f_{max} are roughly proportional to g_m/C_{gs} where g_m is the transconductance and C_{gs} is the gate-source capacitance (that includes contributions from both the top gate and the interribbon gate), see, e.g., Equations (3) and (4) in [5].

Figure 11 shows how our calculated cutoff frequencies for the single-channel and multiple-channel GNR MOSFETs compare to the cutoff frequencies simulated by other groups for GNR MOSFETs and GFETs and to the best reported cutoff frequencies of experimental GFETs. Experimental GNR MOSFETs could not be included in Figure 11 since the RF performance of such transistors has not been reported so far.

As can be seen, our simulated cutoff frequencies for 50-nm gate GNR MOSFETs are lower than those calculated for GFETs with the same gate length by Chauhan *et al.* [47] and Paussa *et al.* [48], both taking phonon scattering into account. This is reasonable since carrier transport in GNRs is degraded compared to that in gapless large-area graphene. Comparing our simulated f_T data with the calculated cutoff frequencies for GNR MOSFETs from [25], where phonon scattering has been taken into account, and those from [21,22] is more difficult since in [21,22,25] transistors with much shorter gates have been considered. Figure 11 shows, however, that our cutoff frequencies are by trend lower than those simulated in [21,22,25]. Although the approach used in the present work is engineering-style and involves less physics than the simulations from [21,22,25] we believe that our results are reasonable. The GNR channels considered in [22,25] have been assumed to be 10 nm wide and have a gap of 0.14 eV only compared to 1.4 eV in our more narrow GNRs. This means that carrier transport in the 10-nm wide ribbons is less degraded than in our GNRs. The simulations in [21,22], on the other hand, assume ballistic transport and therefore are expected to overestimate transistor performance.

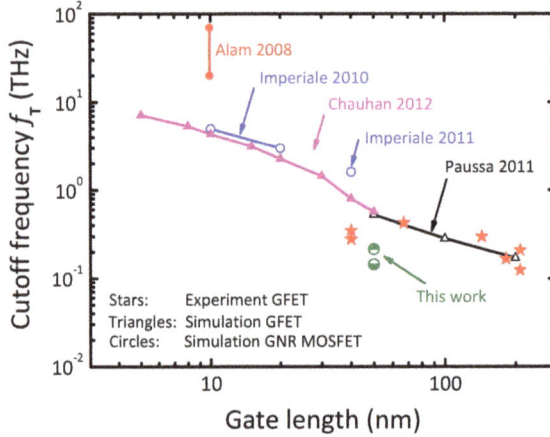

Figure 11. Simulated cutoff frequency of GNR MOSFETs and GFETs as a function of gate length. Data taken from the literature [21,22,47,48] and from the present work. f_T data of experimental GFETs and data taken from the compilations in are also shown [5,6]. The two data points designated by this work correspond to the cutoff frequency of the single-channel GNR MOSFETs, see also Figure 6a, and of the multiple-channel GNR MOSFET with a GNR separation of $5 \times w$.

4. Conclusions

An engineering approach to simulate the steady-state and small-signal behavior of GNR MOSFETs based on a classical 2D device simulator is presented. Modifications implemented in a commercial simulator enable taking the 1D DOS and the material properties of GNRs into account and allow the correct reproduction of the quantum capacitance of GNR MOS structures and of the effects of interribbon gates in multiple-channel GNR MOSFETs. Exemplarily, 50-nm gate ac $N = 7$ GNR MOSFETs in both single-channel and multiple-channel configurations have been investigated in detail. It is shown that multiple-channel GNR transistors show higher normalized drain currents and transconductances compared to their single-channel counterparts. On the other hand, the interribbon gates cause additional gate capacitance components whose effects cannot fully be compensated by the enhanced transconductance. Moreover, GNR MOSFETs show lower cutoff frequencies than

GFETs due to the degraded mobility in narrow GNRs. At the same time, however, the maximum frequency of oscillation of GNR MOSFETs is significantly higher compared to that of GFET due to the semiconducting nature of the GNR channel.

Acknowledgments: This work has been supported by Excellence Research and Intra-Faculty Research Grants of Technische Universität Ilmenau, Germany, and by DFG under contract numbers SCHW 729/16-1 and PE 624/11-1.

Author Contributions: Cedric Nanmeni Bondja developed the approach to account for the 1D density of states in ATLAS, developed the transport model, and performed the simulations; Zhansong Geng performed simulations; Ralf Granzner defined the simulated device structures and analyzed the simulation results; Jörg Pezoldt contributed to the definition of the simulated device structures and to the interpretation of the results; Frank Schwierz conceived the simulation study, compared the simulation results with experimental data and simulation results from the literature, and wrote the paper.

Conflicts of Interest: The authors declare no conflict of interest.

Appendix

In the following, the expressions for the calculation of the carrier sheet density (in units of cm^{-2}) and the quantum capacitance for the 3D (bulk) case [42], the 2D case, and the 1D case [42], which have been used in Section 2.3, are summarized.

3D case

$$n_{sh}^{3D} = t_c \int_0^\infty g_{3D}\left(\varepsilon\right) f\left(\varepsilon - q\varphi_c\right) d\varepsilon \tag{A1}$$

$$g_{3D}\left(\varepsilon\right) = \frac{\nu\, m_{eff}\sqrt{2m_{eff}\varepsilon}}{\pi^2 \hbar^3} \tag{A2}$$

$$f\left(\varepsilon - q\varphi_c\right) = \frac{1}{1 + \exp\left(\dfrac{\varepsilon - q\varphi_c}{k_B T}\right)} \tag{A3}$$

$$C_q^{3D} = \frac{\nu\, q^2 t_c m_{eff}\sqrt{2m_{eff}}}{4k_B T \pi^2 \hbar^3} \int_0^\infty \sqrt{\varepsilon} \cdot \cosh^{-2}\left(\frac{\varepsilon - q\varphi_c}{2k_B T}\right) d\varepsilon \tag{A4}$$

2D case

$$n_{sh}^{2D} = \sum_i \int_0^\infty g_{2D}^i f\left(\varepsilon, E_i\right) d\varepsilon \tag{A5}$$

$$g_{2D}^i = \frac{\nu_i\, m_{effi}}{\pi \hbar^2} \tag{A6}$$

$$f\left(\varepsilon, E_i\right) = \frac{1}{1 + \exp\left(\dfrac{\varepsilon + E_i - q\varphi_c}{k_B T}\right)} \tag{A7}$$

$$C_q^{2D} = \frac{q^2}{\pi \hbar^2} \sum_i \frac{\nu_i \, m_{\text{eff}i}}{1 + \exp\left(\dfrac{E_i - q\varphi_c}{k_B T}\right)} \tag{A8}$$

1D case

$$n_{\text{sh}}^{1D} = \frac{1}{w} \sum_i \int_0^\infty g_{1D}^i f\left(\varepsilon, E_i\right) d\varepsilon \tag{A9}$$

$$g_{1D}^i\left(\varepsilon\right) = \frac{\nu_i}{\pi \hbar} \sqrt{\frac{2m_{\text{eff}i}}{\varepsilon}} \tag{A10}$$

$$C_q^{1D} = \frac{q^2}{w\sqrt{2}hk_B T} \sum_i \nu_i \sqrt{m_{\text{eff}i}} \int_0^\infty \frac{1}{\sqrt{\varepsilon}} \cosh^{-2}\left(\frac{\varepsilon + E_i - q\varphi_s}{2k_B T}\right) d\varepsilon \tag{A11}$$

where φ_c is the channel potential given by $-(E_C - E_F)/q$, q is the elementary charge, ε is the kinetic energy of the electrons, g is the density of states (g^i is the density of states in the ith subband), t_c is the GNR thickness, ν is the valley degeneracy factor, m_{eff} is the density of states effective mass, f is the Fermi–Dirac distribution function, E_i is the position of the ith subband with respect to the conduction band edge, k_B is the Boltzmann constant, and T is the temperature. Note that the expression for the quantum capacitance for the 2D case, *i.e.*, Equation (A8), does not contain an integral since for the expression of the sheet density, *i.e.*, Equation (A5), an analytical solution can be derived.

In the ATLAS simulations, the basic semiconductor equations, *i.e.*, Poisson's equation, the current equations for electrons and holes (using Equation (4) with the parameters given below Table 1), and the continuity equations, are solved self-consistently.

References

1. Geim, A.K.; Novoselov, K.S. The rise of graphene. *Nat. Mater.* **2007**, *6*, 183–191.
2. Wu, Y.; Jenkins, K.A.; Valdes-Garcia, A.; Farmer, D.B.; Zhu, Y.; Bol, A.A.; Dimitrakopoulos, C.; Zhu, W.; Xia, F.; Avouris, P.; *et al.* State-of-the-art graphene high-frequency electronics. *Nano Lett.* **2012**, *2*, 3062–3067.
3. Cheng, R.; Bai, J.; Liao, L.; Zhou, H.; Chen, Y.; Liu, L.; Lin, Y.-C.; Jiang, S.; Huang, Y.; Duan, X. High-frequency self-aligned graphene transistors with transferred gate stacks. *Proc. Natl. Acad. Sci. USA* **2012**, *109*, 11588–11592.
4. Schwierz, F. Graphene transistors. *Nat. Nanotechnol.* **2010**, *5*, 487–496.
5. Schwierz, F. Graphene transistors: Status, prospects, and problems. *Proc. IEEE* **2013**, *101*, 1567–1584.
6. Lemme, M.C.; Li, L.-J.; Palacios, T.; Schwierz, F. Two-dimensional materials for electronic applications. *MRS Bull.* **2014**, *39*, 711–718.

7. Castro, E.V.; Novoselov, K.S.; Morozov, S.V.; Peres, N.M.R.; Lopes-dos-Santos, J.M.B.; Nilsson, J.; Guinea, F.; Geim, A.K.; Castro-Neto, A.H. Biased bilayer graphene: Semiconductor with a gap tunable by the electric field effect. *Phys. Rev. Lett.* **2007**, *99*, 216802.

8. Szafranek, B.N.; Fiori, G.; Schall, D.; Neumaier, D.; Kurz, H. Current saturation and voltage gain in bilayer graphene field effect transistors. *Nano Lett.* **2012**, *12*, 1324–1328.

9. Han, M.Y.; Özyilmaz, B.; Zhang, Y.; Kim, P. Energy band-gap engineering of graphene nanoribbons. *Phys. Rev. Lett.* **2007**, *98*, 206805.

10. Linden, S.; Zhong, D.; Timmer, A.; Aghdassi, N.; Franke, J.H.; Zhang, H.; Feng, X.; Müllen, K.; Fuchs, H.; Chi, L.; *et al.* Electronic structure of spatially aligned graphene nanoribbons on Au(788). *Phys. Rev. Lett.* **2012**, *108*, 216801.

11. Liang, X.; Jung, Y.-S.; Wu, S.; Ismach, A.; Olynick, D.L.; Cabrini, S.; Bokor, J. Formation of bandgap and subbands in graphene nanomeshes with sub-10 nm ribbon width fabricated via nanoimprint lithography. *Nano Lett.* **2010**, *10*, 2454–2460.

12. Berrada, S.; Nguyen, V.H.; Querlioz, D.; Saint-Martin, J.; Alarcon, A.; Chassat, C.; Bournel, A.; Dollfus, P. Graphene nanomesh transistor with high on/off ratio and good saturation behavior. *Appl. Phys. Lett.* **2013**, *103*, 183509.

13. Raza, H.; Kan, E.C. Armchair graphene nanoribbons: Electronic structure and electric-field modulation. *Phys. Rev. B* **2008**, *77*, 245434.

14. Yang, L.; Park, C.-H.; Son, Y.-W.; Cohen, M.L.; Louie, S.G. Quasiparticle energies and band gaps in graphene nanoribbons. *Phys. Rev. Lett.* **2007**, *99*, 186801.

15. Gunlycke, D.; White, C.T. Tight-binding energy dispersions of armchair-edge graphene nanostripes. *Phys. Rev. B* **2008**, *77*, 115116.

16. The International Technology Roadmap for Semiconductors. Available online: http://www.itrs.net (accessed on 15 October 2015).

17. Li, X.; Wang, X.; Zhang, L.; Lee, S.; Dai, H. Chemically derived, ultrasmooth graphene nanoribbon semiconductors. *Science* **2008**, *319*, 1229–1232.

18. Wang, X.; Ouyang, Y.; Li, X.; Wang, H.; Guo, J.; Dai, H. Room-temperature all-semiconducting sub-10-nm graphene nanoribbon field-effect transistors. *Phys. Rev. Lett.* **2008**, *100*, 206803.

19. Bai, J.; Duan, X.; Huang, Y. Rational fabrication of graphene nanoribbons using a nanowire etch mask. *Nano Lett.* **2009**, *9*, 2083–2087.

20. Liao, L.; Bai, J.; Cheng, R.; Lin, Y.-C.; Jiang, S.; Huang, Y.; Duan, X. Top-gated graphene nanoribbon transistors with ultrathin high-*k* dielectrics. *Nano Lett.* **2010**, *10*, 1917–1921.

21. Alam, K. Gate dielectric scaling of top gate carbon nanoribbon on insulator transistors. *J. Appl. Phys.* **2008**, *104*, 074313.

22. Imperiale, I.; Gnudi, A.; Gnani, E.; Reggiani, S.; Baccarani, G. High-frequency analog GNR-FET design criteria. In Proceedings of the 2011 European Solid-State Device Research Conference (ESSDERC), Helsinki, Finland, 12–16 September 2011; pp. 303–306.

23. Harada, N.; Sato, S.; Yokoyama, N. Theoretical investigation of graphene nanoribbon field-effect transistors designed for digital applications. *Jpn. J. Appl. Phys.* **2013**, *52*, 094301.

24. Liang, G.; Neophytou, N.; Lundstrom, M.S.; Nikonov, D.E. Ballistic graphene nanoribbon metal-oxide-semiconductor field-effect transistors: A full real-space quantum transport simulation. *J. Appl. Phys.* **2007**, *102*, 054307.

25. Imperiale, I.; Bonsignore, S.; Gnudi, A.; Gnani, E.; Reggiani, S.; Baccarani, G. Computational study of graphene nanoribbon FETs for RF applications. In Proceedings of the 2010 IEEE International Electron Devices Meeting (IEDM), San Francisco, CA, USA, 6–8 December 2010; pp. 732–735.

26. Fiori, G.; Iannaccone, G. Simulation of graphene nanoribbon field-effect transistors. *IEEE Electron Device Lett.* **2007**, *8*, 760–762.

27. Goharrizi, A.Y.; Pourfarth, M.; Fathipour, M.; Kosina, H. Device performance of graphene nanoribbon field-effect transistors in the presence of edge-line roughness. *IEEE Trans. Electron Devices* **2012**, *59*, 3527–3532.

28. Kliros, G.S. Gate capacitance modeling and width-dependent performance of graphene nanoribbon transistors. *Microelctron. Eng.* **2013**, *112*, 220–226.

29. Bruzzone, S.; Iannaccone, G.; Marzari, N.; Fiori, G. An open-source multiscale framework for the simulation of nanoscale devices. *IEEE Trans. Electron Devices* **2014**, *61*, 48–53.

30. ATLAS User's Manual—Device Simulation Software, Silvaco. Available online: http://dynamic.silvaco. com/dynamicweb/jsp/downloads/DownloadManuals Action.do?req=silentmanuals&nm=atlas (accessed on 15 October 2015).

31. Schwierz, F.; Pezoldt, J.; Granzner, R. Two-dimensional materials and their prospects in transistor applications. *Nanoscale* **2015**, *7*, 8261–8283.

32. Ruffieux, P.; Cai, J.; Plumb, N.; Patthey, L.; Prezzi, D.; Ferretti, A.; Molinari, E.; Feng, X.; Müllen, K.; Pignedoli, C.A.; *et al.* Electronic structure of atomically precise graphene nanoribbons. *ACS Nano* **2012**, *6*, 6930–6935.

33. Chen, Y.-C.; de Oteyza, D.G.; Pedramrazi, Z.; Chen, C.; Fischer, F.R.; Crommie, M.F. Tuning the band gap of graphene nanoribbons synthesized from molecular precursors. *ACS Nano* **2013**, *7*, 6123–6128.

34. Fang, T.; Konar, A.; Xing, H.; Jena, D. Carrier statistics and quantum capacitance of graphene sheets and ribbons. *Appl. Phys. Lett.* **2007**, *91*, 092109.

35. Granzner, R.; Polyakov, V.M.; Schwierz, F.; Kittler, M.; Luyken, R.J.; Rösner, W.; Städele, M. Simulation of nanoscale MOSFETs using modified drift-diffusion and hydrodynamic models and comparison with Monte Carlo results. *Microelectron. Eng.* **2006**, *83*, 241–246.

36. Szabo, A.; Rhyner, R.; Luisier, M. Ab-initio simulations of MoS_2 transistors: From mobility calculation to device performance evaluation. In Proceedings of the 2014 IEEE International Electron Devices Meeting (IEDM), San Francisco, CA, USA, 15–17 December 2014; pp. 725–728.

37. Cao, W.; Kang, J.; Sarkar, D.; Liu, W.; Banerjee, K. Performance evaluation and design considerations of 2D semiconductor based FETs for sub-10 nm VLS. In Proceedings of the 2014 IEEE International Electron Devices Meeting (IEDM), San Francisco, CA, USA, 15–17 December 2014; pp. 729–732.

38. Liu, L.; Lu, Y.; Guo, J. On monolayer MoS_2 field-effect transistors at the scaling limit. *IEEE Trans. Electron Devices* **2013**, *60*, 4133–4139.

39. Ancona, M.G. Electron transport in graphene from a diffusion-drift perspective. *IEEE Trans. Electron Devices* **2010**, *57*, 681–689.

40. Betti, A.; Fiori, G.; Iannaccone, G. Drift velocity peak and negative differential mobility in high field transport in graphene nanoribbons explained by numerical simulations. *Appl. Phys. Lett.* **2011**, *99*, 242108.

41. Caughey, D.M.; Thomas, R.E. Carrier mobilities in silicon empirically related to doping and field. *Proc. IEEE* **1967**, *52*, 2192–2193.

42. Granzner, R.; Thiele, S.; Schippel, C.; Schwierz, F. Quantum effects on the gate capacitance of trigate SOI MOSFETs. *IEEE Trans. Electron Devices* **2010**, *57*, 3231–3237.

43. Unluer, D.; Tseng, F.; Ghosh, A.W.; Stan, M.R. Monolithically patterned wide-narrow-wide all-graphene devices. *IEEE Trans. Nanotechnol.* **2011**, *10*, 931–939.

44. Schwierz, F.; Liou, J.J. *Modern Microwave Transistors*; John Wiley & Sons: Hoboken, NJ, USA, 2003.

45. Schwierz, F. *Microwave Transistors: State of the Art in the 1980s, 1990s, 2000s, and 2010s. A Compilation of 1500 Top References*; TU Ilmenau: Ilmenau, Germany, 2015; unpublished.

46. Kranti, A.; Raskin, J.-P.; Armstrong, G.A. Optimizing FinFET geometry and parasitics for RF applications. In Proceedings of the IEEE International SOI Conference, New Paltz, NY, USA, 6–9 October 2008; pp. 123–124.

47. Chauhan, J.; Liu, L.; Lu, Y.; Guo, J. A computational study of high-frequency behavior of graphene field-effect transistors. *J. Appl. Phys.* **2012**, *111*, 094313.

48. Paussa, A.; Geromel, M.; Palestri, P.; Bresciani, M.; Esseni, D.; Selmi, L. Simulation of graphene nanoscale RF transistors including scattering and generation/recombination mechanisms. In Proceedings of the 2011 International Electron Devices Meeting, Washington, DC, USA, 5–7 December 2011; pp. 271–274.

Electrical Compact Modeling of Graphene Base Transistors

Sébastien Frégonèse, Stefano Venica, Francesco Driussi and Thomas Zimmer

Abstract: Following the recent development of the Graphene Base Transistor (GBT), a new electrical compact model for GBT devices is proposed. The transistor model includes the quantum capacitance model to obtain a self-consistent base potential. It also uses a versatile transfer current equation to be compatible with the different possible GBT configurations and it account for high injection conditions thanks to a transit time based charge model. Finally, the developed large signal model has been implemented in Verilog-A code and can be used for simulation in a standard circuit design environment such as Cadence or ADS. This model has been verified using advanced numerical simulation.

Reprinted from *Electronics*. Cite as: Frégonèse, S.; Venica, S.; Driussi, F.; Zimmer, T. Electrical Compact Modeling of Graphene Base Transistors. *Electronics* **2015**, *4*, 969–977.

1. Introduction

The physical properties of graphene are of highest interest for electronic applications and its properties have been used by several research groups to develop radio frequency (RF) and microwave Graphene Field Effect Transistors (GFET) [1–3]. Unfortunately, the lack of energy bandgap in graphene induces poor DC electrical characteristics and GFETs are still under evaluation [4] and optimization. Also, new transistor concepts are explored such as the Graphene Barristor [5] or the hot electron graphene base transistor (GBT) [6,7]. As explained by the inventors [7], compared to the GFET where the carrier transport is within the plane of the graphene sheet, *"the GBT is based on a vertical arrangement of emitter (E), base (B), and collector (C), just like a hot electron transistor or a vacuum triode"*. This vertical stack considers an emitter-base and a base-collector energy barrier that are controlled by the graphene base and the collector potentials. In the off-state, the carriers face a large barrier potential, while, in the on-state, this barrier vanishes when a sufficient positive bias is applied to the base and collector. The transistor concept has been demonstrated in [7] and it is under optimization to improve the GBT electrical performance [8].

Venica *et al.* and Driussi *et al.* [9,10] have developed physics based device simulators with different level of accuracy in order to improve the understanding of the GBT operation, and to optimize the transistor as a single element [11]. A small signal model has been proposed in [6]. In order to evaluate this transistor in circuit configuration, a large signal compact model is necessary.

In this paper, we propose a large signal compact model, whose verification is done by means of comparison with numerical simulation [9,11]. The paper is organized in two parts: the developed transistor compact model is described in the first part; then the second part compares compact model results to numerical simulation data.

2. Compact Model

The GBT transistor structure is presented in Figure 1. The vertical stack comprises the emitter, the emitter-base region EBi, the graphene base, the base-collector region BCi and the collector. The barrier potential height of the EBi and BCi regions controls the carrier transport mode such as tunneling or thermionic transport. Hence, the choice of the material used for the EBi and BCi region is of major importance. EBi and BCi can be either an insulator material such as SiO_2 or high-k dielectrics to exploit a tunneling transport [6] or a semiconductor such as Ge or Si to foster a thermionic current [12].

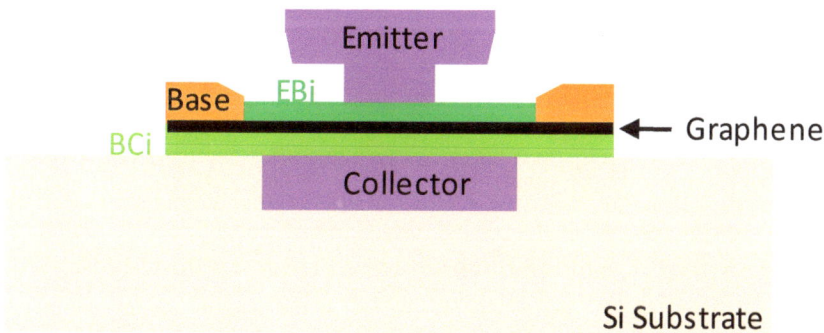

Figure 1. Schematic representation of the transistor structure.

According to the transistor structure (see Figure 1), the following equivalent circuit is proposed (see Figure 2). The extrinsic circuit is composed of three access resistances R_B, R_C, R_E. Concerning the intrinsic part, the core of the model is based on the self-consistent calculation of the internal base potential V_{Bi}, which is a function of the charge in the emitter, the collector and within the graphene layer. These charges are modeled through different capacitances. First, C_Q capacitance models the quantum capacitance of the graphene layer, while C_{BE0} and C_{BC0} describe the EBi and BCi capacitances, respectively. Finally, C_{DC} and C_{DE} are diffusion capacitances that take into account the additional charge due to carrier transport in the EBi and BCi regions. These capacitances are of interest for medium to high injection conditions [9]. Finally, two diodes I_{BE} and I_{BC} are modeling the base-emitter and base-collector current, respectively, and one voltage controlled current source I_{CE} is introduced for the collector-emitter transfer current. Each element is described in the next section.

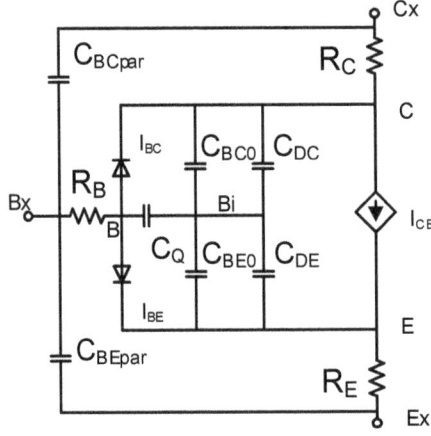

Figure 2. Equivalent circuit of the Graphene Base Transistor (GBT) electrical compact model.

2.1. Self-Consistent Calculation of the Internal Base Potential

The charge in the graphene layer can be computed by combining the specific density of states of graphene with the Fermi approach for the carrier distribution. Assuming that the potential drop into the graphene base $|V_{BBi}| \gg kT/q$ (k is the Boltzmann constant, T is the absolute temperature and q is the elementary charge), the total charge density can be approximated as follows [13]:

$$Q_G \approx q \left(-\frac{q^2}{\pi \left(\hbar v_f \right)^2} |V_{BBi}| V_{BBi} \right) = \frac{1}{2} C_Q V_{BBi} \tag{1}$$

with $\hbar$ the reduced Planck constant, v_f the Fermi velocity and $C_Q = \kappa |V_{BBi}|$, $\kappa = -2 \frac{q^2}{\pi (\hbar v_f)^2}$.

Introducing the graphene charge in the equivalent circuit, the internal base potential can be calculated by solving the following equation:

$$Q_G + C_{BC} V_{BiC} + C_{BE} V_{BiE} = 0 \tag{2}$$

where $V_{BiC} = V_{Bi} - V_C$, $V_{BiE} = V_{Bi} - V_E$, $C_{BC} = C_{BC0} + C_{DC}$, $C_{BE} = C_{BE0} + C_{DE}$ (see Figure 2) and $C_{BC0} = \frac{\varepsilon_{BC} A_E}{e_{BC}}$, $C_{BE0} = \frac{\varepsilon_{BE} A_E}{e_{BE}}$ are the oxide capacitances. e_{BC} and e_{BE} are the insulator thicknesses of EBi and BCi regions, respectively, and ε_{BE} and ε_{BC}

are the associated permittivity. A_E is the emitter area. The diffusion capacitances C_{DE} and C_{DC} will be described in Section II-C.

Substituting Equation (1) in Equation (2), the V_{BBi} potential is introduced in the equation:

$$\frac{1}{2}\kappa |V_{BBi}| V_{BBi} + C_{BC} (-V_{BBi} + V_{BC}) + C_{BE} (-V_{BBi} + V_{BE}) = 0 \tag{3}$$

Equation (3) is a second order polynomial that can be solved and gives the following solution [6]:

$$V_{BBi} = \frac{-(C_{BC} + C_{BE}) + \sqrt{(C_{BC} + C_{BE})^2 \mp 2\kappa (C_{BC} V_{BC} + C_{BE} V_{BE})}}{\pm \kappa} \tag{4}$$

2.1.1. Description of Diodes and Transfer Current Source

As described above, the EBi and BCi material can be either insulator or semiconductor and the charge transport can be dominated by tunneling emission or thermionic transport. In order to obtain a versatile compact model, the diode and transfer current source equations will be based on flexible and simple relationships.

For the base-emitter and base-collector tunnel or semiconductor-graphene diodes, an exponential equation is used (see [14]) and modified to gain in adaptability:

$$I_{BE} = A_E J_{SBE} \exp\left(\frac{V_{BiE} - \phi_{BE}}{B_{BE}}\right) \tag{5}$$

$$I_{BC} = A_E J_{SBC} \exp\left(\frac{V_{BiC} - \phi_{BC}}{B_{BC}}\right) \tag{6}$$

where J_{SBE} and J_{SBC} are the corresponding saturation currents, ϕ_{BE} and ϕ_{BC} are the barrier heights and B_{BE} and B_{BC} are fitting parameters of the slope of the exponential function.

For the transfer current source, a modified Landauer based equation is used [15].

$$I_{CE} = A_E J_{SF} \left[\begin{array}{c} \ln\left(1 + \exp\left(\frac{V_{BiE} - \phi_{BE}}{B_{BE}}\right)\right) \\ -\ln\left(1 + \exp\left(\frac{V_{BiC} - \phi_{BC}}{B_{BC}}\right)\right) \end{array} \right] \times f_{CE} \tag{7}$$

$$f_{CE} = \begin{cases} 0 \, if \, V_{CE} < \phi_{CE} \\ 1 \, if \, V_{CE} > \phi_{CE} \end{cases}$$

J_{SF} is a saturation current and the f_{CE} parameter allows zeroing the current if the collector-emitter barrier ϕ_{CE} is too high to be crossed by the carrier at low V_{CE}. ϕ_{CE} is used as a fitting parameter for the low V_{CE} bias regime.

2.1.2. Medium to High Current Injection Effects

At medium to high current conditions, the charge injected through the transfer current in the EBi and BCi junctions needs to be taken into account. This charge will modify the charge equilibrium in the intrinsic transistor and will induce a shift of the internal base potential. Hence, as suggested in Figure 2, diffusion capacitances are included in the equivalent circuit and this will affect the potential drop V_{BBi} through the Equation (4).

The diffusion charge within the two regions is computed by considering a transit time approach. At medium injection level, the charge can be approximated by

$$Q_{DC} = \tau_{CB}I_{CE}, Q_{DE} = \tau_{EB}I_{CE} \tag{8}$$

τ_{CB}, τ_{EB} are the transit times of the BCi and EBi region at medium injection, respectively. At high injection, an additional charge ΔQ_K [9,12], with respect to Q_{DE} appears when the transfer current I_{CE} overpass the critical current I_{CK}:

$$\Delta Q_K = \Delta \tau_K \left(\frac{I_{CE}}{I_{CK}}\right)^{\gamma} I_{CE} \tag{9}$$

$\Delta\tau_K$ is the additive transit time at high injection and γ is a fitting parameter.

Combining Equations (8) and (9) and deriving the equation, the related capacitance can be deduced:

$$C_{DC} = \frac{dQ_{DC}}{dV_{CE}}, \quad C_{DE} = \frac{d(Q_{DE} + Q_K)}{dV_{BiE}} \tag{10}$$

These elements can be either derived analytically or directly computed using a derivative function in Verilog language.

2.2. Comparison to Numerical Simulation

In order to validate our model and to demonstrate its physical basis, a comparison between physics-based numerical simulations and our compact model is provided. The 1-D numerical model of [9] solves the electrostatics of the GBT self-consistently with the calculated tunneling current and estimates the transit frequency f_T. Concerning the currents, since the physical origin of the base current is still unclear and debated [7,16], a perfectly transparent graphene layer is assumed and the base current is neglected. Hence, we assume a priori that the collector current is the current due to electrons injected from the emitter and, consistently, crossing the whole device.

The simulated device assumes that the EBi and BCi layers are made of two insulators [11]: the EBi region has a 2 nm high permittivity insulator ($\varepsilon_r = 25$)

while the BCi region has a 12 nm oxide with $\varepsilon_r = 2.5$. Only the intrinsic device is simulated and one would need to consider parasitic elements to have a realistic circuit simulation.

First, a comparison between numerical simulations and compact model simulations of the calculated graphene base charge is provided in Figure 3. Despite a deviation at large I_{CE} mainly due to different modeling approaches for the high injection effects (model in [9] solves the potential along the device self-consistently with the traveling electrons), a fairly good agreement is found between the two models. This verifies the adequate calculation of the intrinsic base potential V_{Bi} by the compact model (see Figure 4).

The transfer characteristics of the device are simulated in Figure 5. At low injection condition, a similar behavior is observed despite a small disagreement. A good agreement is observed at medium to high electron injection levels. Figure 6, instead, shows the associated transconductance, which is of major importance for RF circuit simulation. A very good agreement is observed for low and medium bias, a reasonable agreement can be found for high bias. Finally, Figure 7 compares the output curves confirming a good matching between the two models. It should be underlined that the slope g_{CE} (see Figure 8) is properly modeled; this is mandatory to correctly model the voltage gain of an amplifier circuit.

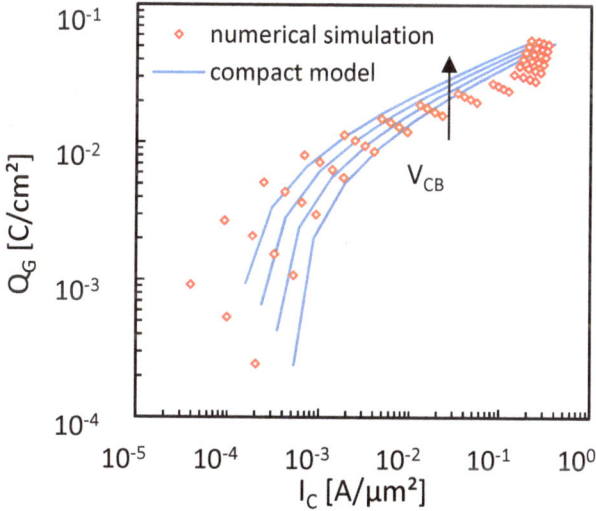

Figure 3. Charge Q_G *versus* I_{CE} curves for different V_{CB} (2, 3, 4, 5 V), numerical simulation and compact model simulation.

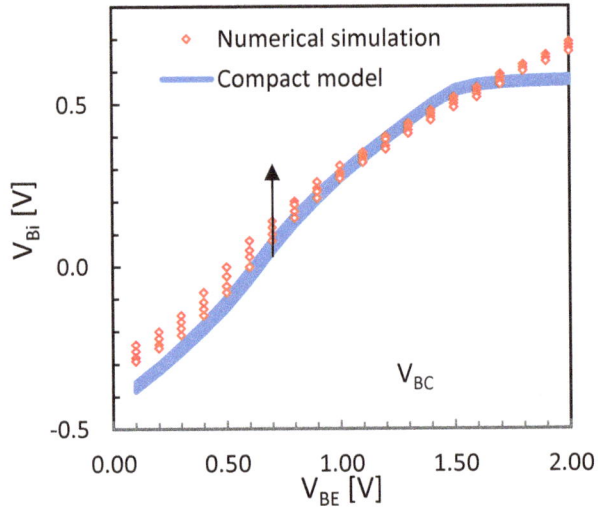

Figure 4. Graphene Fermi potential *versus* V_{BE} curves for different V_{CB} (2, 3, 4, 5 V), numerical simulation and compact model simulation.

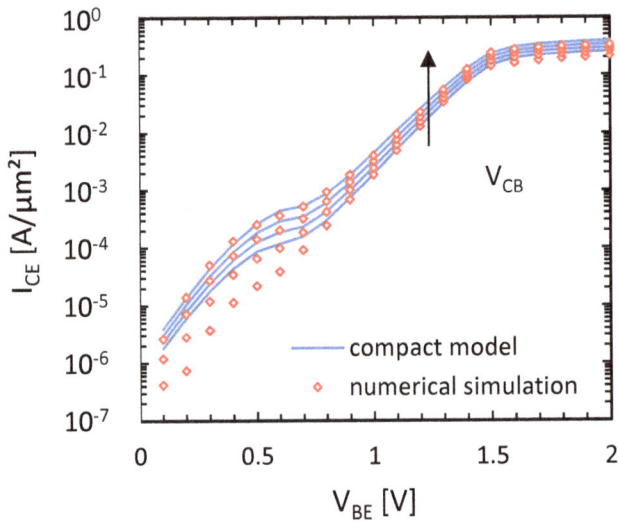

Figure 5. I_{CE} *versus* V_{BE} curves for different V_{CB} values (2, 3, 4, 5 V) simulated with the numerical model and the compact model.

214

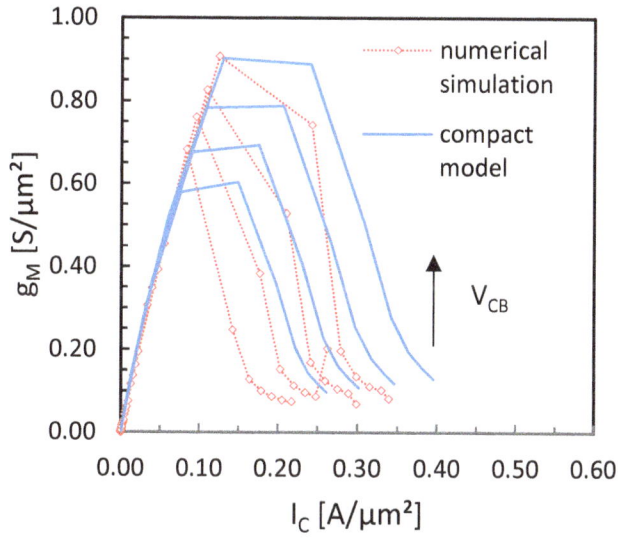

Figure 6. g_M *versus* I_C curves for different V_{CB} (2, 3, 4, 5 V) simulated with the numerical model and the compact model.

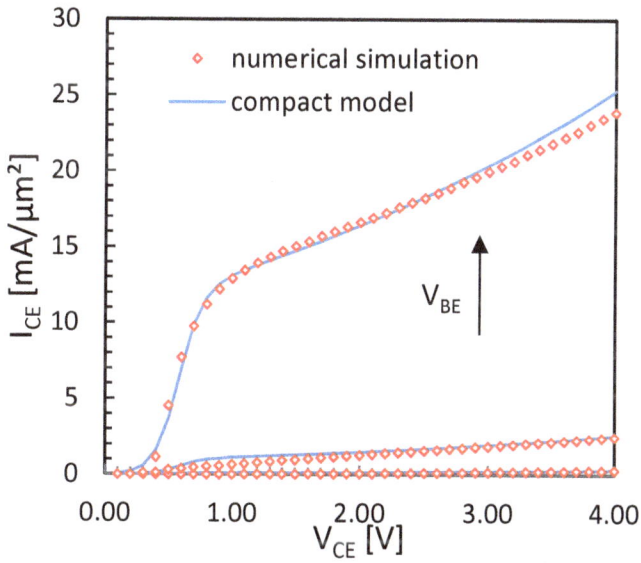

Figure 7. I_{CE} *versus* V_{CE} curves for different V_{BE} values (0.75, 1, 1.25 V), numerical simulation and compact model simulation.

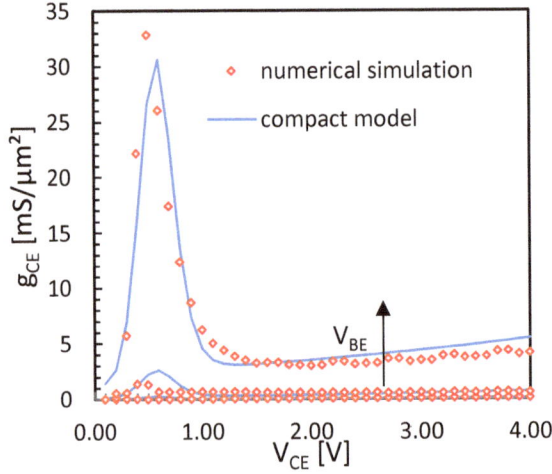

Figure 8. g_{CE} *versus* V_{CE} curves for different V_{BE} values (0.75, 1, 1.25 V), numerical simulation and compact model simulation.

In addition, S parameter simulations have been performed to extract the transit frequency; f_T is then compared to numerical simulation results (see Figure 9). A good agreement is observed up to peak f_T, which is the optimum bias condition for circuit applications. At higher current levels, a deviation is observed. Again this can be due to the different modeling strategies adopted to model the high current effects in the GBT.

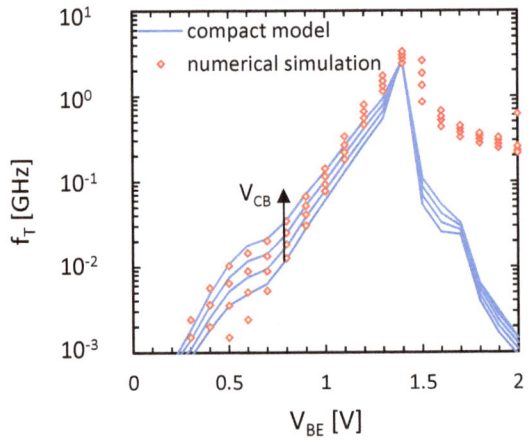

Figure 9. f_T *versus* V_{BE} curves for different V_{CB} (2, 3, 4, 5 V), numerical simulation and compact model simulation.

Table 1 summarizes the used compact model parameters. e_{BC}, e_{BE}, ε_{BE} and ε_{BC} are those used in the physics based simulations, while the others have been extracted by fitting in Figures 3–8.

Table 1. Parameters used in the compact model simulation.

Parameter Name/Unit	Parameter Value
J_{SF} (mA/μm^2)	437
B_{BE} (mV)	35.3
B_{BC} (mV)	94.2
Φ_{CE} (V)	0.57
e_{BE} (nm)	2
e_{BC} (nm)	12
ε_{BE}	25
ε_{BC}	2.5
τ_{BE} (fs)	0.35
τ_{BC} (fs)	6
Φ_{BE} (V)	0.893
Φ_{BC} (V)	1.2
γ	1.55
κ (μF/cm^2)	25
I_{CK} (A/μm^2)	0.7
$\Delta\tau_K$ (fs)	8×10^{-3}

3. Conclusions

We have developed a compact large signal model for GBT devices. Our model represents a good trade-off in terms of physics soundness and implementation complexity. Its accuracy relies on a self-consistent base potential calculation and a physics based charge model associated with an empirical and versatile transfer current equation. The model has been directly implemented in Verilog-A code. Hence, this model can be used to predict circuit performances based on GBT devices. Finally, the compact model accuracy has been verified by comparison with a physics-based electrical model, showing a good agreement with the numerical simulations.

Acknowledgments: This work is part of the GRADE project (317839) supported by the European Commission through the Seventh Framework Program for Research and Technological Development. The authors thank P. Palestri for fruitful discussions.

Author Contributions: S.F. and T.Z. developed the compact model equations. S.V., F.D. developed and performed the physics based simulations. All authors discussed the data and interpretation, and contributed during the writing of the manuscript. All authors have given approval to the final version of the manuscript.

Conflicts of Interest: The authors declare no conflict of interest.

References

1. Lin, Y.-M.; Farmer, D.B.; Jenkins, K.A.; Wu, Y.; Tedesco, J.; Myers-Ward, R.; Eddy, C.; Gaskill, D.; Dimitrakopoulos, C.; Avouris, P. Enhanced Performance in Epitaxial Graphene FETs with Optimized Channel Morphology. *Electron. Device Lett. IEEE* **2011**, *32*, 1343–1345.

2. Wu, Y.Q.; Farmer, D.B.; Valdes-Garcia, A.; Zhu, W.J. Record High RF Performance for Epitaxial Graphene Transistors. In Proceedings of the IEEE International on Electron Devices Meeting (IEDM), Washington, DC, USA, 5–7 December 2011.

3. Liao, L.; Lin, Y.; Bao, M.; Cheng, R.; Bai, J.; Liu, Y.; Qu, Y.; Wang, K.L.; Huang, Y.; Duan, X.; *et al.* High-speed graphene transistors with a self-aligned nanowire gate. *Nature* **2010**, *467*, 305–308.

4. Schwierz, F. Graphene Transistors: Status, Prospects, and Problems. In Proceedings of the IEEE, Ilmenau, Germany, 22 May 2013; Volume 101, pp. 1567–1584.

5. Yang, H.; Heo, J.S.; Park, S.; Song, H.J.; Seo, D.H.; Byun, K.-E.; Kim, P.; Yoo1, K.; Chung, H.-J.; Kim, K. Graphene Barristor, a Triode Device with a Gate-Controlled Schottky Barrier. *Science* **2012**, *336*, 1140–1143.

6. Mehr, W.; Abrowski, J.; Scheytt, C.; Lippert, G.; Xie, Y.-H.; Lemme, M.C.; Ostling, M.; Lupina, G. Vertical graphene base transistor. *IEEE Electron Device Lett.* **2012**, *33*, 691–693.

7. Vaziri, S.; Lupina, G.; Henke, C.; Smith, A.D.; Östling, M.; Dabrowski, J.; Lippert, G.; Mehr, W.; Lemme, M.C. A Graphene-Based Hot Electron Transistor. *Nano Lett.* **2013**, *13*, 1435–1439.

8. Vaziri, S.; Belete, M.; Litta, E.D.; Smith, A.D.; Lupina, G.; Lemme, M.; Östling, M. Bilayer Insulator Tunnel Barriers for Graphene-Based Vertical Hot-electron Transistors. *Nanoscale* **2015**, *7*, 13096–13104.

9. Venica, S.; Driussi, F.; Palestri, P.; Esseni, D.; Vaziri, S.; Selmi, L. Simulation Of DC and RF performance of the graphene base transistor. *IEEE Trans. Electron Devices* **2014**, *61*, 2570–2576.

10. Driussi, F.; Palestri, P.; Selmi, L. Modelling, simulation and design of the vertical Graphene Base Transistor. *Microelectron. Eng.* **2013**, *109*, 338–341.

11. Venica, S.; Driussi, F.; Palestri, P.; Selmi, L. Graphene Base Transistors with optimized emitter and dielectrics. In Proceedings of the Microelectronics, Electronics and Electronic Technology Conference, Opatija, Croatia, 26–30 May 2014; pp. 39–44.

12. Di Lecce, V.; Grassi, R.; Gnudi, A.; Gnani, E.; Reggiani, S.; Baccarani, G. Graphene-base heterojunction transistor: An attractive device for terahertz operation. *IEEE Trans. Electron Devices* **2013**, *60*, 4263–4268.

13. Xu, H.; Zhang, Z.; Peng, L.-M. Measurements and microscopic model of quantum capacitance in graphene. *Appl. Phys. Lett.* **2011**, *98*, 1–3.

14. Vaziri, S.; Belete, M.; Litta, E.D.; Smith, A.D.; Lupina, G.; Lemme, M.C.; Östlinga, M. Bilayer insulator tunnel barriers for graphene-based vertical hot-electron transistors. *Nanoscale* **2015**, *7*, 13096–13104.

15. Ferry, D.K.; Goodnick, S.M. *Transport in Nanostructures*; Cambridge University Press: Cambridge, UK, 1997.
16. Zeng, C.; Song, E.B.; Wang, M.; Lee, S.; Torres, C.M.; Tang, J.; Weiller, B.H.; Wang, K.L. Vertical graphene-base hot-electron transistor. *Nano Lett.* **2013**, *13*, 1435–1439.

Theoretical Analysis of Vibration Frequency of Graphene Sheets Used as Nanomechanical Mass Sensor

Toshiaki Natsuki

Abstract: Nanoelectromechanical resonator sensors based on graphene sheets (GS) show ultrahigh sensitivity to vibration. However, many factors such as the layer number and dimension of the GSs will affect the sensor characteristics. In this study, an analytical model is proposed to investigate the vibration behavior of double-layered graphene sheets (DLGSs) with attached nanoparticles. Based on nonlocal continuum mechanics, the influences of the layer number, dimensions of the GSs, and of the mass and position of nanoparticles attached to the GSs on the vibration response of GS resonators are discussed in detail. The results indicate that nanomasses can easily be detected by GS resonators, which can be used as a highly sensitive nanomechanical element in sensor systems. A logarithmically linear relationship exists between the frequency shift and the attached mass when the total mass attached to GS is less than about 1.0 zg. Accordingly, it is convenient to use a linear calibration for the calculation and determination of attached nanomasses. The simulation approach and the parametric investigation are useful tools for the design of graphene-based nanomass sensors and devices.

Reprinted from *Electronics*. Cite as: Natsuki, T. Theoretical Analysis of Vibration Frequency of Graphene Sheets Used as Nanomechanical Mass Sensor. *Electronics* **2015**, *4*, 723–736.

1. Introduction

Graphene sheets (GSs) have attracted great attention due to their extraordinary mechanical, electrical and thermal properties [1–3]. These fascinating carbon nanomaterials have many potential applications, such as reinforced materials, solar cells, molecule sensors and nanomechanical resonators [4–9]. Jang and his group [10–12] performed molecular dynamics (MD) simulation to investigate the mass sensitivity and resonant frequency of graphene nanomechanical resonators (GNMRs), the temperature-dependent scaling transition in the quality factors of GNMRs, and the effect of polar surfaces on the quality (Q)-factors of zinc oxide (ZnO) nanoresonators. The basic principle of nanoresonator sensors is the detection of a shift of the resonant frequency or wave velocity in the nanosensors caused by attached nanoparticles, including atoms or molecules. The sensitivity of the sensors and their applicability in distinguishing distinct types of atoms/molecules have been discussed by Wand and Arash [13], who simulated the detection of gas atoms based on wave velocity shifts in single-walled carbon nanotubes (SWCNTs) and explored the efficiency of nanotube-based sensors [14].

For structural applications of GSs, knowing their macroscopic properties is required. Therefore, both experimental and theoretical studies of GSs are important issues to design optimal materials and devices. At present, GSs are extensively investigated for applications in optoelectronic devices, high-performance hybrid supercapacitors, and various types of high performance sensors. Nanoelectromechanical systems (NEMS) are emerging as strong candidates for a host of important applications in semiconductor-based technology and fundamental science [15]. Graphene-based NEMS resonators could provide higher sensitivity as nanomechanical mass sensor. The operation of a NEMS mass sensor relies on monitoring how the resonance frequency of a nanomechanical resonator changes when an additional nanomass is adsorbed on its surface [16–18]. There is a fast-growing interest in GSs for use in NEMS resonators, given that lightness and stiffness are the essential characteristics sought after in NEMS resonators for sensing applications [19]. NEMS resonators can be fabricated based on single-layered (SLGSs) or multi-layered graphene sheets (MLGSs) by mechanically exfoliating thin sheets from graphite and putting them over trenches in a silicon oxide layer. Vibrations with fundamental resonant frequency in the megahertz range were actuated, either optically or electrically, and detected optically by interferometry [20].

Current efforts in graphene synthesis include micromechanical cleavage, liquid-phase exfoliation, chemical vapor deposition (CVD), and carbon segregation [21–25]. Double-layered graphene sheets (DLGSs) consist of two single GSs coupled by van der Waals (vdW) interaction forces, and can be synthesized on a silicon carbide substrate [26–28]. DLGSs are of considerable interest because of their unique electronic bands and mechanical properties. Different from SLGSs, the bandgap of DLGSs can be controlled externally by a gate bias and gaps up to 250 meV can be opened, which reveals the large potential of DLGSs for applications in NEMS [29–32]. On the other hand, DLGSs have higher stiffness and natural frequency than SLGSs because of the vdW interaction forces [33–35]. Thus, the resonance frequency and the frequency shift of DLGSs used as NEMS mass sensors can be measured more obviously.

NEMS mass sensors with a resonator can be fabricated from GSs, and the sensing mechanism is based on the fact that the resonant frequency is sensitive to changes in the attached mass [36]. Recently, various theoretical studies have been carried out utilizing the molecular dynamics (MD) method [37,38], molecular structural mechanics [39], or continuum mechanics [35,40–43] to investigate SLGSs for use in NEMS mass sensors. Arash *et al.* studied the vibration properties of SLGSs with five noble gas atoms (He, Ne, Ar, Kr, and Xe) attached using MD simulation to evaluate the applicability of graphene mass sensors [37]. The results indicate that the resolution of a mass sensor made of a square-shaped graphene sheet with a size of 10 nm could achieve an order of 10^{-24} kg and the mass sensitivity could be enhanced by decreasing the size of the SLGSs. Sakhaee-Pour et al. investigated the vibration behavior of defect-free SLGSs using molecular structural mechanics [39]. The principal frequency was highly sensitive to an added nanomass of the order of 10^{-24} kg, corresponding to Reference [37]. Based on a continuum elastic model, Lei et al. [39] analyzed the sensitivity of frequency shift of an atomic-resolution nanomechanical mass sensor modeled by a circular SLGS with attached nanoparticles. The sensitivity of such mass sensor could reach 10^{-27} kg. The effects of the nonlocal parameters and the attached mass on the frequency shifts of GSs were investigated using a nonlocal continuum elastic model [42,43]. The obtained results show that the frequency shift of the SLGS became smaller when the effect of nonlocal parameters was taken into account. Shen et al. [42] did the comparison between the continuum mechanics and the finite element method (FEM) and found a good agreement. This means that the SLGSs with an attached nanoparticle can be simulated accurately based on continuum mechanics. At present, the studies of the mass detection using the graphene-based nanomechanical sensor are focused solely on SLGSs. Because the resonance frequency and its frequency shift of DLGSs resonant could be measured more obviously, DLGSs based mass sensors need to be studied [44].

Ultrasonic vibration and attenuation are important properties related to the design and performance of the sensor devices. GSs appear to be excellent element materials for nanomechanical resonators because they can generate vibrations in the terahertz range. So far, only little has been reported on the resonant frequency analysis for DLGSs used as nanomass sensor, and it is the purpose of the present study to remedy this deficiency. We explore the potential of DLGSs used as a NEMS mass sensor, considering that DLGSs have higher strength and vibration frequency than SLGSs. Based on a nonlocal continuum theory, the influences of the mass and position of attached nanoparticles, of the dimensions of DLGSs, and of nonlocal parameters on the vibration response of DLGS sensor are investigated in detail.

2. Experimental Approach

2.1. Nonlocal Elasticity Theory

According to the general elasticity model, the stress components at any point depend only on the strain component at the same point. In the nonlocal elasticity theory, the stress field at a certain point is considered to be a function of the strain distribution over a certain representative volume of the material centered at that point. The nonlocal elasticity theory [45–47] can be used to study several phenomena related to nanoscale materials. For nonlocal linear elastic solids, the constitutive equation of motion is given by

$$t_{ij,j} + f_i = \rho \ddot{u}_i \qquad (1)$$

where ρ and f_i are the mass density and the applied body forces, respectively, u_i is the displacement vector, and t_{ij} is the stress tensor of the nonlocal elasticity expressed as

$$t_{ij} = \int_V \alpha \left(|x - x'|, e_0 a \right) \sigma_{ij} \left(x' \right) dV \left(x' \right) \qquad (2)$$

where x is a reference point in the body, e_0 is a constant appropriate to the material and has to be determined for each materials independently by experiments or atomistic simulation, a is the internal characteristic length (e.g., lattice parameter, granular size, distance between atoms bound), $|x - x'|$ is the Euclidean distance, and V is an integral region occupied by the body. The nonlocal kernel function $a \left(|x - x'|, e_0 a \right)$ incorporates the nonlocal effects into the constitutive equation and is given as

$$\alpha \left(|x - x'|, e_0 a \right) = \frac{1}{2\pi \left(e_0 a \right)^2} K_0 \left(\frac{\sqrt{x \cdot x}}{e_0 a} \right) \qquad (3)$$

where K_0 is the modified Bessel function.

In Equation (2), σ_{ij} is the local stress tensor of the classical elasticity theory at any point x' in the body and satisfies the constitutive relations

$$\sigma_{ij} \left(x' \right) = \lambda \varepsilon_{kk} \left(x' \right) \delta_{ij} + 2\mu \varepsilon_{ij} \left(x' \right) \qquad (4)$$

where λ and μ are the Lame constants and ε_{ij} is the strain tensor.

For two-dimensional nonlocal elastic beam theory, the stress–strain relation can be expressed as

$$\sigma_{ij} - \left(e_0 a \right)^2 \nabla^2 \sigma_{ij} = \lambda \varepsilon_{kk} \left(x' \right) \delta_{ij} + 2\mu \varepsilon_{ij} \left(x' \right) \qquad (5)$$

where ∇ is the Laplace operator, which is given as $\partial^2/\partial x_1^2 + \partial^2/\partial x_2^2$ in a two-dimensional rectangular coordinate system (x_1, x_2).

The nonlocal elasticity model has been widely adopted for tackling various problems of linear elasticity and micro- or nanostructural mechanics.

2.2. Single-Layer Graphene Sheets

Based on nonlocal continuum mechanics, we consider the dynamic behavior of a single-layer graphene sheet (SLGS) with an attached concentrated mass m_c located at an arbitrary position (x_0, y_0) as shown in Figure 1. The origin is taken at one corner of the mid-plane of the graphene sheet. The x- and y-axes are taken along the length L_a and width L_b of the SLGS, respectively, and the z-axis is taken along the thickness h of the SLGS.

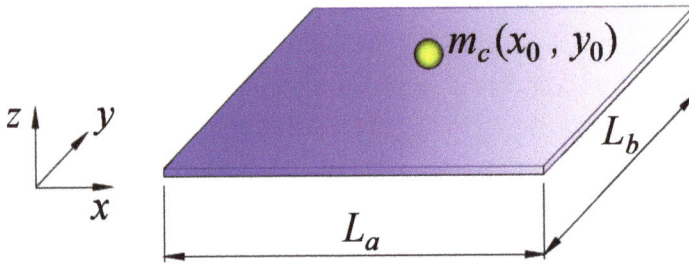

Figure 1. Schematic illustration showing a single-walled GS with an attached mass.

Using the nonlocal elasticity theory, the two-dimensional nonlocal constitutive equations of SLGS are given as

$$\sigma_{xx} - (e_0 a)^2 \left(\frac{\partial^2 \sigma_{xx}}{\partial x^2} + \frac{\partial^2 \sigma_{xx}}{\partial y^2} \right) = \frac{E}{1-v^2} \left(\varepsilon_{xx} + v\varepsilon_{yy} \right) \tag{6a}$$

$$\sigma_{yy} - (e_0 a)^2 \left(\frac{\partial^2 \sigma_{yy}}{\partial x^2} + \frac{\partial^2 \sigma_{yy}}{\partial y^2} \right) = \frac{E}{1-v^2} \left(\varepsilon_{yy} + v\varepsilon_{xx} \right) \tag{6b}$$

$$\tau_{xy} - (e_0 a)^2 \left(\frac{\partial^2 \sigma_{xy}}{\partial x^2} + \frac{\partial^2 \sigma_{xy}}{\partial y^2} \right) = G\gamma_{xy} \tag{6c}$$

where E, G and v are the elastic modulus, the shear modulus, and Poisson's ratio of the GSs, respectively. The internal characteristic length a is the distance between two atoms in a C-C bond, which is 0.142 nm.

The flexural moments of SLGS are obtained from

$$M_{xx} = \int_{-h/2}^{h/2} z\sigma_{xx} dz \tag{7a}$$

$$M_{yy} = \int_{-h/2}^{h/2} z\sigma_{yy} dz \tag{7b}$$

$$M_{xy} = \int_{-h/2}^{h/2} z\tau_{xy} dz \tag{7c}$$

224

When neglecting the displacements of the middle surface in the x and y directions, the relationship between strain and displacement fields is expressed as

$$\varepsilon_{xx} = -z\frac{\partial^2 w}{\partial x^2} \tag{8a}$$

$$\varepsilon_{yy} = -z\frac{\partial^2 w}{\partial y^2} \tag{8b}$$

$$\gamma_{xx} = -2z\frac{\partial^2 w}{\partial x \partial y} \tag{8c}$$

where w is the displacement along the thickness of GSs.

Substituting Equation (6) into Equation (7) and using Equation (8), we have

$$M_{xx} - (e_0 a)^2 \left(\frac{\partial^2 M_{xx}}{\partial x^2} + \frac{\partial^2 M_{xx}}{\partial y^2}\right) = -D\left(\frac{\partial^2 w}{\partial x^2} + v\frac{\partial^2 w}{\partial y^2}\right) \tag{9a}$$

$$M_{yy} - (e_0 a)^2 \left(\frac{\partial^2 M_{yy}}{\partial x^2} + \frac{\partial^2 M_{yy}}{\partial y^2}\right) = -D\left(\frac{\partial^2 w}{\partial y^2} + v\frac{\partial^2 w}{\partial x^2}\right) \tag{9b}$$

$$M_{xy} - (e_0 a)^2 \left(\frac{\partial^2 M_{xy}}{\partial x^2} + \frac{\partial^2 M_{xy}}{\partial y^2}\right) = -D(1-v)\frac{\partial^2 w}{\partial x \partial y} \tag{9c}$$

where D is the flexural rigidity of SLGS, expressed as

$$D = \frac{Eh^3}{12(1-v^2)} \tag{9d}$$

The governing equation for the flexural vibration of SLGS carrying a nanoparticle can be given as

$$\frac{\partial^2 M_{xx}}{\partial x^2} + 2\frac{\partial^2 M_{xy}}{\partial x \partial y} + \frac{\partial^2 M_{yy}}{\partial y^2} = [\rho h + m_c \delta(x - x_0)\delta(y - y_0)]\frac{\partial^2 w}{\partial t^2} \tag{10}$$

where ρ is the mass density of SLGS, m_c is the mass of nanoparticles attached at the position (x_0, y_0), and δ is the Dirac delta function denoted as

$$\delta(x) = \begin{cases} +\infty, & x = 0 \\ 0, & x \neq 0 \end{cases} \tag{11}$$

Substituting Equation (9) into Equation (10), the governing equation can be written in terms of w as

$$D\left(\frac{\partial^4 w}{\partial x^4} + 2\frac{\partial^4 w}{\partial x^2 \partial y^2} + \frac{\partial^4 w}{\partial y^4}\right) + \left[1 - (e_0 a)^2 \left(\frac{\partial^2 w}{\partial x^2} + \frac{\partial^2 w}{\partial y^2}\right)\right] [\rho h + m_c \delta(x - \xi)\delta(x - \eta)]\frac{\partial^2 w}{\partial t^2} = 0 \quad (12)$$

The harmonic solution of Equation (12) can be expressed as

$$w(x, y, t) = Y(x, y)\, e^{i\omega t} \quad (13)$$

where $Y(x,y)$ is the shape function of deflection and ω is the resonant frequency of the SLGS.

Substituting Equation (13) into Equation (12), we obtain

$$\left(\frac{\partial^4 Y}{\partial x^4} + 2\frac{\partial^4 Y}{\partial x^2 \partial y^2} + \frac{\partial^4 Y}{\partial y^4}\right) - \frac{\omega^2}{D}[1 - (e_0 a) 2 \left(\frac{\partial^2 Y}{\partial x^2} + \frac{\partial^2 Y}{\partial y^2}\right)] [\rho h + m_c \delta(x - \xi)\delta(x - \eta)]\frac{\partial^2 Y}{\partial x^2} = 0 \quad (14)$$

Note that the boundary conditions of SLGS with simply supported edges are

$$w = 0, \quad \frac{\partial^2 w}{\partial x^2} = 0, \quad \frac{\partial^2 w}{\partial y^2} = 0 \text{ on } x = 0,\, L_a \text{ and } y = 0,\, L_b \quad (15)$$

Therefore, the shape function (Y) in Equation (13) can be expressed as

$$Y(x, y) = A_{mn} \sin\frac{m\pi x}{L_a}\sin\frac{n\pi y}{L_b} \quad (16)$$

where A_{mn} is the vibration amplitude of oscillation and m and n indicate the mode numbers in the periodic directions.

Substituting the shape function of Equation (16) into Equation (14), then multiplying both sides of Equation (14) by $\sin\frac{m\pi x}{L_a}\sin\frac{n\pi y}{L_b}$ and integrating over the whole region with respect to x and y with the limits $x = 0$ to $x = L_a$ and $y = 0$ to $y = L_b$, after some simplifications, we obtain the following frequency equation

$$\int_0^b\int_0^a A_{mn} D\pi^4 \left(\frac{m^2}{L_a^2} + \frac{n^2}{L_b^2}\right)^2 \sin^2\frac{m\pi x}{L_a}\sin^2\frac{n\pi y}{L_b}\, dx dy$$

$$- \omega^2 \int_0^b\int_0^a A_{mn} \left[1 + (e_0 a)^2 \pi^2 \left(\frac{m^2}{L_a^2} + \frac{n^2}{L_b^2}\right)\right] [\rho h + m_c \delta(x - \xi)\delta(x - \eta)]\sin^2\frac{m\pi x}{L_a}\sin^2\frac{n\pi y}{L_b}\, dx dy = 0 \quad (17)$$

All roots of the above equation are the desired resonant frequencies corresponding to a given shape function. The coefficient of A_{mn} should be zero for the non-trivial solution. Thus, the resonant frequency of mass sensor can be determined from

$$\omega_{mn}^2 = \frac{D\pi^4 \left(\frac{m^2}{L_a^2} + \frac{n^2}{L_b^2} \right)^2}{\left[1 + (e_0 a)^2 \pi^2 \left(\frac{m^2}{L_a^2} + \frac{n^2}{L_b^2} \right) \right] \left(\rho h + \frac{4m_c}{L_a L_b} \sin^2 m\pi\varsigma \sin^2 n\pi\eta \right)} \tag{18}$$

When the nonlocal parameter ($e_0 a$) is assumed to be zero, the resonant frequency of a SLGS with attached nanoparticles can also be obtained from classical elasticity theory.

2.3. Double-Layer Graphene Sheets

DLGSs are composed of two single layers of GSs, interacting with each other by vdW forces. To the upper sheet of DLGSs nanoparticles are attached while no nanoparticles are attached to the lower graphene sheet. Thus, the governing equations for the vibration of the DLGS are given as the two coupled equations

$$D\nabla^4 w_1 + \left[1 - (e_0 a)^2 \nabla^2 \right] \left[\rho h + m_c \delta (x - x_0) \delta (y - y_0) \right] \frac{\partial^2 w_1}{\partial t^2} = p_{12} \tag{19a}$$

$$D\nabla^4 w_2 + \rho h \left[1 - (e_0 a)^2 \nabla^2 \right] \frac{\partial^2 w_2}{\partial t^2} = p_{21} \tag{19b}$$

where $w_j (x, y, t)$, $j = 1, 2$, are the flexural deflections of the upper ($j = 1$) and lower ($j = 2$) sheet, p_{12} ($p_{21} = -p_{12}$) is the transverse pressure caused by the vdW forces between the two layers of DLGSs, ρ is the mass density of the GSs, h is the thickness of each layer in DLGSs, t is the time, and δ is the Dirac delta function. The bi-harmonic operator and the flexural rigidity are given by

$$\nabla^4 = \frac{\partial^4}{\partial x^4} + \frac{\partial^4}{\partial x^2 \partial y^2} + \frac{\partial^4}{\partial y^4}, \quad D = \frac{Eh^3}{12(1 - v^2)} \tag{20}$$

where E is the elastic modulus of GSs and v is Poisson's ratio.

In Equation (19), the distributed transverse pressure acting on the upper and lower layers of DLGSs can be given by

$$p_{12} = c (w_2 - w_1) \tag{21}$$

227

where c is the vdW interaction coefficient between the upper and lower layers, which can be obtained from the Lennard–Jones pair potential [48], given as:

$$c = b \left(\frac{4\sqrt{3}}{9a}\right)^2 \frac{24\varepsilon}{\sigma^2} \left(\frac{\sigma}{a}\right)^8 \left[\frac{3003\pi}{256} \sum_{k=0}^{5} \frac{(-1)^k}{2k+1} \begin{pmatrix} 5 \\ k \end{pmatrix} \left(\frac{\sigma}{a}\right)^6 \frac{1}{(\bar{z}_1 - \bar{z}_2)^{12}} \right] \tag{22}$$

Here, $\varepsilon = 2.968\ meV$ and $\sigma = 0.34\ nm$ are parameters determined by the physical properties of GSs, $\bar{z}_j = z_j/a$, $(j = 1, 2)$, where z_j is the coordinate of the jth layer in the direction of thickness with the origin at the midplane of the GSs, and is the coordinate of the jth layer in the direction of thickness with the origin at the midplane of the GSs, and $a = 1.42\ nm$ is the C-C bond length.

In Equation (19), we define the following function:

$$\mu(x, y) = \rho h + m_c \delta(x - x_0)\delta(y - y_0) \tag{23}$$

Thus, μ is the variable mass distribution function of the upper sheet.

To obtain the vibration frequency for the governing equations Equation (19) of DLGS, we can introduce the substitution

$$w_j(x, y, t) = Y_j(x, y)\, e^{i\omega t}, \quad j = 1, 2 \tag{24}$$

where $Y_j(x, y)$, $j = 1, 2$ is the shape function of deflection in the upper and lower sheets and ω is the resonant frequency of the DLGS sensor.

Substituting Equation (24) into Equation (19), the coupled governing equations of the vibration in DLGSs are written in following matrix form:

$$\begin{bmatrix} D\nabla^4 + \mu\omega^2 (e_0 a)^2 \nabla^2 + c - \mu\omega^2 & -c \\ -c & D\nabla^4 + \rho h \omega^2 (e_0 a)^2 \nabla^2 + c - \rho h \omega^2 \end{bmatrix} \begin{Bmatrix} Y_1 \\ Y_2 \end{Bmatrix} = \begin{Bmatrix} 0 \\ 0 \end{Bmatrix} \tag{25}$$

Algebraic manipulation of Equation (25) reduces it to a single equation:

$$\nabla^8 Y + \frac{(\mu + \rho h)\,\omega^2 (e_0 a)^2}{D} \nabla^6 Y + \frac{(2c - \rho h \omega^2 - \mu\omega^2)\,D + \mu\rho h \omega^4 (e_0 a)^4}{D^2} \nabla^4 Y +$$
$$\frac{[c(\rho h + \mu) - 2\rho h \omega^2]\,\omega^2 (e_0 a)^2}{D^2} \nabla^2 Y + \left[\frac{\rho h \mu\omega^4 - c(\rho h + \mu)\,\omega^2}{D^2} \right] Y = 0 \tag{26}$$

where $Y = Y_1$, or $Y = Y_2$.

For a simply supported boundary condition, the shape function of deflection of DLGS in Equation (26) can be expressed as

$$Y(x, y) = C_{mn} \sin\frac{m\pi x}{L_a} \sin\frac{n\pi y}{L_b} \tag{27}$$

where C_{mn} is the vibration amplitude of oscillation and m and n indicate the mode numbers.

In the analysis of SLGS, the effect of the nonlocal parameter $(e_0 a)$ on the vibration characteristics has been investigated. For the following analysis of DLGSs, we only focus on the discussion of the layer numbers, attached mass and vibration frequencies.

Substituting Equations (23) and (27) into Equation (26), and putting $e_0 a = 0$, yields

$$
\pi^8 \left(\frac{m^2}{L_a^2} + \frac{n^2}{L_b^2} \right)^4 Y + \frac{(2c - \rho h \omega^2)}{D} \pi^4 \left(\frac{m^2}{L_a^2} + \frac{n^2}{L_b^2} \right)^2 Y + \left[\frac{(\rho h)^2 \omega^4 - 2c\rho h \omega^2}{D^2} \right] Y
$$
$$
= \left\{ \frac{\pi^4}{D} \left(\frac{m^2}{L_a^2} + \frac{n^2}{L_b^2} \right)^2 \omega^2 - \frac{\rho h}{D^2} \omega^4 + \frac{c}{D^2} \omega^2 \right\} m_c \delta (x - x_0) \delta (y - y_0) Y
$$

(28)

After multiplying both sides of Equation (28) by $\sin \frac{m \pi x}{L_a} \sin \frac{n \pi y}{L_b}$ and integrating over the whole region with respect to x and y within the limits from $x = 0$ to $x = L_a$ and $y = 0$ to $y = L_b$, after some simplifications, we obtain the following polynomial expression of the frequency

$$
r_0 \omega^4 - r_1 \omega^2 + r_2 = 0
$$

(29)

where the coefficients r_0, r_1, and r_2 are

$$
r_0 = \left(\frac{\rho h}{D} \right)^2 + \frac{4 \rho h}{D^2 L_a L_b} m_c \sin^2 m \pi \xi \sin^2 \eta
$$

(30a)

$$
r_1 = \frac{2 \rho h \pi^4}{D} \left(\frac{m^2}{L_a^2} + \frac{n^2}{L_b^2} \right)^2 + \frac{2 \rho h c}{D^2} + \left[\frac{4 \pi^4}{abD} \left(\frac{m^2}{L_a^2} + \frac{n^2}{L_b^2} \right)^2 + \frac{c}{D^2} \right] m_c \sin^2 m \pi \xi \sin^2 m \pi \eta
$$

(30b)

$$
r_2 = \pi^8 \left(\frac{m^2}{L_a^2} + \frac{n^2}{L_b^2} \right)^4 + \frac{2c \pi^4}{D} \left(\frac{m^2}{L_a^2} + \frac{n^2}{L_b^2} \right)^2
$$

(30c)

where and $\xi = x_0/L_a$ and $\eta = y_0/L_b$ define the non-dimensional position of attached nanoparticles.

The solution of Equation (29) yields the resonate frequency of DLGSs. The high frequency of DLGSs with anti-phase mode, in which the upper and lower layers of DLGSs moves in the opposite direction, can be obtained from

$$
\omega_{mn}^2 = \frac{r_1 + \sqrt{r_1^2 - 4 r_0 r_2}}{2 r_0}
$$

(31)

3. Analytical Results and Discussion

In our simulations, the nanomechanical resonator is considered to simply consist of supported GSs loaded with a nanoparticle. The vibration mode is taken to be the fundamental frequency of $m = n = 1$, and the anti-phase mode in which the deflection of the upper and lower layers in DLGSs occurs in the opposite direction. In order to investigate the vibration behavior of a realistic graphene-based nanoscale mechanical mass sensor, the geometrical dimensions and the material constants were taken from the recent literature. The aspect ratio of the DLGSs is defined as the side length to thickness $2h$ratio, where the effective thickness h of each layer is 0.127 nm. The equivalent Young's modulus E and density ρ of the GSs are taken to be 2.81 TPa and 2300 kg/m^3, respectively [49].

Figures 2 and 3 show the resonant frequency and the frequency shift of SLGSs as a function of the attached nanomass for different nonlocal parameters. Here, the frequency shift is defined as the difference between the natural frequency of a GS with and without attached nanoparticles, *i.e.*, $\Delta\omega = \Delta(\omega) - \Delta(\omega + m_c)$. The nanoparticles are attached on the center ($\xi = \eta = 0.5$) of the GSs with $L_a/2h = L_b/2h = 20$. The resonant frequency shown in Figure 2 decreases with increasing mass of attached nanoparticles. It is observed form Figure 3 that the frequency shift ($\Delta\omega$) of GSs carrying attached nanoparticles is positive because the attached particles increase the overall mass of the GS resonator, and the value of the shift increases with increasing attached mass. For the nonlocal parameter e_0a used in this simulation, we take that e_0a is 0, 1.0 nm and 2.0 nm [50–52]. The adopted value of the scaling parameter e_0 depends on the crystal structure in lattice dynamics, and the internal characteristics length a = 1.42 nm (C-C bond length) for the graphene structure. The nonlocal parameter affects the vibration simulation results of the GSs due to the size-dependent mechanical properties. It is found that the influence of the nonlocal parameter on the frequency shift becomes larger when the GSs carry a relatively small mass of nanoparticles.

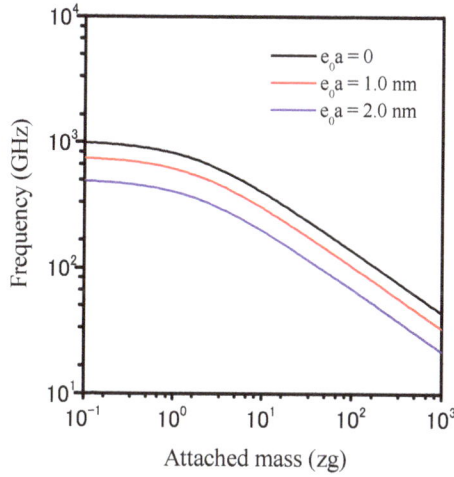

Figure 2. Variation of the resonant frequency in SLGSs as a function of attached mass for different nonlocal parameter ($L_a/2h = L_b/2h = 20$, $\xi = \eta = 0.5$).

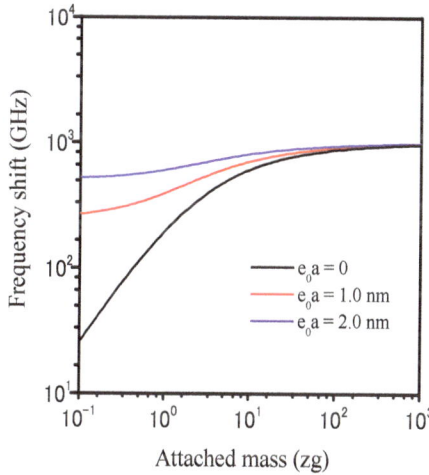

Figure 3. Variation of the frequency shift in SLGSs as a function of attached mass for different nonlocal parameter ($L_a/2h = L_b/2h = 20$, $\xi = \eta = 0.5$).

A comparison of the resonant frequency and the frequency shift between the SLGSs and DLGSs is shown in Figures 4 and 5 as a function of the mass of the nanoparticles attached to the center of sheet. As shown in Figure 4, the value of resonant frequency decreases with increasing mass and the DLGSs have higher vibration frequency than the SLGSs. It is seen in Figure 5 that the frequency shift of the GS resonator increases with increasing nanoparticle mass, especially for DLGSs. This suggests that DLGSs used as nanomechanical resonator can provide higher sensitivity than SLGSs. The relationship between the frequency shift and the attached mass is nearly logarithmic linear for small attached masses. The linear relationship of DLGSs has a wider range than that of SLGSs. Thus, the result indicates that DLGSs can more easily be used to estimate the attached mass than SLGSs according to the changes in the resonant frequency. The double logarithmic linear relationship of the DLGSs is less than about 1.0 zg. According to this simulation, the relationship between the frequency shift ($\Delta\omega$) and the attached mass (m_c) can be well represented by the exponential function of $\Delta\omega = \alpha m_c^\beta$, which has a double logarithmic, linear relationship. The values of the parameters α and β can be determined easily by fitting the experimental results. It is seen that the variation of frequency shifts of the DLGSs agrees qualitatively with the results for circular and square SLGSs reported by Lei *et al.* [40] and Shen *et al.* [42]. The frequency shift appears when the attached mass is less than 1.0 zg, and thus the mass sensitivity of this kind of DLGS resonator can reach the range of atomic mass unit.

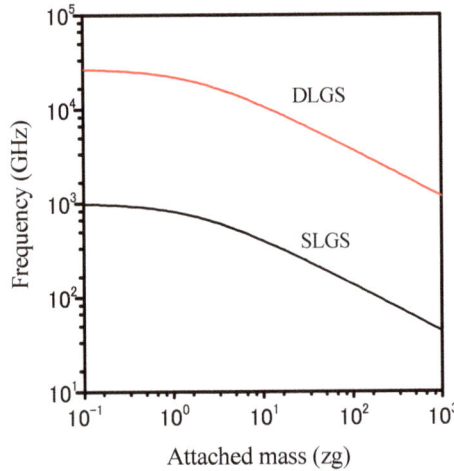

Figure 4. Comparison of the resonant frequency variation for DLGSs- and SLGSs-based mass sensors ($L_a/2h = L_b/2h = 20$, $\xi = \eta = 0.5$).

The effect of the location of attached nanoparticles on the frequency shift of the DLGSs is shown in Figure 6. The attached mass is located either at the center $(\xi, \eta) = (0.5, 0.5)$ of the DLGSs, near to the corner $(\xi, \eta) = (0.25, 0.25)$, or near to the edge $(\xi, \eta) = (0.5, 0.25)$. It can be found that the location of the nanoparticles influences the frequency shift significantly. The value of the frequency shift rises as the mass is close to the center of the DLGSs, but the dependence on location becomes smaller when the attached mass increases. Figure 7 shows the DLGSs aspect ratio (length to width) effect on the frequency shift of DLGSs with an attached nanoparticle at the center. The DLGS dimension significantly affects the frequency shift, which becomes larger as one side length of the DLGS decreases. The double logarithmic relationship between the frequency shift and the attached mass becomes highly linear when the side length of DLGS increases.

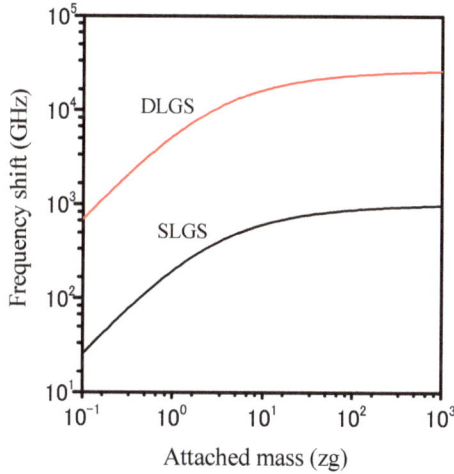

Figure 5. Comparison of the frequency shift for DLGSs- and SLGSs-based on mass sensor ($L_a/2h = L_b/2h = 20$, $\xi = \eta = 0.5$).

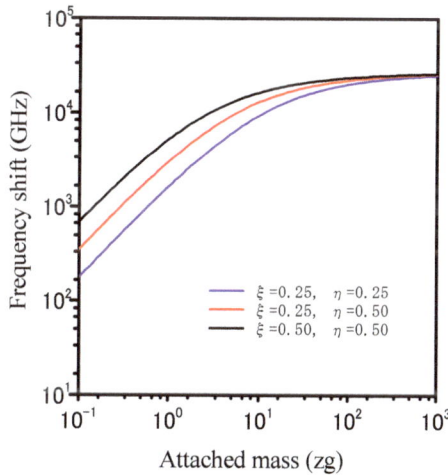

Figure 6. Location effect of the attached nanoparticle on the frequency shift of DLGS resonator ($L_a/2h = L_b/2h = 20$).

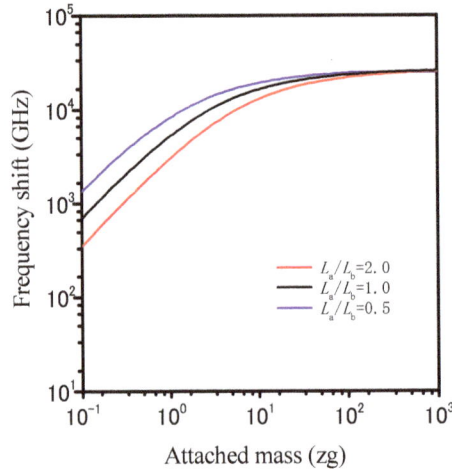

Figure 7. Influence of DLGS dimension on the frequency shift of nanomechanical resonator ($L_b/2h = 20$, $\xi = \eta = 0.5$).

Finally, Figure 8 indicates the relationship between the frequency shift and the aspect ratio of DLGSs with different attached masses. The resonant frequency is sensitive to the side length of DLGSs due to the stiffness dependence of DLGS resonators, especially for small masses of the nanoparticles. It is found that the frequency shift of the DLGS resonator increases with increasing attached mass and decreasing aspect ratio. This suggests that, to provide higher sensitivity, the dimension of the DLGSs used as sensor elements can be changed.

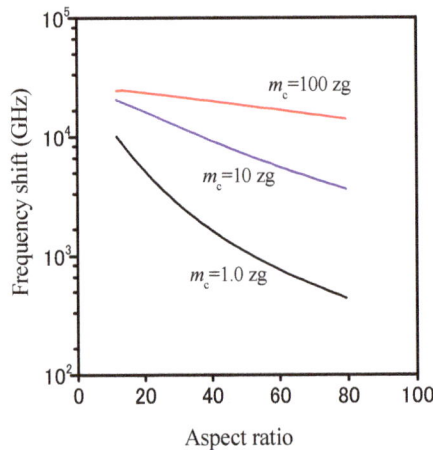

Figure 8. Relationship between the frequency shift and the aspect ratio of DLGSs for different attached mass ($\xi = \eta = 0.5$).

4. Conclusions

Based on nonlocal elasticity theory, we present a vibration analysis of supported DLGSs carrying an attached mass, taking the small-scale effect into account. The DLGSs are regarded as two single-layered GSs coupled by vdW interaction forces. The mass of nanoparticles attached to the DLGSs can be derived by measuring the frequency change of the GS resonator. According to the present analytical model, the influences of the nonlocal parameters, the attached mass, and the position of the nanoparticles on the frequency shift of GSs are analyzed and discussed in detail. A logarithmically linear relationship exists between the vibration frequency and the attached mass when the total mass is less than 1.0 zg for DLGSs. Moreover, the small-scale effects considerably influence the frequency shift of GSs, especially for small masses of attached nanoparticles. The results indicate that GS resonators could provide high sensitivity as a nanomechanical mass sensor. The analytical method developed here can serve as a useful design approach for graphene-based mass sensor.

Conflicts of Interest: The author declares no conflict of interest.

References

1. Frank, I.W.; Tanenbaum, D.M.; van der Zande, A.M.; McEuen, P.L. Mechanical properties of suspended graphene sheets. *J. Vac. Sci. Technol. B* **2007**, *25*, 2558–2561.
2. Novoselov, K.S.; Geim, A.K.; Morozov, S.V.; Jiang, D.; Zhang, Y.; Dubonos, S.V.; Grigorieva, I.V.; Firsov, A.A. Electric field effect in atomically thin carbon films. *Science* **2004**, *306*, 666–669.
3. Balandin, A.A. Thermal properties of graphene and nanostructured carbon materials. *Nat. Mater.* **2011**, *10*, 569–581.
4. Du, J.; Cheng, H.M. The fabrication, properties, and uses of graphene/polymer composites. *Macromol. Chem. Phys.* **2012**, *213*, 1060–1077.
5. Chen, C.; Rosenblatt, S.; Bolotin, K.I.; Kalb, W.; Kim, P.; Kymissis, I.; Stormer, H.L.; Heinz, T.F.; Hone, J. Performance of monolayer graphene nanomechanical resonators with electrical readout. *Nat. Nanotechnol.* **2009**, *4*, 861–867.
6. Wang, X.; Zhi, L.; Mullen, K. Transparent, conductive graphene electrodes for dye-sensitized solar cells. *Nano Lett.* **2008**, *8*, 323–327.
7. Schedin, F.; Geim, A.K.; Morozov, S.V.; Hill, E.W.; Blake, P.; Katsnelson, M.I.; Novoselov, K.S. Detection of individual gas molecules adsorbed on grapheme. *Nat. Mater.* **2007**, *6*, 652–655.
8. Arash, B.; Jiang, J.W.; Rabczuk, T. A review on nanomechanical resonators and their applications in sensors and molecular transportation. *Appl. Phys. Rev.* **2015**, *2*, 021301.
9. Jiang, J.W.; Park, H.; Rabczuk, T. MoS2 Nanoresonators: Intrinsically better than graphene? *Nanoscale* **2014**, *6*, 3618–3625.

10. Jiang, J.W.; Wang, B.S.; Park, H.; Rabczuk, T. Adsorbate migration effects on continuous and discontinuous temperature-dependent transitions in the quality factors of graphene nanoresonators. *Nanotechnology* **2014**, *25*, 025501.

11. Jiang, J.W.; Park, H.; Rabczuk, T. Preserving the Q-factors of ZnO nanoresonators via polar surface reconstruction. *Nanotechnology* **2013**, *24*, 405705.

12. Jiang, J.W.; Park, H.S.; Rabczuk, T. Enhancing the mass sensitivity of graphene nanoresonators via nonlinear oscillations: The elective strain mechanism. *Nanotechnology* **2012**, *23*, 47550.

13. Wang, Q.; Arash, B. A review on applications of carbon nanotubes and graphenes as nano-resonator sensors. *Comput. Mater. Sci.* **2014**, *82*, 350–360.

14. Arash, B.; Wang, Q. Detection of gas atoms with carbon nanotubes. *Sci. Rep.* **2013**, *3*.

15. Ekinci, K.L.; Huang, X.M.H.; Roukes, M.L. Ultrasensitive nanoelectromechanical mass detection. *Appl. Phys. Lett.* **2004**, *84*, 4469–4471.

16. Chaste, J.; Eichler, A.; Moser, J.; Ceballos, G.; Rurali, R.; Bachtold, A. A nanomechanical mass sensor with yoctogram resolution. *Nat. Nanotechnol.* **2012**, *7*, 301–304.

17. Eichler, A.; Moser, J.; Chaste, J.; Zdrojek, M.; Wilson-Rae, I.; Bachtold, A. Nonlinear damping in mechanical resonators made from carbon nanotubes and graphene. *Nat. Nanotechnol.* **2011**, *6*, 339–342.

18. Pang, W.; Yan, L.; Zhang, H.; Yu, H.Y.; Kim, E.S.; Tang, W.C. Femtogram mass sensing platform based on lateral extensional mode piezoelectric resonator. *Appl. Phys. Lett.* **2006**, *88*, 243503.

19. Geim, A.K. Graphene: status and prospects. *Science* **2009**, *324*, 1530–1534.

20. Bunch, J.S.; van der Zande, A.M.; Verbridge, S.S.; Frank, I.W.; Tanenbaum, D.M.; Parpia, J.M.; Craighead, H.G.; McEuen, P.L. Electromechanical Resonators from Graphene Sheets. *Science* **2007**, *315*, 490–493.

21. Xu, M.S.; Gao, Y.; Yang, X.; Chen, H.Z. Unique synthesis of graphene-based materials for clean energy and biological sensing applications. *Chin. Sci. Bull.* **2012**, *57*, 3000–3009.

22. Novoselov, K.S.; Jiang, D.; Schedin, F.; Booth, T.J.; Khotkevich, V.V.; Morozov, S.V.; Geim, A.K. Two-dimensional atomic crystals. *Proc. Natl. Acad. Sci. USA* **2005**, *102*, 10451–10453.

23. Wei, D.C.; Liu, Y.Q. Controllable synthesis of graphene and its applications. *Adv. Mater.* **2010**, *22*, 3225–3241.

24. Kim, K.S.; Zhao, Y.; Jang, H.; Lee, S.Y.; Kim, J.M.; Kim, K.S.; Ahn, J.H.; Kim, P.; Choi, J.Y.; Hong, B.H. Large-scale pattern growth of graphene films for stretchable transparent electrodes. *Nature* **2009**, *457*, 706–710.

25. Sutter, P.W.; Flege, J.I.; Sutter, E.A. Epitaxial graphene on ruthenium. *Nat. Mater.* **2008**, *7*, 406–411.

26. Arghavan, S.; Singh, A.V. Effects of van der Waals interactions on the nonlinear vibration of multi-layered graphene sheets. *J. Phys. D: Appl. Phys.* **2012**, *45*, 455305.

27. Ohta, T.; Bostwick, A.; Seyller, T.; Horn, K.; Rotenberg, E. Controlling the electronic structure of bilayer grapheme. *Science* **2006**, *313*, 951–954.

28. Virojanadara, C.; Yakimova, R.; Zakharov, A.A.; Johansson, L.I. Large homogeneous mono-/bi-layer graphene on 6h-sic(0001) and buffer layer elimination. *J. Phys. D Appl. Phys.* **2010**, *43*, 374010.

29. Zhang, Y.; Tang, T.T.; Girit, C.; Hao, Z.; Martin, M.C.; Zettl, A.; Crommie, M.F.; Shen, Y.R.; Wang, F. Direct observation of a widely tunable bandgap in bilayer grapheme. *Nature* **2009**, *459*, 820–823.

30. Meng, H.; Dai, Y.; Ye, Y.; Luo, J.X.; Shi, Z.J.; Dai, L.; Qin, G.G. Bilayer graphene anode for small molecular organic electroluminescence. *J. Phys. D Appl. Phys.* **2012**, *45*, 245103.

31. Popov, A.M.; Lebedeva, I.V.; Knizhnik, A.A.; Lozovik, Y.E.; Potapkin, B.V. Structure, energetic and tribological properties, and possible applications in nems of argon-separated double-layer grapheme. *J. Phys. Chem. C* **2013**, *117*, 11428–11435.

32. Popov, A.M.; Lebedeva, I.V.; Knizhnik, A.A.; Lozovik, Y.E.; Potapkin, B.V.; Poklonski, N.A.; Siahlo, A.I.; Vyrko, S.A. AA stacking, tribological and electronic properties of double-layer graphene with krypton spacer. *J. Chem. Phys.* **2013**, *139*, 154705.

33. Wang, J.; He, X.; Kitipornchai, S.; Zhang, H. Geometrical nonlinear free vibration of multi-layered graphene sheets. *J. Phys. D Appl. Phys.* **2011**, *44*, 135401.

34. Shi, J.X.; Ni, Q.Q.; Lei, X.W.; Natsuki, T. Nonlocal vibration of embedded double-layer graphene nanoribbons in in-phase and anti-phase modes. *Phys. E* **2012**, *44*, 1136–1141.

35. Natsuki, T.; Shi, J.X.; Ni, Q.Q. Vibration analysis of circular double-layered graphene sheets. *J. Appl. Phys.* **2012**, *111*, 044310.

36. Hill, E.W.; Vijayaraghavan, A.; Novoselov, K. Graphene sensors. *IEEE Sens. J.* **2011**, *11*, 3161–3170.

37. Arash, B.; Wang, Q.; Duan, W.H. Detection of gas atoms via vibration of graphenes. *Phys. Lett. A* **2011**, *375*, 2411–2415.

38. Jiang, J.W.; Park, H.S.; Rabczuk, T. Enhancing the mass sensitivity of graphene nanoresonators via nonlinear oscillations: The effective strain mechanism. *Nanotechnology* **2012**, *23*, 475501.

39. Sakhaee-Pour, A.; Ahmadiana, M.T.; Vafaib, A. Applications of single-layered graphene sheets as mass sensors and atomistic dust detectors. *Solid State Commun.* **2008**, *145*, 168–172.

40. Lei, X.W.; Natsuki, T.; Shi, J.X.; Ni, Q.Q. Vibration analysis of circular double-layered graphene sheets. *J. Appl. Phys.* **2013**, *113*, 154313.

41. Dai, M.D.; Kim, C.W.; Eom, K. Nonlinear vibration behavior of graphene resonators and their applications in sensitive mass detection. *Nanoscale Res. Lett.* **2012**, *7*, 449.

42. Shen, Z.B.; Tang, H.L.; Li, D.K.; Tang, G.J. Vibration of single-layered graphene sheet-based nanomechanical sensor via nonlocal kirchhoff plate theory. *Comp. Mater. Sci.* **2012**, *61*, 200–205.

43. Lee, H.L.; Yang, Y.C.; Chang, W.J. Mass Detection Using a Graphene-Based Nanomechanical Resonator. *Jpn. J. Appl. Phys.* **2013**, *52*, 025101.

44. Natsuki, T.; Shi, J.X.; Ni, Q.-Q. Vibration analysis of nanomechanical mass sensor using double-layered graphene sheets resonators. *J. Appl. Phys.* **2013**, *114*, 094307.

45. Lu, P.; Zhang, P.Q.; Lee, H.P.; Wang, C.M.; Reddy, J.N. Non-local elastic plate theories. *Proc. R. Soc. A* **2007**, *463*, 3225–3240.

46. Polizzotto, C. Nonlocal elasticity and related variational principles. *Inter. J. Solid. Struc.* **2001**, *38*, 7359–7380.

47. Arash, B.; Wang, Q. A review on the application of nonlocal elastic models in modeling of carbon nanotubes and graphenes. *Comput. Mater. Sci.* **2012**, *51*, 303–313.

48. Kitipornchai, S.; He, X.Q.; Liew, K.M. Continuum model for the vibration of multilayered graphene sheets. *Phys. Rev. B* **2005**, *72*, 075443.

49. Shi, J.X.; Natsuki, T.; Lei, X.W.; Ni, Q.Q. Equivalent Young's modulus and thickness of graphene sheets for the continuum mechanical models. *Appl. Phys. Lett.* **2014**, *104*, 223101.

50. Narendar, S.; Gopalakrishnan, S. Nonlocal scale effects on wave propagation in multi-walled carbon nanotubes. *Comput. Mater. Sci.* **2009**, *47*, 526–538.

51. Amiriana, B.; Hosseini-Arab, R.; Moosavia, H. Thermo-mechanical vibration of short carbon nanotubes embedded in pasternak foundation based on nonlocal elasticity theory. *Shock. Vib.* **2013**, *20*, 821–832.

52. Duan, W.H.; Wang, C.M.; Zhang, Y.Y. Calibration of nonlocal scaling effect parameter for free vibration of carbon nanotubes by molecular dynamics. *J. Appl. Phys.* **2007**, *101*, 024305.

Dimensional Quantization and the Resonance Concept of the Low-Threshold Field Emission

Georgy Fursey, Pavel Konorov, Boris Pavlov and Adil Yafyasov

Abstract: We present a brief critical review of modern theoretical interpretations of the low-threshold field emission phenomenon for metallic electrodes covered with carbon structures, taking the latest experiments into consideration, and confirming the continuity of spectrum of resonance states localized on the interface of the metallic body of the cathode and the carbon cover. Our proposal allowed us to interpret the double maxima of the emitted electron's distribution on full energy. The theoretical interpretation is presented in a previous paper which describes the (1 + 1) model of a periodic 1D continuous interface. The overlapping of the double maxima may be interpreted taking into account a 2D superlattice periodic structure of the metal-vacuum interface, while the energy of emitted electrons lies on the overlapping spectral gaps of the interface 2D periodic lattice.

Reprinted from *Electronics*. Cite as: Fursey, G.; Konorov, P.; Pavlov, B.; Yafyasov, A. Dimensional Quantization and the Resonance Concept of the Low-Threshold Field Emission. *Electronics* **2015**, *4*, 1101–1108.

1. Brief Review of Modern Experiments in the Low-Threshold Field Emission

In numerous recent experiments, see for instance [1–8], extremely low-threshold field emission from the carbon nano-clusters was observed for electric fields $(10^4 - 10^5 \text{ V/cm})$. This is a surprisingly strong effect, because the field initiating a noticeable emission from these materials is 2–3 orders less than the field required for the field-emission from the traditional metals and semiconductors. Despite the obviously unusual nature of the effect, numerous authors, see for instance [6,8], have attempted to explain the low-threshold phenomenon trivially with use of the classical Fowler-Nordheim machinery, based on the enhancing of the field at the micro-protrusions. They assume that the local field F_s near the emitting center is calculated as $F_0 = \gamma F_0$, where γ is the field enhancement coefficient, defined by the micro-geometry, and F_0 is the field of the equivalent flat capacitor. This completely classical explanation of the low -threshold emission phenomenon is not universal, and certainly non valid for carbon-covered cathodes. This has been considered in our recent papers [2,3,7] because the surface of the carbon flakes, obtained by the detonation synthesis are perfectly smooth with rare and relatively small protrusions. These protrusions are able to lower the threshold by a factor of 5, while a lowering factor of 10^2 is observed in our experiments. We suggested in [2,3,7] an alternative explanation of the threshold lowering (field enhancement) based on the dimensional quantization in the under-surface space-charge region on the metal-graphene interface.

Recent experiments done by our group confirm the continuity of spectrum 2D-size quantization and allow us to estimate the effective mass m^* and the de Broglie wavelength in space-charge region depending on electron's concentration $n_{ex} = \frac{1}{q}\int_0^\tau J(t)dt$, where $j = I/S$, $I = 80$ A is the current and $S \approx 0.75$ cm^2—the area of the cathode, $\tau \approx 2 \times 10^{-9}$ s. Monitoring of current density allows us to estimate the experimental density of charges $Q = 2.4 \times 10^{-7}$ Coul/cm^2, which corresponds to density of electrons $n \approx 1.3 \times 10^{12}$ cm^{-2}.

On another hand, from the size-quantization theory [9] the 2D density of electrons is estimated as

$$n_{2D} = \frac{m^*kT}{2\pi\hbar^2} \ln\left(1 + exp\frac{E_0 - E_F}{kT}\right) \tag{1}$$

Here m^* is the effective electron mass, E_0—the size-quantization level, E_F—the Fermi level. In our case

$$E_0 \approx E_F, \, n_{exp} = n_{2D} = \frac{m^*kT}{2\pi\hbar^2} \ln 2 \tag{2}$$

The de Broglie electron wavelength [9] in graphene flakes is

$$\lambda = \frac{2\pi\hbar}{\sqrt{2m^* (kT)}} \tag{3}$$

Based on the preceding Equations (2) and (3), we estimate the electron effective mass and the de Broglie electron

$$m^* = \frac{(2\pi\hbar)^2}{2(kT)} \tag{4}$$

At room temperature we have $\lambda \approx 18$ nm, $m^* \leq 10^{-2} \, m_0$. We should notice that the experiment based estimates yield the values $m^* \leq 10^{-2} \, m_0$. The above value of the electron effective mass yield the values one order higher than ones for graphene, obtained from independent experiments, see [10].

Another important result obtained by our group is discovery of the second maximum on the dependence of distribution of electron on total energy. This result does not have yet an adequate interpretation.

The classical Fowler-Nordheim machinery used to calculate the transmission coefficient for simple rectangular potential barrier, see [11], gives an exponentially small value of the transmission coefficient $T \approx e^{-qa}$ with $q = \sqrt{v - 2mE\hbar^{-2}}$ for the under - barrier tunneling with $v \gg 2mE\hbar^{-2}$, and the width of the barrier is equal to a.

In our papers [2,7] we assume that the spectrum of the size quantization is discrete and reduced to several discrete levels. Based on this assumption, we developed a resonance version of the classical Fowler-Nordheim machinery, considering the complex levels as resonances. The role of field enhancing factor in our interpretation was played by the small effective mass of electron in the carbon structure. Indeed, the field is measured by the steepness of the slope of the potential. However, the effective steepness is calculated with respect to the de Broglie wavelength which is m/m_e times bigger than the conventional de Broglie wavelength at the same energy. The corresponding formula for the transmission coefficient was derived [7] for the general 1D model of the space-charge region, with complex discrete spectrum of the surface levels.

2. The Resonance Interpretation

Presence of these resonance details in the barrier may result in much larger values of the transmission coefficient T for electrons with certain energy. In [2] we emulated T by delta-barrier supplied with inner structure defined by a differential operator on a finite interval. We realize that this differential operator does not have any physical meaning itself, but serves just as a mathematical detail of the model. Besides, fitting of the corresponding model, with this differential operator, presents some technical difficulties, in particular solving the relevant inverse problem. In [7] we base our conclusions upon a simpler model, substituting inner structure by a finite matrix, which is fit based on experimental data on size-quantization. Similarly to [2] we emulate the barrier in [7] by the generalized Datta and Das Sarma boundary condition, see [7,12,13]. The 1D solvable model of the contact zone of the emitter is constructed hereafter based on division of the normal coordinate into three layers; the metal (1) $\infty < x < 0$, the vacuum (2), $a < x < \infty$ and carbon deposit (3), $0 \leq x \leq a$., with the portions of the wave-function denoted correspondingly.

The components x_s, $s = 1, 2, 3$ of the wave-function of the electron satisfy on the domains the Schrödinger equations, for instance

$$-\frac{\hbar^2}{2m_s} \frac{d^2 u^s}{dx_s^2} + q_s(x_s)u^s = Eu^s, \, s = 1, 2 \tag{5}$$

Here $q_1 = q_2 = 0$, end q_3 is a rectangular barrier hight q.

241

$$u(x) = \begin{cases} u_1(x) = & e^{-ipx} + Te^{ipx}, & -\infty < x < 0 \\ u_3(x) = & Be^{\sqrt{\frac{2m(q-E)}{\hbar^2}}x} + Be^{-\sqrt{\frac{2m(q-E)}{\hbar^2}}x}, & 0 < x < a, \\ u_2(x) = & Te^{-ipx}, & a < x < \infty \end{cases} \qquad (6)$$

Following [2,7], we model the component u^3 of the wave-function on the barrier—the solution of a differential equation on a finite interval, by the finite vector and, correspondingly, substitute the barrier by a zero-range model with an inner structure, see [14], defined by the finite Hermitian matrix A. The corresponding boundary form is represented in terms of the corresponding "boundary values" $\zeta_\pm^u$ as

$$J^s(u,v) = \frac{\hbar^2}{2m_s} \left[\frac{du^s}{dx^s} \bar{v}^s - u^s \frac{d\bar{v}}{dx^s} \right], s = 1,2$$

The corresponding boundary forms of the inner structure in terms of the adequate boundary values is

$$J^3(u,v) = \zeta_+^u \bar{\zeta}_-^v - \zeta_-^u \bar{\zeta}_+^v$$

The zero-range model of the resonance field emission is defined by the boundary condition imposed on the data, which annihilate the sum of the boundary forms $J^1(u,v) + J^2(u'v) + J^3(u,v)$. When discussing the zero-range model of the interface lattice, we assume, that the deposit zone is substituted by the model barrier, but the zone inherits the small effective mass $m_1 \leq m_2$ from the deposit on the contact interface. In particular the sum is vanishing while the boundary data are imposed to the generalized Datta-Das Sarma boundary condition at the contact of the deposit and vacuum (that is exactly on the barrier). This boundary condition is defined, similarly to Datta-Das Sarma, [12], by the vector parameter $\vec{\beta} = (\beta_1, \beta_2, \beta_3)$ as:

$$\frac{u^1}{\beta_1} = \frac{u^2}{\beta_2} = \frac{\zeta_-^u}{\beta_1}, \frac{\hbar^2}{2m_1} u'_1 \vec{\beta}_1 + \frac{\hbar^2}{2m_2} u'_2 \vec{\beta}_2 + \zeta_+^u \vec{\beta} = 0 \qquad (7)$$

The quantum-mechanical meaning of the similar parameter $\vec{\beta}$ in the case of T-junction is revealed in [12]. In our case, the parameter is defined by the geometry of the contact zone. We assume here that it can be fitted based on the experimental data. The boundary values $\zeta_\pm^u$ of the component of the wave-function on the barrier are connected via the corresponding Weyl-Titchmarsh function $\mathcal{M}(E) = P_N \frac{I+EA}{A-EI} P_N$, and selected "deficiency" subspace N, defining the connection of the inner space (the space of the size-quantization on the barrier), with the vacuum and the deposit. The energy E plays a role of the spectral parameter, so that the dimension of the operator A is energy. Hereafter, we assume that the Weyl function $\mathcal{M}(E)$ is scalar,

but our model can be easily extended to the general case with multidimensional Weyl function. Assuming that the wave-function of the electron in the deposit and in vacuum is a scattered wave, we represent the components of it in the deposit and in vacuum as $u_1 = e^{ipx} + e^{-ipx}R_1$, $u_2 = T_2 e^{-ipx}$. Substituting this scattering Ansatz into the above boundary conditions, we obtain an expression for the transmission coefficient T from the deposit into vacuum :

$$T(\lambda) = \frac{\bar{\beta}_1 \beta_2 \, m_1^{-1/2}}{|\beta_1^2|m_1^{-1/2} + |\beta_2^2|m_2^{-1/2} + i|\beta_3|^2 \left[\hbar \mathcal{M} \sqrt{2E}\right]^{-1}} \tag{8}$$

In the non-resonance situation, $\mathcal{M} \approx$ Const, the Datta-Das Sarma parameter $(1,1,e^{qa/2})$ defines the exponential small transmission rate $T \approx e^{-aq}$. Then, in the resonance situation, $\mathcal{M} = \mathcal{M}(\lambda)$ the transmission is exponentially small on the complement of the set of poles of $\mathcal{M}$, but is essentially greater at the poles λ_p, $\mathcal{M}(\lambda_p) = \infty$, where $\mathcal{M}^{-1}(\lambda_p) = 0$. With regard of $m_1 \ll m_2$ we have at the poles, that

$$T = \frac{\bar{\beta}_1 \beta_2 \, m_1^{-1/2}}{|\beta_1|^2 m_1^{-1/2} + |\beta_2|^2 m_2^{-1/2}} \approx \frac{\beta_2}{\beta_1}$$

which can be essentially greater than exponential estimate $T \approx e^{-2av}$. Then, in the resonance situation, $\mathcal{M} = \mathcal{M}(\lambda)$ based on $\beta_3 \approx e^{vd}$ we see the peak of the transmission coefficient at the eigenvalues of the matrix A which play a role of the levels of the size - quantization on the barrier (with special boundary conditions on the contact of the barrier with the inner part of the deposit and the vacuum. This condition is compatible with unitarity of the full scattering matrix on the interface deposit-vacuum, if the weights m_1^{-1}, m_2^{-1} are taken into account.

In the ballistic regime, the current j is proportional to the weighted integral of the transmission coefficient T. In particular, for linear spectrum $E = \hbar v_F k$ the integration is performed on the energy variable

$$j = -\frac{2ev_F}{\hbar^2} \int T(E) \left[f(E) - f(E + eV)\right] E dE$$

Here, eV is the voltage drop, $f(E)$ is the Fermi distribution, and v_F is the Fermi velocity. For low temperature the integrand with the Fermi distribution is reduced to

$$[f(E) - f(E + eV)] \approx -eV\delta(E - E_F)$$

In our case we have generally:

$$j = -\frac{2ev_F}{\hbar^2} \int \frac{\bar{\beta}_1 \beta_2 \, m_1^{-1/2} \, [f(E) - f(E + eV)] \, E dE}{|\beta_1|^2 m_1^{-1/2} + |\beta_2|^2 m_2^{-1/2} + i|\beta_2|^2 [\hbar\sqrt{2E}\mathcal{M}]^{-1}}$$

with $\mathcal{M} = \mathcal{M}(E)$, and, for low temperature, just the value of T at the Fermi level, with a trivial coefficient.

3. Discussion: The Role of the Interface Structure

The above model of resonance tunneling through the size - quantization level presents a qualitative interpretation the low-threshold emission. The magnitude of the threshold measured in experiment (see [2,3]) is $qV = (20 - 100)$ meV. These values of the threshold may be qualitatively explained based on resonance tunneling through the size-quantization level. Up to choice of the statistical distribution, this magnitude corresponds to $qV = E_0 - E_F = (2 - 3)kT$, for room temperature.

The reason for this may have various causes. First of all: the one-dimensional version of the electron transmission through the interface barrier (even with regard to size-quantization levels). Another cause is absence of theoretical equivalent of the multilayer structure in the accepted model of the cathode, usually substituted by the two-layer structure of the contact of graphene and the metal surface, without taking into account mutual influence of the components. Finally, the naive model does not take into account the two-dimensional structure of the graphene-metal interface, which may play an essential role for both classical Fowler-Nordheim or resonance transmission scenario of the field emission. In any case, the enhanced values of the measured electron mass may be interpreted as evidence of the emission from narrow spectral bands, rather than one from the discrete resonance levels. Even considering the simplest $(1 + 1)$ model of the graphene-metal interface, we discover that the electron mass may be greater than in pure graphene due to possible level crossings arising from the dispersion curves of the materials involved—the graphene and the metal. Even a naive $(1 + 1)$ model of the graphene-metal interface shows the levels crossings and relevant super-narrow spectral bands and gaps with various values of effective electron mass.

Indeed, let us observe the influence of periodicity of the graphene-metal interface in the simplest case $(1 + 1)$ of 1D of the periodic structure of the interface. Using a discrete model of a 1D periodic graphene-metal superlattice, we obtain the dispersion function of the lattice, with relevant Weyl-Titchmarsh function $M(\lambda)$, taking into account an interaction only between the nearest neighbors in the lattice:

$$\mathcal{M}(\lambda) - B \cos p = 0 \qquad (9)$$

Here, the Weyl-Titchmarsh function of the graphene-metal superlattice is composed of the functions of component lattices and the interaction is taken into account by 2×2 matrix B. The effective mass of electron in the resulting interface lattice does not coincide with the mass in each of component lattices, but depends on the interaction β. Generally, it may be greater than the mass of electron in graphene.

The Weyl-Titchmarsh function has a positive imaginary part in the upper half-plane of energy $\Im\lambda \geq 0$ and real on the spectral gaps of the lattice. Then the effective mass of the electron on the lattice is calculated as $\left[\frac{d\lambda^2}{dp^2}\right]$, which yields, at the ends of the spectral bands, where $\left[\frac{d\lambda}{dp}\right] = 0$

$$m = -\frac{\frac{d\mathcal{M}}{d\lambda}}{\beta \cos p} = -\frac{\frac{d\mathcal{M}}{d\lambda}}{\mathcal{M}(\lambda)} \tag{10}$$

Based on effective mass we may judge on density of states which allows to estimate the emission current based on above formulae for the transmission coefficient Equation (8). While calculating the transmission coefficient with regard of 1D periodic structure of the graphene-metal interface,

$$T(\lambda) = \frac{2\pi i}{2\pi i + \mathcal{M}(\lambda)}$$

we notice that the transmission coefficient may be large on the gaps of the interface-lattice and relatively small on the open spectral bands of the interface. For the 2D interface, two spectral bands may overlap, yielding the resulting picture with two maxima in the electron's distribution on full energy, see [1]. We plan to continue our study on the corresponding phenomena in our upcoming publications, based on 2D mathematical tools prepared in our preceding papers, see for instance [13].

4. Conclusions

The paper provides a brief overview of the main results of an experimental study of field emission from the graphene-like structures, which implies that the emissions are of a low-threshold character.

One-dimensional theoretical model of the issue, based on the resonant tunneling effect, is analyzed. The advantages and disadvantages of the one-dimensional model are discussed.

Acknowledgments: We are grateful to V. Bogevolnov, who attracted our attention to this cheap and highly interesting material and supplied us with a sufficient amount of it and to M. Polyakov for experimental dates of pulse studies.

Author Contributions: The article is written by authors in the creative equivalent to participation in the creation of the text in the presentation and discussion of the results.

Conflicts of Interest: The authors declare no conflict of interest.

References

1. Fursey, G.N.; Egorov, N.V.; Zakirov, I.I.; Yafyasov A.M.; Antonova L.I,; Trofimov V.V. Peculiarities of the total energy distribution of the field emission electrons from graphene-like structures. *Radiotech. Electron.* **2016**, *61*, 79–82, in press. (In Russian)

2. Yafyasov,A; Bogevolnov, V.; Fursey, G.; Pavlov, B.; Polyakov, M.; Ibragimov, A. Low-threshold emission from carbon nano-structures In: *Ultramicroscopy* **2011**, *111*, 409–414.

3. Fursey, G.N.; Polyakov, M. A.; Kantonistov, A.A.,; Yafyasov, A. M.; Pavlov, B. S.; Bogevolnov, V. B. Field and Explosive Emission from Graphene-Like Structures. *Tech. Phys.* **2013**, *58*, 845–851.

4. Fursey, G.N. *Field Emission in Vacuum Microelectronics*; Springer, Kluwer Academic/Plenum: New York, NY, USA, 2005; pp.161–168.

5. Fursey, G.N. Field emission in vacuum micro-electronics. *Appl. Surf. Sci.* **2003**, *215*, 1113–1134.

6. Forbes,G.; Xanthakis, J.P. Field penetration into amorphous-carbon films: Consequences for field-induced electron emission. *Surf. Interface Anal.* **2007**, *39*, 139–145.

7. Fursey, G.N.; Polyakov, M.; Pavlov, B.; Yafyasov, A.; Bogevolnov, V. Exceptionally low Threshold of Field Emission from carbon nano-clusters. Extended abstract. In Proceedings of the 24th International Vacuum Nanoelectronics Conference, Bergishe Universitat, Wuppertal, Germany, 18–22 July 2011.

8. Modinos, A.; Xanthakis, J.P. Electron emission from amorphous carbon nitride film. *Appl. Phys. Lett.* **1998**, *73*, 28–37.

9. Davies, J.H. *The Physics of Low-Dimensional Semiconductor*; An Introduction; Cambridge University Press: Cambridge, UK, 1998; p. 438.

10. Novoselov, K.S.; Geim, A.K.; Morozov, S.V.; Jiang, D.; Katsnelson, M.I.; Grigorieva, I.V.; Dubonos, S.V.; Firsov, A.A. Two-dimensional gas of massless Dirac electrons in graphene. *Nature* **2005**, *438*, 197–200.

11. Flugge, S. *Practical Quantum Mechanics I*. Springer Verlag, Heidelberg: New York, NY, USA, 1974.

12. Harmer, M.; Pavlov, B.; Yafyasov, A. Boundary condition at the junction. *J. Comput. Electron.* **2007**, *6*, 153–157.

13. Bagraev, N.; Martin, G.; Pavlov, B.; Yafyasov, A.; Goncharov, L.; Zubkova, A. *The Dispersion Function of a Quasi-2D Periodic Sandwich via DN-Map*. In Proceedings of the 11th International Conference on Computational Structures Technology, Dubrovnik, Croatia, 4–7 September 2012; Barry, H.N., Ed.; Topping, Heriot-Watt University: Edinburgh, UK, 2012; p. 53.

14. Pavlov, B. The theory of extensions and explicitly-soluble models. *Russian Math. Surv.* **1987**, *42*, 127–168.

MDPI AG

St. Alban-Anlage 66

4052 Basel, Switzerland

Tel. +41 61 683 77 34

Fax +41 61 302 89 18

http://www.mdpi.com

Electronics Editorial Office

E-mail: electronics@mdpi.com

http://www.mdpi.com/journal/electronics

www.ingramcontent.com/pod-product-compliance
Lightning Source LLC
Chambersburg PA
CBHW051922190326

41458CB00026B/6378